NONLINEAR DISTORTION IN WIRELESS SYSTEMS

NONLINEAR DISTORTION IN WIRELESS SYSTEMS

MODELING AND SIMULATION WITH MATLAB®

Khaled M. Gharaibeh

Yarmouk University, Jordan

IEEE PRESS

A John Wiley & Sons, Ltd., Publication

Chapters 3, 4 and 5 contain some material taken from the following sources: 1. Khaled M. Gharaibeh, Kevin Gard and M. B. Steer, "In-band Distortion of Multisines," IEEE Transaction on Microwave Theory and Techniques, Vol. 54, pp. 3227–3236, Aug. 2006. 2. Khaled. M. Gharaibeh and M. B. Steer, "Modeling distortion in multichannel communication systems," IEEE Trans. Microwave Theory and Tech., Vol. 53, No. 5, pp. 1682–1692, May 2005. 3. Khaled Gharaibeh and M. B. Steer, "Characterization of Cross Modulation in Multi-channel Power Amplifiers Using a Statistically Based Behavioral Modeling Technique," IEEE Transaction on Microwave Theory and Techniques, Dec. 2003. Reproduced by permission 2003, 2005, 2006 of © IEEE.

Chapter 6 contains some material taken from the following source: Khaled M. Gharaibeh, Kevin Gard and M. B. Steer, "Estimation of Co-Channel Nonlinear Distortion and SNDR in Wireless System," IET Microwave Antenna and Propagation, 2007, Volume 1, Issue 5, pp. 1078–1085. Reproduced by permission of © 2007 IET.

Chapter 7 contains some material taken from the following sources: 1. Reprinted from AEU – International Journal of Electronics and Communications, **64**, Khaled M. Gharaibeh, "On the relationship between the noise-to-power ratio (NPR) and the effective in-band distortion of WCDMA signals," 273–279, © 2010, with permission from Elsevier. 2. Khaled M. Gharaibeh, 2009. Simulation of Noise Figure of Nonlinear Amplifiers Using the Orthogonalization of the Nonlinear Model. International J. RF Microwave Comput-Eng. 19, 502–511. Reproduced with permission from John Wiley & Sons, Ltd. 3. Khaled M. Gharaibeh, K. Gard and M. B. Steer, 2004. Accurate Estimation of Digital Communication System Metrics – SNR, EVM and Rho in a Nonlinear Amplifier Environment. In Proc. of the 42 Automatic RF Techniques Group (ARFTG) Orlando, FL, pp. 41–44.

MATLAB® and Simulink® are trademarks of The MathWorks, Inc. and are used with permission. The MathWorks does not warrant the accuracy of the text or exercises in this book. This book's use or discussion of MATLAB® and Simulink® software or related products does not constitute endorsement or sponsorship by The MathWorks of a particular pedagogical approach or particular use of the MATLAB® and Simulink® software.

Library of Congress Cataloging-in-Publication Data

Gharaibeh, Khaled M.
  Nonlinear distortion in wireless systems : modeling and simulation with MATLAB® / Khaled M. Gharaibeh.
     p. cm.
  Includes bibliographical references and index.
  ISBN 978-0-470-66104-8 (hardback)
  1. Signal processing–Computer simulation.   2.   Electric distortion–Computer simulation.   3.   Nonlinear systems–Computer simulation.   4.   Wireless communication systems–Computer simulation.   5.   MATLAB®.   I.   Title.
   TK5102.9.G48 2012
   621.382′2028553–dc23

                                                                                    2011029722

A catalogue record for this book is available from the British Library.
  ISBN: 9780470661048 (H/B)
  ISBN: 9781119961727 (ePDF)
  ISBN: 9781119961734 (oBook)
  ISBN: 9781119964117 (ePub)
  ISBN: 9781119964124 (Mobi)

Set in 10/12 Times by Laserwords Private Limited, Chennai, India
Printed in Malaysia by Ho Printing (M) Sdn Bhd

To my wife Rania and my sons Mohammed, Ibrahim and Abdullah . . .

Contents

Preface

Modeling and simulation of nonlinear systems provide communication system designers with a tool to predict and verify overall system performance under nonlinearity and complex communication signals. Traditionally, RF system designers use deterministic signals (discrete tones), which can be implemented in circuit simulators, to predict the performance of their nonlinear circuits/systems. However, RF system designers are usually faced with the problem of predicting system performance when the input to the system is real-world communication signals which have a random nature. In this case, system distortion cannot be quantified and simulated using traditional approaches which do not take into account the random nature of real-world communication signals. Many books which discuss modeling and simulation of nonlinear system exist. However, and up to the knowledge of the author, very few of them have targeted modeling and simulation of nonlinear distortion from a stochastic point of view.

This book describes the principles of modeling and simulation of nonlinear distortion in single and multichannel wireless communication systems using both deterministic and stochastic signals. Models and simulation methods of nonlinear amplifiers explain in detail how to analyze and evaluate the performance of data communication links and how to establish the performance metrics under nonlinear transformations. The book relies extensively on using random process theory to develop simulation tools for predicting system performance. The analysis presented in the book provides a linkage between deterministic and stochastic views of nonlinear distortion which enables system and circuit designers to understand the nonlinear phenomena and hence, to be able to design wireless communication systems efficiently.

This book also addresses the problem of how to embed models of nonlinear distortion in system-level simulators such as MATLAB® and Simulink® where practical techniques that professionals can use immediately on their projects are presented. It provides MATLAB® simulation modules and a comprehensive reference of models needed for the simulation of nonlinear distortion in wireless communication systems. The book explores simulation and programming issues and provides a comprehensive reference of simulation tools for nonlinearity in wireless communication systems. Together, these provide a powerful resource for students, professors and engineers who are working on the design and verification of nonlinear systems such as High-Power Amplifiers (HPA), Low-Noise Amplifiers (LNA), mixers, etc. in the context of a wireless communication system design.

The book is divided into three major parts totaling ten chapters. The first part consists of three chapters and provides the basics needed to understand the nonlinear phenomena and discusses the basics of modeling nonlinearity in wireless systems. Chapter 1 is an introduction to wireless communications systems and nonlinearity and it serves as an introduction to the problem of modeling nonlinear distortion. Chapter 2 is an introduction to wireless system standards and signal models; and Chapter 3 presents various models of nonlinearity and their parameter extraction. This includes analytical models such as Volterra based models as well empirical model such as block models.

The second part consists of three chapters and discusses major techniques used for characterizing nonlinear distortion in wireless communication systems. Two major approaches are presented; the first is based on modeling deterministic signal distortion and the second utilizes random signals to characterize nonlinear distortion in real-world communication systems. Chapter 4 provides the reader with the deterministic view of nonlinear distortion where it discusses the analysis of nonlinear distortion using single and multiple tones. Closed form expressions that relate signal distortion to nonlinear system characteristics are presented for various nonlinear models such as power series, Volterra model and block models. Chapter 5 discusses analysis of the response of nonlinear system to a random input signals which represent real-world communication signals. This chapter demonstrates the probabilistic view of nonlinear distortion and provides the basic mathematical tools needed to analyze nonlinear distortion which include autocorrelation function analysis and nonlinear spectral analysis. It also discusses the analysis of nonlinear distortion in multichannel systems using random signals. Chapter 6 presents the concept of the orthogonalization of the behavioral model which is used to identify in-band distortion components responsible for the degradation of wireless system performance. Finally, Chapter 7 uses the concepts developed in the previous chapters and presents the derivation of communication system figures of merit in terms of nonlinearity. These include Adjacent Channel Power Ratio (ACPR), Signal-to-Noise and Distortion Ratio (SNDR), Noise-to-Power Ratio (NPR), Error Vector Magnitude (EVM), Noise Figure (NF) and Bit Error Rate (BER).

The last part provides the reader with techniques for implementing various models of nonlinearity and nonlinear distortion in MATLAB® and Simulink®. Chapter 8 provides an introduction to the simulation of communication systems in MATLAB® and in Simulink®, where the basics of simulations of modern wireless communication systems are presented. Chapter 9 explains how to use MATLAB® to simulate various types of nonlinearity and how to analyze, predict and evaluate the performance of data communication systems under nonlinearity. Finally, Chapter 10 explains how to use Simulink® to analyze, predict and evaluate the performance of wireless communication systems related to nonlinear distortion and provides a comprehensive reference of models for simulation of nonlinear distortion.

To complement the material of the book chapters, three appendices are included which serve as supporting material. Appendix A provides the basics of signal and system analysis which includes time and frequency representation of signals and linear system analysis. Appendix B provides an introduction to random variables and random processes on which, the bulk of the material of the book is based, and Appendix C provides an introduction to MATLAB® and MATLAB® simulations.

For more information, please visit the companion website, www.wiley.com/go/ gharaibeh_modeling.

Khaled M. Gharaibeh, Irbid, Jordan

List of Abbreviations

AMPS	Advanced Mobile Phone System
ADS	Advanced Design System
ADC	Analog-to-Digital Converter
AM–AM	Amplitude Modulation–Amplitude Modulation
AM–PM	Amplitude Modulation–Phase Modulation
ACPR	Adjacent-Channel Power Ratio
ASK	Amplitude Shift Keying
AWGN	Additive White Gaussian Noise
BER	Bit Error Rate
BPF	Band Pass filter
CCPR	Co-Channel Power Ratio
CDMA	Code Division Multiple Access
CW	Continuous Wave
DAC	Digital-to-Analog Conversion
DPCH	Dedicated Physical Channel
DCS	Digital Cellular System
DUT	Device Under Test
DSP	Digital Signal Processing
DS-SS	Direct Sequence Spread Spectrum
DECT	Digital European Cordless Telephone
DL	Down-Link
dB	Decibel
EVM	Error Vector Magnitude
EDGE	Enhanced Data rates for GSM Evolution
ETSI	European Telecommunications Standards Institute
FSK	Frequency Shift Keying
FDMA	Frequency Division Multiple Access
GSM	Global System for Mobile
GMSK	Gaussian Minimum Shift Keying
GHz	Gigahertz
GPS	Generalized Power Series
HB	Harmonic Balance

ISI	Inter Symbol Interference
ICI	Inter Channel Interference
IF	Intermediate Frequency
IMD	Intermodulation Distortion
IMR	Intermodulation Ratio
IIP3	Input Third-Order Intercept Point
LAN	Local Area Network
LNA	Low Noise Amplifier
LTE	Long Term Evolution
LPF	Low Pass Filter
LOS	Line-of-Sight
MHz	MegaHertz
MISO	Multiple Input Single Output
NPR	Noise-to-Power Ratio
NF	Noise Figure
NNF	Nonlinear Noise Figure
NBGN	Narrow band Gaussian Noise
NII	National Information Infrastructure
NLOS	Non-Line-Of-Sight
OIP3	Output Third-Order Intercept Point
OFDM	Orthogonal Frequency Division Multiplexing
OSC	Oscillator
PA	Power Amplifier
PCS	Personal Communication System
PSK	Phase Shift Keying
PSD	Power Spectral Density
PAR	Peak-to-Average Ratio
QAM	Qudrature Amplitude Modulation
RF	Radio Frequency
SNR	Signal-to-Noise Ratio
SNDR	Signal-to-Noise and Distortion Ratio
THD	Total Harmonic Distortion
UMTS	Universal Mobile Telephone System
UL	Uplink
VGA	Variable Gain Amplifier
VNA	Vector Network Analyzer
VSG	Vector Signal Generator
VSA	Vector Signal Analyzer
WLAN	Wireless Local Area Network
WMAN	Wireless Metropolitan Network
WCDMA	Wide Band Code Division Multiple Access
WSS	Wide Sense Stationary

List of Figures

List of Tables

Acknowledgements

The author would like to thank a number of people who helped making this book a reality. First and foremost, many thanks go to my wife Rania for her patience and her support. Secondly, I would like to thank the technical staff of Wiley-Blackwell for their continuous support and cooperation. Lastly, I would like to thank Prof. Michael B Steer and Dr. Kevin Gard from North Carolina State University for their continuous support, motivation and for providing some of the measurements presented in this book.

1

Introduction

A fundamental design concern for system designers of wireless communication systems is modeling distortion introduced by the nonlinear behavior of the devices incorporated in the design. Nonlinearity in wireless communication systems usually exists in the RF front ends and is produced by nonlinear devices such as power amplifiers, low-noise amplifiers, mixers, etc. Nonlinearity is responsible for introducing signal components that contribute to degrading system performance in a similar fashion to system noise, interference or channel impairments.

In order to understand the effects of nonlinear distortion on the performance of wireless communication systems, it is important to understand the architecture of these systems and the components that are responsible for introducing nonlinear distortion. On the other hand, it is also important to understand how these systems can be modeled and simulated. Modeling and simulation of nonlinear systems is an important step towards the efficient design of modern communication systems.

In this chapter, an overview of nonlinearity and nonlinear distortion in wireless systems is given. The common sources of nonlinearity in wireless communication system are presented. Then, the concept of nonlinear distortion produced by the nonlinear behavior and its relationship to the performance of wireless systems is given in the subsequent sections. In the last section of this chapter, an overview of the most common modeling and simulations approaches of nonlinear systems and circuits are presented that will serve as an introduction to subsequent chapters that discuss modeling and simulation of nonlinear distortion.

1.1 Nonlinearity in Wireless Communication Systems

Nonlinearity in wireless systems is originated in the nonlinear devices incorporate in the design of the transmitter and the receiver. The main blocks that introduces nonlinear distortion are mixers, power amplifiers in wireless transmitters and low-noise amplifiers in wireless receivers.

Nonlinear Distortion in Wireless Systems: Modeling and Simulation with MATLAB®, First Edition.
Khaled M. Gharaibeh.
© 2012 John Wiley & Sons, Ltd. Published 2012 by John Wiley & Sons, Ltd.

1.1.1 Power Amplifiers

Power Amplifiers (PAs) are devices that are used at the end of the transmitter chain in order to produce a signal with a power suitable for transmission through an antenna. In a wireless transmitter, a baseband signal sent to a PA that is connected to an antenna through a matching circuit after being modulated and up-converted. The PA performs power amplification by multiplying the signal by a gain factor that results in an amplified signal whose power is much higher than the input signal. Ideally, a PA has constant gain across all input powers, however, a practical PA has a maximum output power that is determined by the DC input power. Hence, as this limit is approached gradually, the apparent gain of the PA decreases with increasing the input power.

Therefore, depending on the input operating power, the PA is considered linear if it is operated in a power range within the linear amplification range of its characteristics. If the amplifier is operated close to or within the saturation region of its characteristics, then it is considered nonlinear. It is usually desirable to operate the PA near its saturation region in order to obtain maximum power efficiency; however, this means that nonlinear distortion is introduced at the output of the PA, which is undesirable specially when the input signal has a varying amplitude.

PAs are characterized by their gain versus input power curves for single-tone input that are known as Amplitude Modulation–Amplitude Modulation (AM–AM) conversion characteristics. In a typical AM–AM characteristics, the gain of the PA remains constant (output power increases linearly) with increasing input power up to a saturation point where the gain drops (output power remains constant with increasing input power). This is due to the input/output characteristics of the active device (transistor) incorporated in the PA design. Saturation of the PA characteristics is a manifestation of the PA nonlinearity where the output power does not follow the input power by a constant gain. Another manifestation of nonlinearity is the phase characteristics of the PA where the phase of the output signal deviates from the input phase by an angle that depends on the input signal power. These phase characteristics are called the Amplitude Modulation–Phase Modulation (AM–PM) characteristics and they measure the phase distortion introduced by a PA. AM–PM characteristics are mainly due to voltage dependent collector capacitance (caused by a varying depletion layer width) (Cylan, 2005).

The AM–AM and AM–PM characteristics can be formulated by considering a single tone signal of the form

$$x(t) = R \cos(2\pi f_0 t + \psi) \tag{1.1}$$

where R is the signal amplitude, f_0 is the tone frequency and ψ is its phase. The output of nonlinearity can be expressed as

$$y(t) = F[R] \cos(2\pi f_0 t + \psi + \Phi(R)) \tag{1.2}$$

where $F[.]$ is the amplitude distortion as a function of the input amplitude and represent the AM–AM characteristics of the nonlinear device and $\Phi[.]$ is the phase distortion as a function of the input amplitude and represents the AM–PM characteristics. Figure 1.1 shows a typical AM–AM and AM–PM characteristics of a PA.

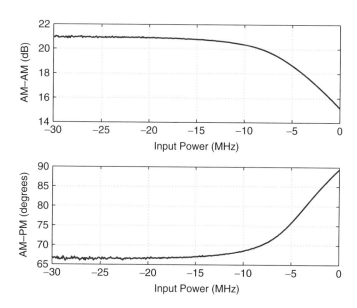

Figure 1.1 AM–AM and AM–PM conversions. Reproduced by permission of © 2002 Artech House.

1.1.1.1 Memory Effects

The AM–AM and AM–PM characteristics define the PA linearity at a single frequency. However, the PA characteristics may differ at different frequencies. This is due to the memory effects of the PA that are usually caused by the time constants in the biasing circuit or are due to the impedance mismatch with the amplifier. Figure 1.2 shows the

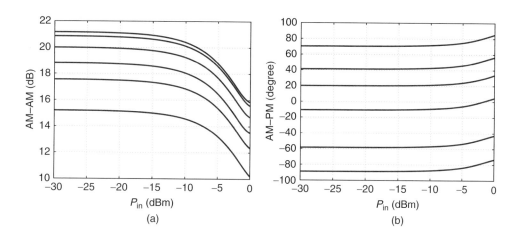

Figure 1.2 AM–AM conversion at different frequencies.

AM–AM and AM–PM of a PA at different frequencies. In fact, the existence of the AM–PM characteristics indicates that the PA has memory since a memoryless PA has only AM–AM characteristics. However, if the time constants of the memory are smaller than the maximum signal envelope frequency, the system is called a quasi-memoryless system. In general, if the bandwidth of the PA is much larger than the modulation bandwidth of the signal, the PA can be considered as memoryless. Memory effects can manifest themselves as hysteresis in the time domain, which causes the intermodulation products of a two-tone test to be asymmetrical in the frequency domain.

The most common approaches to modeling the wideband behavior (i.e. memory effects) of nonlinear amplifiers are those based on Volterra series analysis. Volterra series analysis represents an analytical approach to modeling nonlinearity since it represents nonlinearity in a similar way in which Taylor series does for analytic functions. A general Volterra series model of a nonlinear system is described by the following functional expansion of continuous functions (Lunsford, 1993):

$$y(t) = \sum_{n=1}^{\infty} F_n(x(t)) = \sum_{n=0}^{\infty} y_n(t) \tag{1.3}$$

where $F_n(x(t))$ is the Volterra functional and is defined as

$$F_n(x(t)) = \int_{-\infty}^{\infty} \ldots \int_{-\infty}^{\infty} h_n(\lambda_1, \ldots, \lambda_n) \prod_{i=1}^{n} x(\lambda_i) d\lambda_i \tag{1.4}$$

where $h_n(\lambda_1, \ldots, \lambda_n)$ is the n-dimensional Volterra kernel that can be symmetric without loss of generality and leads to a unique set of Volterra kernels. Other models for nonlinear systems with memory are discussed in Chapter 3.

1.1.1.2   PA Classes

PA designs fall into a number of classes according to their biasing conditions. Different classes have different nonlinear characteristics and hence, different power efficiencies. The most common PA classes are Class A, B, AB, C, D, E and F, where Class A, B and AB being linear (with low power efficiency) whereas the others are nonlinear (with high power efficiency). The PA class is distinguished by its biasing conditions that are chosen to give a desired PA linearity at the expense of its efficiency (Briffa, 1996; Cripps, 2000).

Higher positive bias voltages provide a higher linear region of operation. However, positive bias voltages (such as class A operations) lead to higher conduction angles, which means low power efficiency. On the other hand, negative bias voltages (such as with class C operation) lead to lower conduction angles but this requires a higher input drive level to maintain power efficiency. This makes the PA operate near saturation, which means nonlinear operation. Figure 1.3 shows typical RF output power and power efficiency of a PA versus conduction angle where different classes are defined (Cripps, 2000).

1.1.2   Low-Noise Amplifiers (LNAs)

The low-noise amplifier is the first block in any RF receiver and is responsible for the amplification of the received signal, which usually has very low power. An LNA in an

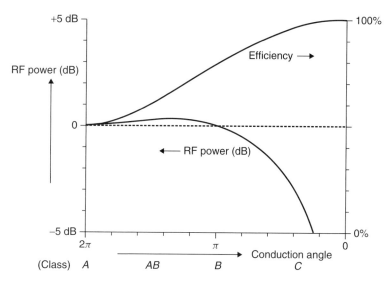

Figure 1.3 PA efficiency vs. conduction angle (Cripps, 2000). Reproduced by permission of © 2000 Artech House.

RF receiver, therefore, plays an important role in the quality of the reception process, and hence the design of LNAs represents a limiting factor in the performance of the overall communication system.

Since an LNA in an RF wireless receiver is required to amplify weak signals in the presence of noise, the design of an LNA requires a tradeoff of a number of factors such as linearity, noise performance, stability, power consumption and complexity. Linearity and noise performance are responsible for determining the dynamic range of the amplifier where the minimum and maximum allowed signals levels are determined. An LNA is designed with its gain compression (due to nonlinearity) determined by the maximum received signal level expected in a certain application. On the other hand, an LNA is designed with a noise performance such that its added noise is below the minimum expected received signal level. Therefore, the main difference between LNAs and PAs is that in PA design noise performance is not an issue since PAs usually operate with input signal powers that are much higher than the inherent noise of the PA circuit. This means that the LNA nonlinearity is linked to the noise performance of its circuit where a tradeoff must be made. On the other hand, the interaction of received signal with channel noise by the nonlinear amplification in an LNA represents a source of distortion introduced inside the band of the received signal (Chabrak, 2006; Gharaibeh, 2009).

Noise performance of an LNA is quantified in terms of the Noise Figure (NF) that measures the amount of noise that is added by the LNA. NF of an LNA is defined as the ratio of the Signal-to-Noise Ratio at the input (SNR_i) to the Signal to Noise Ratio at the output (SNR_o). However, the nonlinearity of LNAs hinders the simple evaluation of their NF. The interaction of signal and noise by nonlinearity results in many output signal components that may be correlated to each other and this complicates the evaluation of the SNR_o. On the other hand, since nonlinear distortion adds to system noise, NF of nonlinear

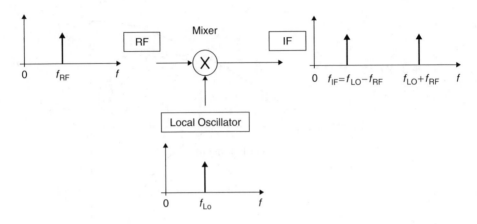

Figure 1.4 A conceptual diagram of an ideal mixer (Cripps, 2000).

amplifiers is highly dependent on the input power because signal distortion increases as the amplifier is driven close to saturation.

1.1.3 Mixers

A mixer is a product device used for frequency up- and down-conversion process at both the transmitter and receiver. A mixer is therefore, inherently nonlinear for normal operation, however, this nonlinearity may result in unwanted effects. In general, the main challenge in the design of mixers is to maintain a second-order nonlinearity needed for the mixing process especially with direct conversion transmitters or receivers. However, this is not easy to achieve as mixers usually exhibit higher-order nonlinearity in addition to the desired second order nonlinearity (Sheng *et al.*, 2003; Terrovitis and Meyer, 2000). Figure 1.4 shows a conceptual diagram of an ideal mixer that operates as downconverter. An ideal mixer is a three port system where the Radio Frequency port (RF) and the Local Oscillator port (LO) are its inputs, while the Intermediate Frequency port (IF) is its output. In this mode of operation and with single-tone inputs, the mixer multiplies the RF-signal with the LO-signal where two sinusoides are generated at the IF output port; one at the sum frequency and another at the difference frequency. A mixer is non ideal if it comprises higher-order nonlinearities (above 2), which means that its output will consist of other spectral components than the two produced by the ideal mixer (Vandermot *et al.*, 2006).

1.2 Nonlinear Distortion in Wireless Systems

As discussed in the previous sections, nonlinearity in wireless communication systems is originated in the PAs, LNAs and mixers. This nonlinearity limits the delivered output power because of the compression nonlinear characteristics and also introduces unwanted signal components at the output of the nonlinear device. These unwanted signal components are called "nonlinear distortion" that is manifested as harmonics at multiples of the fundamental frequencies when the input signal consists of discrete tones and, as spectral regrowth when the input signal spectrum has a finite bandwidth.

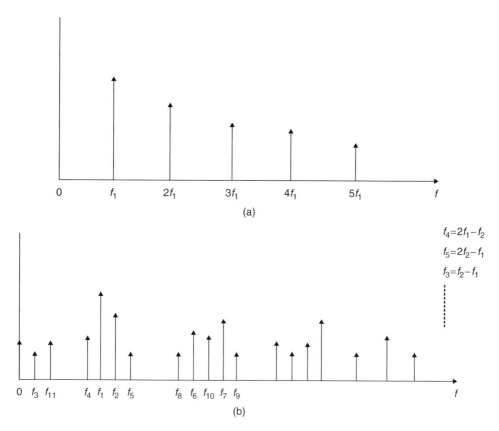

Figure 1.5 Spectra of the nonlinear response to (a) single-tone signal and (b) two-tone signal (Rhyne and Steer, 1987).

Understanding nonlinear distortion is usually based on studying the response of nonlinearity to single or multiple sinusoids (tones) (Gharaibeh *et al.*, 2006; Pedro and de-Carvalho, 2001; Rhyne and Steer, 1987; Vandermot *et al.*, 2006). Figure 1.5 shows the nonlinear response to single- and two-tone signals (Rhyne and Steer, 1987). With a single-tone signal, nonlinearity results in compression of the signal amplitude and the introduction of harmonics at multiples of the fundamental frequency. With two-tone signals nonlinearity results in additional signal components called intermodulation products (IP) that fall very close to the fundamental tones and represent the interaction between the input sinusoids by nonlinearity. In the case of multitone signals, nonlinear distortion is manifested as a large number of intermodulation products that fall within and outside the bandwidth of the input signal. The distortion components that fall inside the bandwidth of the input signal are termed "in-band distortion" and those that fall outside the bandwidth of the signal are termed "out-of-band distortion". The nonlinear distortion of multiple tone signals is usually used as an approximation of the nonlinear distortion of finite-bandwidth signals which can be approximated by a large number of tones with different amplitudes, phases and frequencies.

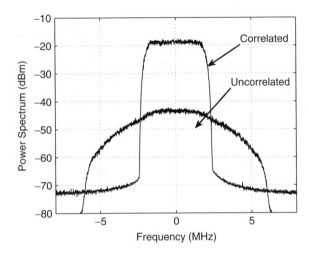

Figure 1.6 Spectrum of the nonlinear response to a digitally modulated signal partitioned into linear and spectral regrowth components.

With digitally modulated signals, nonlinearity results in two main impairments to the output spectrum. The first is gain compression that results in reduced output power relative to linear operation of the nonlinear device, and the second is called spectral regrowth that consists of distortion components that appear inside and outside the bandwidth of the input signal. Figure 1.6 shows the power spectrum at the output of a nonlinearity partitioned into linear with gain compression and spectral regrowth components. The in-band portion of spectral regrowth results in the degradation of system performance while the out-of-band component results in the degradation of the performance of other systems that use adjacent frequency channels.

In the following subsections, the main impairments that nonlinear distortion cause to wireless communications systems are presented.

1.2.1 Adjacent-Channel Interference

Adjacent-Channel Interference (ACI) is a well-known manifestation of nonlinear behavior in wireless systems. ACI is an irreducible out-of-band component of spectral regrowth that is responsible for introducing interference in the adjacent frequency channels in systems that use Frequency Division Multiplexing (FDM) as shown in Figure 1.7. In old cellular systems, such as analogue systems and Global System for Mobile (GSM) systems, frequency planning was required to cope with interference so that the same frequency channels are not used in neighboring cells. Modern wireless systems use Code Division Multiple Access (CDMA) technology where the concept of clusters is not used, and hence the same channels are reused in adjacent cells. That is, channels adjacent to a main channel are used for communication in the same cell. Distortion introduced in an adjacent channel is therefore more significant than with other systems.

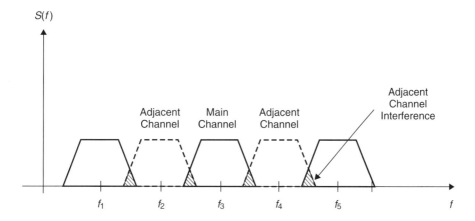

Figure 1.7   Adjacent channel interference in FDM systems (Rhyne and Steer, 1987).

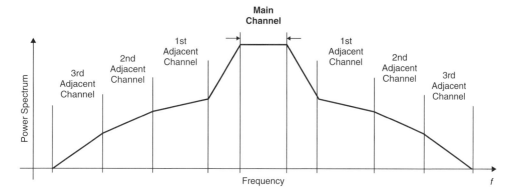

Figure 1.8   Definition of ACPR.

Therefore, it is necessary for the system design to impose a maximum level on the amount of distortion that can be introduced in adjacent cells. Since it is not possible to design for absolute distortion levels, relative levels of power in an adjacent channel to power radiated in the main channel is specified. The most commonly used measure of this distortion is called Adjacent Channel Power Ratio (ACPR) that refers to the ratio of the power in the main channel to the power in one of the adjacent channels (See Figure 1.8). The definition of ACPR is standard dependent and will be discussed in Chapter 7.

1.2.2 Modulation Quality and Degradation of System Performance

As discussed above, in-band distortion is responsible for the degradation of system performance that is manifested as the degradation of Signal-to-Noise Ratio (SNR) and ultimately system Bit Error Rate (BER). For linear modulation schemes, the nonlinear behavior is manifested as compression and rotation of the signal constellation, which results in increasing system probability of error (bit error rate).

A basic approach to quantify nonlinear distortion in digital communication systems is to consider a memoryless nonlinearity which can be characterized by the AM–AM and AM–PM characteristics as in Equation (1.2). Therefore, for a digitally modulated signal of the form

$$x(t) = r(t) \cos(2\pi f_0 t + \psi(t)) \tag{1.5}$$

where $r(t)$ is a time-varying signal amplitude and represents the amplitude modulation of the signal; and $\psi(t)$ is the time-varying phase of the signal that represents phase modulation, the output of the nonlinearity can be expressed as

$$y(t) = F[r(t)] \cos(2\pi f_0 t + \psi(t) + \Phi(r(t))) \tag{1.6}$$

where $F[.]$ is the amplitude distortion as a function of the input amplitude and represents the AM–AM characteristics of the nonlinear device and $\Phi[.]$ is the phase distortion as a function of the input amplitude and represents the AM–PM characteristics. Therefore, with nonlinear amplification, the AM–AM distortion corrupts the envelope of the signal and the AM–PM distortion corrupts its phase. Figure. 1.9(a) shows the constellation diagram of a 16 QAM signal with Raised-Cosine (RC) pulse shaping and Figure. 1.9(b) shows the constellation diagram after amplification by a nonlinear amplifier. It is clear that the constellation diagram becomes distorted and this distortion appears as a compression and rotation of the constellation point. Consequently the received signal will be harder to detect and this results in degradation of the system BER (Briffa, 1996).

It is important to note that systems which use linear modulation schemes (QPSK, QAM, etc.), which have become common in communication systems and standards due to their higher spectral efficiency, are more susceptible to nonlinear amplification than systems which use modulation schemes with constant envelopes such as FSK (frequency modulation) or its variants. If the signal has time-varying amplitude then its instantaneous

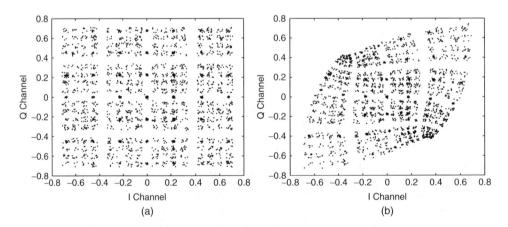

Figure 1.9   Signal constellation (a) before and (b) after nonlinear amplification.

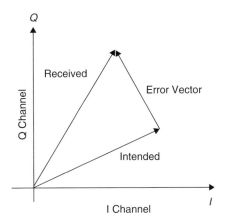

Figure 1.10   Definition of EVM.

input power changes continuously. This means that the signal at the PA output is distorted if the amplifier gain are not linear. If the signal has a constant envelope, then the PA linearity is not an important issue because the instantaneous input power stays constant and hence, there are no gain and phase variations for a specific operating point (Cylan, 2005).

Nonlinear distortion in wireless transmitters is usually quantified by the Error Vector Magnitude (EVM) which is a measure of the departure of signal constellation from its ideal reference because of nonlinearity. EVM is defined as the distance between the desired and actual signal vectors (error vector) as shown in Figure 1.10, normalized to a fraction of the signal amplitude. The actual value of the constellation point can deviate from the ideal value significantly, depending on PA nonlinearity. EVM quantifies in-band distortion which causes high bit error rates during reception of the transmitted data. Therefore, EVM specifications must be fulfilled in order to have a good quality of communication. Another measure of the fidelity of the signal under nonlinear amplification in some systems, like CDMA systems, is the waveform quality factor (ρ). The waveform quality factor is a measure of the correlation between a scaled version of the input and the output waveforms (Aparin, 2001).

1.2.3   Receiver Desensitization and Cross-Modulation

In a mobile receiver, the interaction of multiple signals by nonlinearity results in the problem of receiver desensitization. For example, one of the stringent requirements in second-generation CDMA receiver design (such as the IS-95 system) is the reception of a CDMA channel in the presence of a single-tone jammer (Aparin, 2003; Gharaibeh and Steer, 2005; IS95 Standard, 1993). The single tone jammer models, for example, a narrow-band Advanced Mobile Phone Service (AMPS) signal transmitted from a nearby AMPS base station or any other type of jamming or interference. In this scenario, desensitization of the single tone is a measure of the receiver's ability to receive a CDMA signal at its assigned frequency and in the presence of a single-tone jammer at a given frequency offset

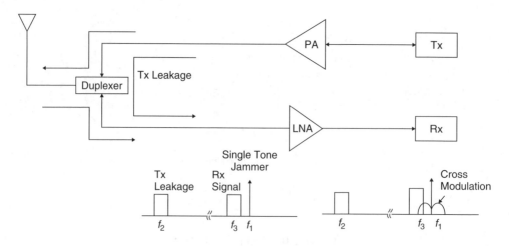

Figure 1.11   Receiver desensitization in reverse link CDMA system (Gharaibeh, 2004).

from the CDMA signal center frequency. This situation is shown in Figure 1.11 where the poor isolation between the transmitter (Tx) and receiver (Rx) ports of the duplexers causes a stronger Tx leakage to the Rx input. The interference introduced by the jammer results from cross-modulation of the jammer and transmitter leakage where Tx leakage modulation is transferred to the jammer by the Rx nonlinearities causing widening the jammer spectrum, as shown in Figure 1.11. This process results in a significant amount of interference in the Rx channels adjacent to the jammer reducing the Rx sensitivity (Aparin, 2003; Gharaibeh and Steer, 2005).

1.3   Modeling and Simulation of Nonlinear Systems

1.3.1   Modeling and Simulation in Engineering

System modeling and simulation are inevitable tools in modern engineering design. In addition to optimizing system design, the role of modeling and simulation is to allow concepts to be explored even before enabling technologies have been brought to maturity. In order to analyze the concept of system modeling and system simulations, it is important to define the concepts of a system, a model and simulation individually.

A system is defined as a collection of objects that interact together to achieve the objective of which the system is designed for. Systems produce outputs that depend on their inputs and on their contents, both of which may change with time and space. A model of a system is a mathematical representation that provides a simplification of the actual system in order to provide understanding and at the same time preserve the ability to predict the system output. The simplification of reality is usually done on two main aspects of the system. The first is the simplification of the details of the system, that is, the components of the system. The second is the simplification associated with the dynamics of the system that represent the cause–effect characteristics of the system that are separated in time and/or space (Jeruchem *et al.*, 2000).

Modeling of systems is usually faced with the question of how much the model represents the actual system given that it is a simplification of reality. The other issue that may arise at the same time is related to the amount of simplification that is allowed in order to develop a good model. The answers to these questions, in fact, depend on the level of understanding that we expect from the model. Too much simplification leads to a weak representation of reality and hence, the model does not provide enough understanding. On the other hand, less simplification means that more details are included in the model (in order to enhance the representation of reality) but this results in a complicated model that also does not promote understanding. Therefore, developing a model is a tradeoff between complexity and accuracy (IThink, 2009).

In general, there are two main types of modeling in engineering: physical modeling and behavioral modeling. A physical model is based on knowledge of the system components that together make up a real system and also, on knowledge of the rules that describe their interaction. Examples of such models are circuit-level models that are used in engineering design processes. On the other hand, behavioral (black box) models are based on the input/output measurements data and hence, their accuracy depends on the quality of measurements (Arabi, 2008; Jeruchem et al., 2000). In general, physical models are more accurate than behavioral models since they involve more details on the system, while behavioral models are simpler and easier to simulate.

Simulation is defined as implementing a system model in a computer program that runs over time to achieve two main goals; the first is to study the interactions among system components; and secondly to study dynamics of the system. In other words, a simulation algorithm aims at studying how the system output changes when it is driven by different inputs and/or studying how the system behaves when the interactions among its components are defined differently. One of the benefits of simulation is that it provides time and space compression of the system where the behavior of the system over a long time period can be viewed with much shorter period by simulation, provided that the model is accurate (Frevert et al., 2005; IThink, 2009; Jeruchem et al., 2000).

Modeling and simulation are interrelated activities in the sense that when a model is used in simulation of a system, the simulator may produce results that indicate that the model is incorrect when compared to reality. This means that the model needs to be updated in order to achieve a closer representation of the actual system. This process continues until a good model and a good simulation algorithm are achieved that eventually lead to the desired representation of the real system. On the other hand, the level of details included in the model has a major influence on the number of computations needed to simulate the system. Therefore, the complexity of the simulation algorithm is always proportional to the complexity of the model since increasing the level of the model details or increasing the model dynamics means that more computations are needed to simulate the system.

Modeling and simulation are engineering disciplines that usually develop through practice rather than by studying their methodology. It is always the combination of the knowledge about the problem and the model developer practice that makes the process of modeling and simulation successful. In most engineering processes, the model developer simulates the model in order to get insight into the validity of the model. The developer then revises the model and simulates it again until an acceptable level of understanding about the system is obtained (Frevert et al., 2005).

1.3.2 Modeling and Simulation for Communication System Design

As discussed in the previous section, modeling and simulation of systems are activities that accompany system design process. The design process is classified into several design levels that start at the "system level" and ends at the "gate level" in digital systems or the "circuit level" in analogue system; a process usually known as the top-down approach, see Figure 1.12 (Frevert *et al.*, 2005). System-level design of modern communication system usually follows the "specification approach" that is based on three main components; a set of specifications an algorithm that models the proposed system and a simulation tool. System-level design using this approach starts by setting up a list of specifications for the intended system and then developing an algorithm based on these specifications that achieves the desired performance criteria. A simulator is then used to evaluate the selected algorithm and to provide insight into which parameters of the algorithm need to be adapted in order to achieve the predefined performance criteria. On the other hand, a simulator provides insight on the mechanics of the proposed algorithm where its behavior is studied over the evolution of time. This process is repeated until an optimal algorithm that meets all the system specifications is reached.

In communication system design these algorithms mainly aim at transmitting data from the signal source at point A to a sink at point B. Algorithms that are specified at this level may be for example (Frevert *et al.*, 2005):

- data structure and protocol;
- Forward Error Correction techniques (FEC);
- modulation techniques (QPSK, QAM, GMSK, OFDM);
- channel equalization and synchronization.

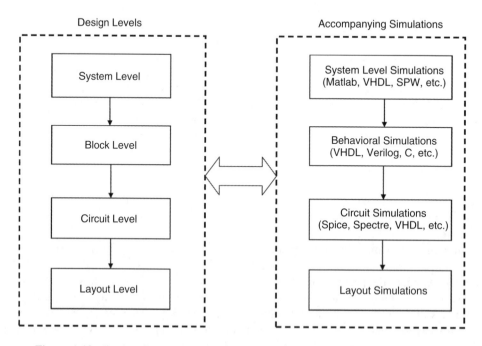

Figure 1.12   Design flow and the accompanying simulations (Frevert *et al.*, 2005).

Examples of system-level simulators are CoCentric System Studio®, MATLAB®, and SPW®. A commonly encountered example of system-level simulation of communication systems is the simulation of the degradation of the postulated system given a communication link budget. System BER is usually simulated versus SNR and distortion parameters that are set according to an initial link budget. If the result of the simulation says that the performance measure cannot be met, then the link budget or the distortion parameters must be modified. This process is repeated until the performance measure is met (Frevert *et al.*, 2005).

System-level design verified by simulation is then used as an input to the following design steps where the system is broken down into many components, each of which is designed individually by the aid of simulations. Therefore, system-level design and simulation enables the building blocks of the system to be identified. The next step towards complete system design is to design these building blocks. This design level is called "block-level" design and consists of detailed descriptions of each block which, altogether, lead to the functionalities identified by the system-level design. In communication systems, these blocks can be analogue or digital or mixed blocks and can be designed individually using block-level simulators that enable the verification of the functionality of each block. Examples of block level simulators are VHDL-AMS, Verilog-AMS and ISS (Frevert *et al.*, 2005).

The next level of system design is called the gate level in digital subsystems or the circuit level in analogue subsystems. In this level, gates or circuits are represented by netlists that contain gates or analogue elements. Simulation at this level is done using circuit level simulators such as VHDL and Verilog for digital circuits or SPICE and Spectre for analogue circuits. Circuit simulation enables the design at the block level to evaluated and modified to meet the performance measures of each of the block (Frevert *et al.*, 2005).

1.3.3 Behavioral Modeling of Nonlinear Systems

Behavioral modeling of nonlinear systems refers to a class of modeling techniques where a nonlinear system is dealt with as a black-box. The black-box model is developed using a set of Input-Output (I/O) measurements and then by using system optimization theory, a function or a set of functions along with their parameters are extracted. Unlike circuit-level modeling, no information is required about the circuit topology, however, intuitive or a priori knowledge about the system helps in developing the model. For example, in modeling nonlinear power amplifiers, a priori knowledge is available since nonlinearity originates from active devices, which is well understood in circuit theory. Therefore, this nonlinearity takes input-output characteristics that are well known a priori and then the modeling problem reduces to estimating system parameters using classical parameter estimation theory such as least squares and its variants.

Different behavioral modeling approaches have been developed in the literature for modeling nonlinear systems. Perhaps the most common are those based on the Volterra and Wiener theory of nonlinear system identification (Bendat, 1990; Chen, 1989; Jeruchem *et al.*, 2000; Schetzen, 1981; Shi and Sun, 1990). A detailed discussion on different modeling approaches for nonlinear systems is presented in Chapter 3.

Table 1.1 Periodic steady-state methods for various nonlinear circuits
(Mayaram, 2000)

Simulation Method	Nonlinear Device
Time Domain	Amplifiers, Mixers, Oscillators
Frequency Domain	Mildly Nonlinear Amplifiers, Mixers, Oscillators
Mixed Time–Freq	Mixers, Switched Capacitor
Envelope	Oscillators, PLLs
Linear Time-Varying	Mixers, Switched Capacitor

1.3.4 Simulation of Nonlinear Circuits

Simulation methods of nonlinear circuits are usually divided into main categories: time domain methods and frequency-domain methods. In time-domain methods, a circuit is simulated by solving the system differential equations numerically with a time discretization method (Kundert *et al.*, 1989). On the other hand, frequency-domain methods use the Fourier analysis to solve the circuit equations in the frequency domain. The two methods produce a periodic steady-state solution to the circuit equations. The main difference between these two methods is that time-domain methods can deal with strongly nonlinear circuits and discontinuities while frequency-domain methods deal with circuit components that can be characterized in the frequency domain.

Other approaches include mixed frequency–time approaches and circuit envelope methods which use a combination of both time and frequency methods to simulate a circuit and aim at reducing the drawbacks of each method in specific applications. In the following, a brief description of these methods is presented. Table 1.1 shows the applications of each of the presented simulation methods.

1.3.4.1 Time-Domain Methods

Time-domain methods are used to find the steady-state solution of the underlying differential equations of a circuit assuming that the solution is periodic. In these methods, a set of nodal voltages are determined such that $\mathbf{v}(T) = \mathbf{v}(0)$, where T is the period and $v(0)$ is the initial condition that forces the solution to be periodic.

The solution of the circuit differential equations is usually done using iterative approaches such as the Newton shooting method that is used by most commercial circuit simulators such as **Spice**® and **Spectre**®. Since this method is iterative, the time needed to reach a solution depends on a prescribed error tolerance.

There are a number of issues associated with time-domain transient simulations. First, time-domain methods have limitations on the size of the circuits to be simulated because large circuits tend to produce large matrices that need to be efficiently manipulated by computers. Therefore, time-domain methods are usually useful for circuits with less than 300 nodes (Mayaram, 2000). Secondly, time-domain methods usually require excessive computational times because they deal with the absolute bandwidth of the signal rather than the baseband bandwidth as in envelope simulations. Finally, time-domain methods have limitations when used for simulation of distortion in nonlinear circuits where special care must be taken when choosing the error tolerance used in obtaining the iterative solution of the circuit differential equations (Mayaram, 2000).

1.3.4.2 Harmonic Balance (HB) Method

Harmonic Balance (HB) is a frequency-domain method for simulation of mildly nonlinear analogue circuits. HB is based on finding the frequency spectrum of the resulting currents in the circuit given a periodic voltage excitation when the circuit is in its periodic steady state. HB methods are useful in computing quantities that define nonlinear distortion such as third-order Intermodulation Distortion (IMD3), third order Input Intercept Point (IIP3) and Total Harmonic Distortion (THD). HB methods can also be used to perform nonlinear noise analysis of analogue circuits and also to compute harmonics, phase noise, and amplitude limits of oscillator (Agilent, 2006). An example of circuit simulators based on HB is Agilent ADS.

In harmonic balance, the system of nonlinear equations is formulated in both time and frequency domains where the linear contributions are calculated in the frequency domain and the nonlinear contributions in the time domain. Therefore, given an input voltage excitation to the circuit in the frequency domain $V(\omega)$, the resulting current $I(\omega) = F(V(\omega))$ in the linear part of the circuit is found by solving circuit equations using the Kirchhoff Current Law (KCL) at each node. KCL is applied for a number of independent frequencies (harmonic) and the current is calculated at those harmonics by solving a system of linear equations. (Gilmore, 1991; Rabaie *et al.*, 1988).

For the nonlinear part of the circuit, the current–voltage calculations are performed in the time domain where the time-domain representation of the input voltage excitation is found using the Fourier series as

$$v(t) = a_0 + \sum_i a_i \cos(\omega_i t) + \sum b_i \sin(\omega_i t) \tag{1.7}$$

where $\omega_i = i\omega_0$ and ω_0 is the fundamental frequency of the input signal. The time-domain waveform is then applied to the nonlinear part of the circuit and the current response $i(t) = f(v(t))$ is determined in the time domain. The time-domain current is then transformed to frequency domain using Fourier transform.

In a next step the current found in the nonlinear part of the circuit is converted to the frequency domain using Fourier transform and then the frequency spectrum of all the currents at a node is balanced at each frequency. This results in a set of linear simultaneous equations that can be solved in the frequency domain. The resulting currents are then converted to the time domain using the inverse Fourier transform.

Harmonic balance simulations represent an acceptable frequency-domain approach for simulating the response of amplifiers, mixers when stimulated by multisine signal sources. However, as the number of input tones increases, the simulation time increases. Hence, HB simulations are not usually practical when simulating non-periodic digitally modulated waveforms since these cannot be represented by discrete tones or need a large number of tones to approximate their continuous spectrum, which means very long simulation times (Yap, 1997).

1.3.4.3 Mixed Frequency–Time Methods

Combinations of both basic methods; the time-domain and frequency-domain methods, result in a new method called the mixed time–frequency method which result from a combination of the circuit envelope method and the quasi-periodic shooting method.

The basic idea of mixed frequency–time methods is that a quasi-periodic signal such as a modulated signal or a multitone signal can be recovered by simulating a finite number of periods of the carrier signal distributed evenly over the period of the modulation signal. For example consider, a two-tone signal of the form (Kundert, 2003)

$$v(t) = \sum_{n=-\infty}^{\infty} \sum_{m=-\infty}^{\infty} V_{nm} e^{j2\pi(nf_1+mf_2)t} \tag{1.8}$$

where f_1 and f_2 are the- fundamental frequencies. Now, if these frequencies are designated as a carrier (f_1) and an envelope of a modulated signal (f_2), then the signal $v(t)$ represents a quasi-periodic signal. The response of the circuit is also quasi-periodic and can then be represented by a Fourier series as (Kundert, 2003)

$$i(t) = \sum_{n=-\infty}^{\infty} \sum_{m=-\infty}^{\infty} I_{nm} e^{j2\pi(nf_1+mf_2)t} \tag{1.9}$$

In mixed frequency–time simulation, the objective is to find the discrete response $i_n = i(nT_1)$ of the circuit instead of the continuous waveform $i(t)$ obtained from sampling $i(t)$ at the carrier frequency, where $T_1 = 1/f_1$. The key idea with mixed frequency–time simulations is that sampling the quasi-periodic two-tone signal at the carrier frequency (f_1) results in a sampled waveform that is periodic at the modulation frequency. This means that if K harmonics are needed to represent the modulation signal (envelope) then the quasi-periodic signal can be recovered by knowing only ($2K + 1$) cycles of the carrier evenly distributed over the period of the modulation (envelope) signal. Hence, the simulation time is determined by the number of harmonics needed to represent the envelope waveform and not the carrier, which means less simulation time than time-domain methods (Kundert, 2003).

One application of mixed frequency–time methods is the distortion analysis of switched-capacitor circuits where the path of the signal is linear but the clock causes switching. This is because the period of the clock is usually orders of magnitude smaller than the time interval of interest and hence, classical circuit simulation algorithms become extraordinarily computationally expensive (Mayaram, 2000).

1.3.4.4 Envelope Methods

Time-domain methods are not suitable for simulation of high-frequency modulated signals because the simulation must follow the fast varying carrier signal. This results in prohibitively large number of time steps and hence, long simulation times and large memory requirements. Due to the fact that in modulated signals a high-frequency carrier is used to carry slowly varying information signals where the carrier does not contain any information, simulation of modulated signals in RF circuits can be done on the envelope level rather than simulating the RF modulated carrier (Berenji et al., 2006).

Circuit envelope simulation methods applies time-domain analysis techniques on top of the harmonic balance simulations where the solution to the circuit equations is represented as the sum of harmonics with time-varying complex envelopes (Yap, 1997):

$$i(t) = \Re \left(\sum_{k=1}^{N} i_k(t) e^{j\omega_k t} \right) \tag{1.10}$$

where $i_k(t)$ represents the time-varying complex envelope of the current waveform that represents the modulation of the carrier. This complex envelope is simulated using time-domain techniques where the time step of the simulation is proportional to the bandwidth of the complex envelope and not to the carrier frequency. This means much lower time steps and hence much lower simulation time. On the other hand, the carrier is simulated either in the frequency domain using HB or in conventional time-domain transient simulators. In this way, the modulation complex envelope is solved for in the time domain instead of simulating multiple separate tones as in the HB where all frequency components must be solved for simultaneously. This means that the matrix sizes used in the HB solution remain reasonable for simulation. The result of circuit envelope simulations is a time-varying frequency spectrum where the time-varying envelope is not represented by additional tones but rather by the time-varying behavior of the spectrum (Berenji *et al.*, 2006; Yap, 1997).

For example, a two-tone signal can be interpreted as a periodically modulated periodic signal by rearranging equation (1.8) as (Kundert, 2003)

$$v(t) = \sum_{n=-\infty}^{\infty} V_n(t) e^{j2\pi n f_1 t} \tag{1.11}$$

where

$$V_n(t) = \sum_{m=-\infty}^{\infty} V_{nm} e^{j2\pi m f_2 t} \tag{1.12}$$

This form is equivalent to representing the two-tone signal by a conventional Fourier series with Fourier coefficients defined for integer multiples of f_1, except that the Fourier coefficients themselves are time-varying. These time-varying coefficients are periodic with period $T_2 = 1/f_2$ and can themselves be represented by a Fourier series (Kundert, 2003).

1.4   Organization of the Book

The remainder of this book consists of 9 chapters. In Chapter 2, an introduction to wireless system standards and signal models is presented. Chapter 3 presents the different models of nonlinearity that will be used in other chapters for modeling and simulation of nonlinear distortion. Chapter 4 provides analysis of nonlinear transformation of deterministic signal including single and multi-tone analysis. Closed-form expressions that relate signal distortion to nonlinear system characteristics are presented for various nonlinear models such as power series, the Volterra model and block models. Chapter 5 discusses the analysis of the response of nonlinear system to random input signals that represent real-world communication signals. This chapter provides the reader with the probabilistic view of nonlinear distortion and presents quantification of distortion using autocorrelation function analysis, nonlinear spectral analysis and the analysis of multichannel nonlinear distortion. Chapter 6 presents the identification of correlated and uncorrelated distortion using the concept of the orthogonalization of the behavioral model. This analysis is used to identify the distortion components responsible for the degradation of wireless system performance. Chapter 7 uses the concepts developed in Chapter 6 and introduces modeling of communication system figures of merit related to nonlinearity such as spectral mask parameters, Signal-to-Noise and Distortion Ratio (SNDR), Noise-to-Power Ratio (NPR), EVM, NF and BER. Chapter 8 provides an introduction to the simulation of communication systems in MATLAB® where the basics of simulation of modern wireless communication systems in Simulink® are presented. Chapter 9 explains how to use MATLAB® to simulate various types of nonlinearity and how to analyze, predict and evaluate the performance of data communication systems under nonlinearity. Chapter 10 explains how to use Simulink® to analyze, predict and evaluate the performance of data communication systems under nonlinearity and provides a comprehensive reference for models of nonlinearity in Simulink®.

1.5 Summary

In this chapter the basics of nonlinearity and nonlinear distortion in wireless communication systems have been presented. The main sources of nonlinearity in wireless systems and their behavior that causes performance degradation have been discussed. It has been shown that nonlinear distortion is one of the most important considerations in the design of wireless systems. It has also been shown that modeling and simulation of nonlinear distortion are important as accompanying activities during the design cycle of any wireless system. This chapter serves as an introduction to the topics that will follow in the subsequent chapters in the sense that all modeling and simulation issues that will be discussed next are justified by the importance of the problem of nonlinear distortion in wireless systems.

2

Wireless Communication Systems, Standards and Signal Models

As shown in the previous chapter, impairments to communication systems due to nonlinearity depend on the type of input signals. Modern wireless standards involve highly complex digital signal processing stages that render the approximation of signals by simple multisine signals or simple bandlimited noise inaccurate in the analysis and simulation of nonlinear distortion. Hence, the study of the effect of nonlinear distortion on the performance of wireless communication systems requires knowledge and understanding of the details of the various blocks in wireless transmitters and receivers that are responsible of signal generation, modulation, multiple access and demodulation. Modern wireless standards involve customized structures for signal processing and multiple access techniques and hence, when the effect of nonlinear distortion on the performance of such systems is to be modeled and simulated, these structures need to be studied and analyzed.

This chapter provides the basics of modern wireless communication systems including transmitter architectures, receiver architectures, digital signal processing and multiple-access techniques. In addition, common wireless and mobile system standards and their signal models are presented in order to provide real-world examples of the effect of nonlinear distortion on wireless communication systems. The material presented in this chapter serves as an introduction to the analysis of wireless communication signals under nonlinear transformation, which will be presented in the following chapters.

2.1 Wireless System Architecture

The basic architecture of a communication system consist of three basic blocks; the transmitter that is responsible for signal conditioning to a form suitable for relaying the signal through a channel, the channel that represents the medium through which the signal propagates to the receiver and the receiver that regenerates the transmitted signal. Figure 2.1 shows the basic blocks in any communication system. The transmitter and

Nonlinear Distortion in Wireless Systems: Modeling and Simulation with MATLAB®, First Edition.
Khaled M. Gharaibeh.
© 2012 John Wiley & Sons, Ltd. Published 2012 by John Wiley & Sons, Ltd.

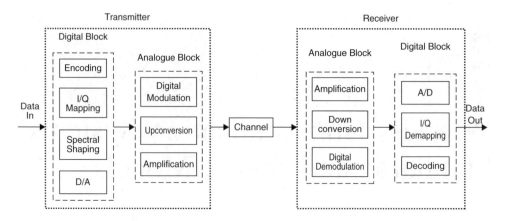

Figure 2.1   Basic blocks in communication systems.

receiver consist of two main blocks: the first is the digital block and the second is the analogue block (Chabrak, 2006). The digital block is responsible of converting the raw data to a form suitable for transmission. In the transmitter side, this includes analogue-to-digital conversion and a variety of encoding processes that converts a raw stream of data to another stream of data that is suitable for transmission and that enhances the transmission process in a noisy channel. In the receiver side, the digital block is responsible for recovering the transmitter raw data after inversion of the data conversion process done at the transmitter. The analogue block is responsible for converting digital data into an analogue signal suitable for transmission through the channel. The channel consists of the medium through which the transmitter signal must propagate to reach the receiver, which may be space, a conductor (such as a copper wire or a transmission line) or an optical fibre that uses light for relaying data.

In wireless systems, the channel is simply the space through which the transmitted signal must propagate to reach the receiver. Therefore, the main task of the transmitter is to convert the signal to a form suitable for transmission through space. This process is executed through the RF stage of the transmitter where modulation, up conversion, amplification and transmission through an antenna take place. At the receiver side, the transmitted signal is first received by an RF stage of the receiver where signal reception through an antenna, amplification, down conversion and demodulation processes are executed. The RF stage in both the transmitter and receiver constitute the bulk of the size, cost and power consumption of the communication system. On the other hand, this stage is usually responsible for the degradation of the system performance and the trade offs in the design of its components represent the bulk of the design process of the overall system.

In the following subsections, a detailed description of the architectures used in the design of RF stages in both the transmitter and the receiver is presented. Discussing the anatomy of these stages will help localize the problem of identifying the blocks responsible for nonlinear distortion, which is the topic of this book.

2.1.1 RF Transmitter Architectures

A digital wireless transmitter consists of three major stages: modulation, up-conversion, power amplification and signal transmission into the air through an antenna. Figure 2.1 shows the main stages of a digital wireless transmitter. The input to the transmitter is generated by a digital signal processor where source coding, channel coding, interleaving, data mapping and pulse shaping take place. The output of the digital signal processor is called a baseband signal that has a low frequency and a low power and is converted to a passband signal that has a high frequency and a high power after the modulation process.

Different classifications of the transmitter architecture of digital wireless transmitters exist (Chabrak, 2006; Cylan, 2005; Razavi, 1998). Wireless transmitters can be classified according to the way the modulator is implemented. In this classification, two architectures exist: the first is the quadrature architecture where I and Q baseband signals are modulated separately and then combined; and the second is the polar architecture where the magnitude and phase of the baseband signal are modulated separately and then combined to produce the modulated signal. Another classification is done according to the way frequency conversion is done. This includes mixer-based architectures and Phase-Locked Loop (PLL) architectures. In the mixer-based approach, two architectures exist: direct conversion and indirect conversion and these differ in complexity and performance. The main difference between these two architectures is that mixer-based transmitters can operate with either linear or nonlinear modulation techniques, while the PLL-based transmitters are limited to constant envelope modulation schemes. However, PLL-based transmitters provide a higher level of integration because they eliminate the need for discrete components. Another classification of wireless transmitters is done according to nonlinearities included in their design. Nonlinearity is mainly related to the PA that is used to produce high-power signals suitable for transmission on an antenna (Cylan, 2005; Razavi, 1998).

The choice of the transmitter architecture depends on many factors, among which circuit complexity, linearity and power efficiency are the most important. Circuit complexity is related to the modulation and up conversion process where baseband signals are converted to passband signals via frequency mixing process. Circuit complexity affects the size and the cost of the transmitter that are limiting factors when mobile hand-held devices are considered. Power efficiency and linearity are usually related to the RF power amplification stage of the transmitter and they are also important criteria to be considered in the design of wireless transmitters, especially for battery-powered hand-held devices which are required to have minimal power consumption in order to expand the battery life. Power efficiency and linearity are always conflicting requirements since high efficacy transmitters are usually not power efficient and vice versa. Therefore, the performance of a wireless transmitter is usually quantified by parameters that are related to power efficiency and linearity, such as the maximum transmitted power and the amount of nonlinear distortion introduced by the nonlinear devices incorporated in the design of the transmitter (Chabrak, 2006; Cylan, 2005; Razavi, 1998).

In the following subsection the most common architectures for linear and nonlinear wireless transmitters are described with respect to their benefits and drawbacks. The choice

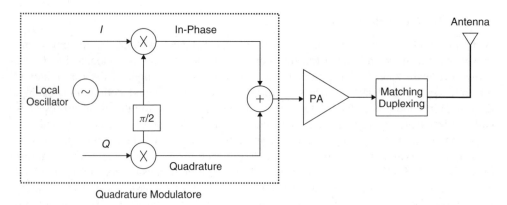

Quadrature Modulatore

Figure 2.2 Zero-IF transmitter architecture (Cylan, 2005). Reproduced by permission of © 2005 Springer.

of the transmitter architecture is important for proper system operation, performance improvement and reduced circuit complexity.

2.1.1.1 Mixer-Based Transmitters

Direct Conversion (Homodyne) Architecture

Direct conversion transmitters generate the high-frequency signal by the modulation process in one step as shown in Figure 2.2. The baseband signal, which consists of In-phase (I) and Quadrature (Q) baseband data, modulates a carrier generated by an RF Local Oscillator (LO). The modulated high-frequency signal is then power amplified and transmitted to an antenna through a matching circuit. The modulation process is carried out by two mixers and an adder where the I channel modulates a carrier with frequency f_c and the Q channel modulates a phase shifted version of the same carrier by 90 degrees. The resulting modulated signals are then added together to form the passband signal. In some systems, the resulting modulated signal is bandpass filtered in order to suppress noise and distortion that exist within the receiver band. Homodyne transmitters usually have the disadvantage of LO pulling by the power amplifier. This problem can be solved by the proper shielding of the LO and by using offset LO technique, however, this imposes additional circuit complexity on the transmitter design. Other disadvantages of homodyne transmitters include carrier leakage, phase-gain mismatch and the need for gain control at RF frequencies. The main advantage of the direct conversion transmitter is its simplicity with respect to the amount of circuitry required in the design. On the other hand, direct conversion transmitters are suitable for high-level integration because they need a low number of passive elements (Cylan, 2005).

Examples of systems that use homodyne transmitters are the Philips Digital Enhanced Cordless System (DECT) and Lucent GSM system.

Two-Step Conversion (Heterodyne) Architecture

In a heterodyne transmitter the baseband I and Q signals are first modulated at an Intermediate Frequency (IF) and the resulting signal is then up converted to the transmit RF

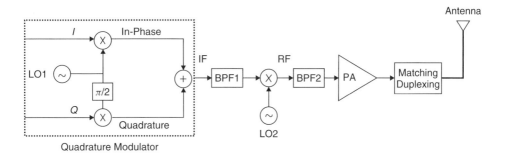

Figure 2.3   Heterodyne transmitter architecture (Cylan, 2005; Razavi, 1998).

frequency. Therefore, in this design, two LOs and two mixers are required. In addition, and two bandpass filter are needed after each of the mixing processes in order to eliminate harmonics and unwanted sidebands that result from the mixing process. Figure 2.3 shows a block diagram of a heterodyne transmitter. This transmitter enjoys the advantage of having a carrier frequency at the PA output that is different from the LO frequency used in quadrature modulation, which means that LO pulling does not occur. On the other hand, this architecture allows gain control to be done at a much Lower Frequency (IF) than the case of homodyne transmitters where gain control is done at the RF frequency.

Two-step conversion architectures have many disadvantages. The main problem with these architectures is the increased circuit complexity in comparison to homodyne transmitters since more circuitry is employed in the design, especially bandpass filters, which are required to have high out of band suppression. These filters are usually difficult to achieve at RF and hence they are usually implemented off-chip, which means that heterodyne transmitters allow only a low-level integration (Cylan, 2005; Razavi, 1998).

2.1.1.2   Linear and Nonlinear Transmitters

Nonlinearity in an RF transmitter usually refers to the nonlinearity of the PA, hence a transmitter that uses a nonlinear PA is called a "nonlinear transmitter" and the one that uses a linear PA is called a "linear transmitter". Nonlinear PAs are highly power efficient, however, they produce nonlinear distortion, which is undesirable in any communication system. Nonlinear distortion is usually amplitude dependent and hence, nonlinear amplifiers are more suitable for systems that use constant envelope modulation such as Frequency Shift Keying (FSK), Minimum Shift Keying (MSK) and Gaussian Minimum Shift Keying (GMSK). An example of such systems is the GSM system that uses GMSK which is a constant envelope modulation. If the modulated signal envelope varies with time, such as M-ary Phase Shift Keying (PSK), Quadrature Amplitude Modulation (QAM), etc., which are used in systems like EDGE, WCDMA and CDMA2000, then a linear PA is required. A detailed analysis of nonlinear distortion will be presented in the next section.

Nonlinear Transmitters

In a nonlinear transmitter the modulated signal is passed through a bandpass filter and a limiter amplifier before power amplification, as shown in Figure 2.4. The bandpass

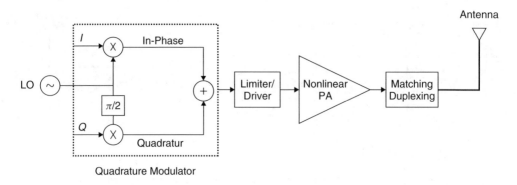

Figure 2.4 Nonlinear transmitter architecture (Cylan, 2005).

filter suppresses harmonics that result from the modulation process that may cause further harmonics and intermodulation products to be introduced if amplified by the PA. The limiter amplifier eliminates any envelope fluctuations in order to maintain the operation of the PA within the high-efficiency region of operation. One important feature of nonlinear transmitters is that the output power of the PA is not adjusted by the power of the input signal but rather by a control voltage which can be applied to the PA. Changing the control voltage will cause the bias of the PA to change and hence a change in the PA characteristics (including an increase in the output power) is obtained (Chabrak, 2006; Cylan, 2005; Razavi, 1998).

Linear Transmitters

A linear transmitter uses a linear PA. Linear PAs are usually operated with some power back-off in order to deliver linear output power (Chabrak, 2006; Cylan, 2005; Razavi, 1998). The linearity of the PA in a linear transmitter means that the PA does not produce harmonics or intermodulation products, which means that the need for a BandPass Filter (BPF) after the modulator is eliminated. On the other hand, a linear PA has a constant gain and hence, its output power can only be adjusted by adjusting its input power. Therefore, the limiter amplifier used in a nonlinear transmitter is replaced by a Variable Gain Amplifier (VGA) which is used to adjust the operating input power of the PA that controls its output power (Chabrak, 2006; Cylan, 2005; Razavi, 1998). Figure 2.5 shows a block diagram of a linear RF transmitter which uses a direct conversion architecture.

2.1.2 Receiver Architecture

A digital wireless system receiver usually consists of three main stages: an RF front end, an Analogue-to-Digital Converter (ADC) and a digital front end, as shown in Figure 2.6. The RF front end is responsible for amplification, filtering and down conversion of the received RF signal to some Intermediate Frequency (IF). The design of this stage depends on the required signal quality, which is mainly affected by RF imperfections and noise considerations. The ADC stage is mainly responsible for demodulation of the resulting IF signal and conversion of the analog signal into the digital domain through sampling and

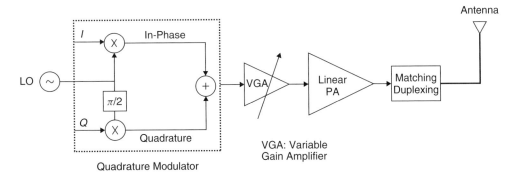

Figure 2.5   Linear transmitter architecture (Cylan, 2005).

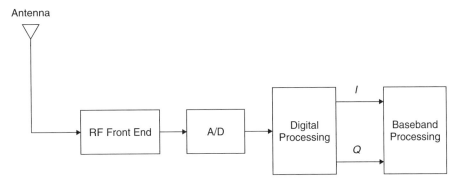

Figure 2.6   Wireless receiver architecture (Cylan, 2005).

quantization processes. The digital front end constitutes the stage where digital processing of the received data is performed in order to reproduce the digital baseband data.

Wireless receivers are classified according to their IF stage. Perhaps the most common and the oldest architecture is the heterodyne concept that includes two mixing and filtering stages. However, advances in silicon technology made possible the development of zero-IF architectures, which provides reduced complexity, lower power consumption and a high level of integration (Chabrak, 2006; Cylan, 2005).

2.1.2.1   Heterodyne Receiver

A heterodyne receiver performs two main tasks, the first is down conversion of the received signal to baseband and the second is the selection of the desired frequency channel in an FDMA system. Figure 2.7 shows a block diagram of a superheterodyne receiver where an RF stage selects the RF band of interest that is then amplified by a LNA and then fed to an IF stage where down conversion to an IF frequency is performed. The RF stage partially filters out undesired signals that fall outside the RF band of interest. The IF stage consists of a down conversion mixer, an IF filter and an IF amplifier. The IF

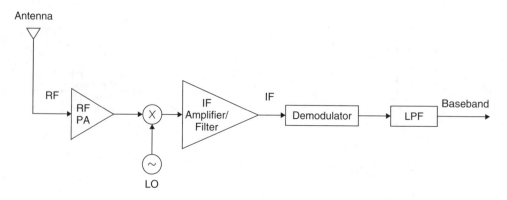

Figure 2.7 Superheterodyne receiver architecture (Cylan, 2005; Razavi, 1998).

filter is responsible for the selection of the desired frequency channel at a fixed frequency which is the IF. The resulting signal is the frequency channel of interested modulated at the IF frequency. This signal is then demodulated to baseband using either coherent (which requires another mixing stage) or non coherent demodulation using either envelope detection or the use of nonlinear devices (Chabrak, 2006; Cylan, 2005).

The main feature of this architecture is that channel selection is done at a single frequency (the IF) instead of tuning the filter that performs the channel selection at RF. This is because it is usually much easier and more cost-effective to implement a band pass filter with good selectivity at a fixed center frequency, rather than to improve the selectivity of the RF stage.

The selection of the IF frequency depends on the band of frequency to which the desired channel belongs that determines the bandwidth of the RF stage. This is due to the image rejection problem that results from the mixer characteristics where a channel at the (image frequency) is down converted to the same IF band. The image frequency is the carrier frequency centered at $f_c + 2 f_{IF}$ where f_c is the desired channel frequency. Therefore, the RF stage must filter out the interfering channels at the image frequency in order to prevent interference at IF. Hence, depending on the bandwidth of the RF stage, the IF frequency is selected in order to ensure that the image frequency is rejected.

In superheterodyne receivers it is possible to use more than one IF stage in order to relax the specifications of filters incorporated in the design resulting in multi stage heterodyne receivers. However, the image frequency problem is still the same in these receivers and hence, image rejection filters are needed between multiple IF stages.

An example of a wireless system that uses a superheterodyne receiver is the *IEEE 802.11* WLAN system.

2.1.2.2 Zero-IF Receiver

A zero-IF receiver (also known as a homodyne receiver) performs the frequency down conversion of the received signal in one step to baseband. Figure 2.8 shows a block diagram of a zero-IF receiver where the received consists of a single down conversion step using one mixing stage. The benefits of having zero-IF are mainly related to the

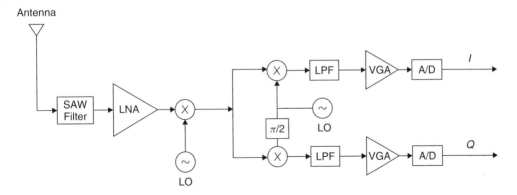

Figure 2.8 Zero-IF receiver (Chabrak, 2006).

simplicity of the design which results in a high-level of integration required for modern wireless systems. In a zero-IF receiver, the need for image rejection filters and mixers is eliminated; and this simplifies the circuit design. On the other hand, IF filters are replaced by low-pass filters that are more suitable for monolithic integration and low power consumption applications. The main advantage of zero-IF receivers with regard to the LNA design is that the LNA is not required to be matched to 50 Ω loads when dealing with off chip image rejection filters (Chabrak, 2006; Cylan, 2005).

Despite the simplicity of zero-IF receivers, there are a number of disadvantages associated with performing the down-conversion in one step. The first is the effect of DC offset voltages that can corrupt the received signal and saturate the following stage of the receiver. These DC offset voltages result mainly from the second order intermodulation products in the mixer that lie at DC and have a time varying nature. Another disadvantage is the I–Q mismatch, when the zero-IF receiver is used with quadrature modulation, which degrades the received signal constellation, resulting in higher system error rates (Chabrak, 2006; Razavi, 1998).

An example of a wireless system that uses zero-IF receivers is the UMTS mobile system.

2.1.2.3 Low-IF Receiver

In low-IF receivers, the IF frequency is selected to be very low in order to allow down conversion to baseband by digital mixing. Figure 2.9 shows the structure of a low-IF receiver where the LNA is followed by a mixing stage for down conversion to a low IF frequency. Then the signal is down converted to baseband and low-pass filtered in the digital domain after ADC. The image frequency problem of superheterodyne receiver is dealt with in Digital Signal Processing (DSP) where image reject down conversion does not produce self-mixing problems. Therefore, if bandpass filtering is applied after the first mixing stage, I/Q mismatch and DC-offsets are eliminated (Chabrak, 2006). Low-IF receivers enjoy all the advantages of direct conversion receivers and at the same time, enjoy all the weaknesses of direct conversion receivers. However, since image rejection is done in

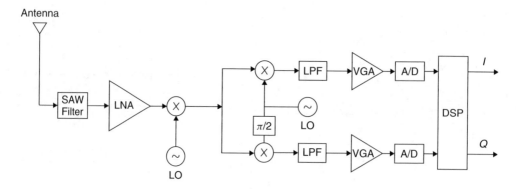

Figure 2.9   Low-IF receiver (Chabrak, 2006).

the digital domain, low-IF receivers can outperform direct conversion receivers if high-performance ADC and low IQ phase and gain mismatch are employed. On the other hand, since the IF signal has to be digitized (bandwidth of at least twice that of the baseband signal), the ADC has to have higher bandwidth capabilities, which makes low-IF receiver more suitable for narrowband systems.

An example of a wireless system that uses low-IF receivers is the GSM mobile system.

2.2   Digital Signal Processing in Wireless Systems

Digital wireless transceiver technology has become the dominant technology in most modern wireless system given the advances in hardware and digital signal processing in recent decades. Digital transmission offers several advantages over analogue transmission such as higher spectral efficiency, the ability for data error correction, resistance to channel impairments, the ability for more efficient multiple access strategies, and better data security. On the other hand, DSP hardware has become cheaper, more compact in size and faster than ever, and this has been reflected in the design of the overall wireless system.

The DSP block of a digital wireless transmitter is responsible for a number of operations on the baseband signal that precede the up conversion process. Figure 2.10 shows a block diagram of the DSP functionalities included in most wireless systems (Briffa, 1996). These operations are done using digital circuitry and their type depends on wireless standard specified for a given wireless system. The DSP block of a wireless system usually consists of Analogue-to-Digital (A/D) conversion, data encoding, baseband pulse shaping and digital modulation. The digital modulation block is responsible for mapping information bits to an analogue signal suitable for transmission over a wireless channel. However, given the sophistication of the mobile environment (channel impairments), the

Figure 2.10   DSP functionalities included in most wireless systems.

demand for high data rates and the need for a high spectral efficiency, modern wireless systems employ different combinations (and possibly interchanging roles) of coding and modulation techniques. These techniques include using spread spectrum, space time coding and multi carrier techniques.

2.2.1 Digital Modulation

Digital modulation techniques are classified according to the amplitude variability of the modulated signal into two main categories: linear and nonlinear techniques. In linear modulation the data bits modulate the amplitude or the phase of the carrier (amplitude and phase modulation), while in nonlinear modulation, data bits modulate the frequency of the carrier (frequency modulation). Hence, a linearly modulated signal usually has a non constant envelope, while a nonlinearly modulated signal has a constant envelope.

The choice of the modulation technique depends on many factors among which, spectral efficiency and power efficiency are the most important in wireless and mobile system applications. These are often conflicting requirements in the sense that linear modulation usually has better spectral efficiency than nonlinear modulation since the latter leads to spectral broadening. However, given the non constant envelope of the linearly modulated signals, they are more susceptible to channel impairments such as fading and interference. On the other hand, since information is embedded in the signal envelope, linear modulation requires the use of linear power amplifiers that are more expensive and less power efficient. Another tradeoff factor is the complexity of the receiver with respect to the need for a coherent phase reference with the transmitted signal. Systems that use coherent demodulation are usually more complex than systems that use non coherent demodulation. Thus, a tradeoff has to be made between spectral efficiency, power efficiency, robustness against channel impairments and cost (Goldsmith, 2005).

The power and spectral efficiency of a modulation technique is determined by the constellation size, which determines how many bits are transmitted per signal. This is defined as M-ary modulation where M is the number of states (levels, phases or frequencies) of the transmitted signal and $k = \log_2 M$ is the number of bits per transmitted symbol. The larger values of M means that more bits can be transmitted in a given bandwidth, which means higher spectral efficiency. However, larger values of M makes the transmitted signal more susceptible to noise and fading, which means a higher SNR is required for proper detection at the receiver.

A general model for a digitally modulated signal is given by

$$x(t) = A(t) \cos(\omega_c t + \omega(t)t + \theta(t)) \tag{2.1}$$

where $A(t)$, $\omega(t)$ and $\theta(t)$ represent the modulated amplitude, frequency and phase of a carrier by the information signal. Phase and frequency modulation can be combined in one term as angle modulation. Thus, the modulated signal can be written as:

$$x(t) = A(t) \cos(\omega_c t + \phi(t)) \tag{2.2}$$

where $\phi(t) = \omega(t)t + \theta(t)$. Using some trigonometric identities, the modulated signal $x(t)$ can be written as

$$x(t) = I(t) \cos(\omega_c t) + Q(t) \sin(\omega_c t) \tag{2.3}$$

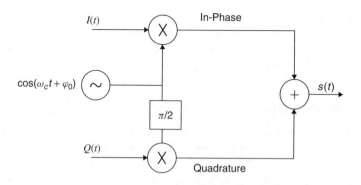

Figure 2.11   Quadrature modulator.

where $I(t) = A(t)\cos(\phi(t))$ is called the in-phase component and $Q(t) = \sin(\phi(t))$ is called the quadrature component. This formulation enables the realization of any modulation technique using quadrature modulators where the input baseband data is split into two data streams and then each stream modulates the amplitude of two carriers having the same frequency but with 90 degrees of phase shift.

Figure 2.11 shows a block diagram of a quadrature modulator where the baseband signals $I(t)$ and $Q(t)$ are up converted to a carrier frequency (ω_c) using two mixers. The first mixer multiplies the LO output by the in-phase signal and the other mixer multiplies the LO output after a 90° phase shift by the quadrature signal. The outputs of the mixers are then summed to yield the modulated signal. The benefit of quadrature modulators is that the I and Q signals are modulated on the same carrier, however, they are orthogonal to each other.

Modulated signals can also be written in terms of their complex envelope. The real signal $x(t)$ in Equation 2.2 can be written as the real part of a complex signal

$$x(t) = \Re[A(t)e^{j\omega_c t}e^{j\phi(t)}] \tag{2.4}$$

$$= \Re[\tilde{x}(t)e^{j\omega_c t}] \tag{2.5}$$

The complex envelope $\tilde{x}(t)$ can therefore, be used to represent the RF signal without reference to the carrier signal. The complex envelope signal is given by

$$\tilde{x}(t) = I(t) + jQ(t)$$

$$= A(t)e^{j\phi(t)} \tag{2.6}$$

where the amplitude and phase of the modulated carrier can be written in terms of the complex envelope of the signal as

$$A(t) = \sqrt{I^2(t) + Q^2(t)}$$

$$\phi(t) = \tan^{-1}\left[\frac{Q(t)}{I(t)}\right]. \tag{2.7}$$

2.2.1.1 Linear Modulation

In linear modulation the amplitude or the phase of a carrier is modulated by baseband data. A quadrature modulator maps the input data stream into two data streams I_i and Q_i, where i denotes the i-th symbol period (T_s) and then a group of bits ($k = \log_2 M$) are mapped into a certain phase or amplitude of the transmitted signal $x(t)$ for $0 \leq t \leq T_s$ (Briffa, 1996). Therefore, the transmitted signal in a symbol period T_s is given by

$$x_i(t) = I_i \cos(\omega_c t) + Q_i(t) \sin(\omega_c t) \tag{2.8}$$

Generally, there are three possible modulation formats that can be generated from this mapping: M-ary Amplitude Shift Keying (MASK) or M-ary Pulse Amplitude Modulation (MPAM), M-ary Phase Shift Keying (MPSK) and M-ary Quadrature Amplitude Modulation (MQAM). In MASK, the quadrature component is set to zero, and hence the transmitted signal can be written as:

$$x_i(t) = A_i \cos(\omega_c t), 0 \leq t \leq T_s \tag{2.9}$$

where $A_i = (2i - 1 - M)d, i = 1, 2, ..., M$ is a set of amplitudes of the transmitted symbols taking M levels where each group of $k = \log_2 M$ bits are transmitted as one symbol in a symbol time equals $\log_2 M T_b$; where T_b is the bit period. Note that the set A_i defines the signal constellation which is one-dimensional in this case as shown in Figure 2.12. The parameter d defines the distance between symbols on the signal constellation that is typically a function of the signal energy. MPSK encodes the information into the phase of the carrier and hence, the amplitude of the carrier remains constant. Thus, the transmitted signal can be written as

$$x_i(t) = A \cos\left(\omega_c t + 2\pi \frac{i-1}{M}\right), 0 \leq t \leq T_s$$
$$= A \cos\left(2\pi \frac{i-1}{M}\right) \cos(\omega_c t) - A \sin\left(2\pi \frac{i-1}{M}\right) \sin(\omega_c t) \tag{2.10}$$

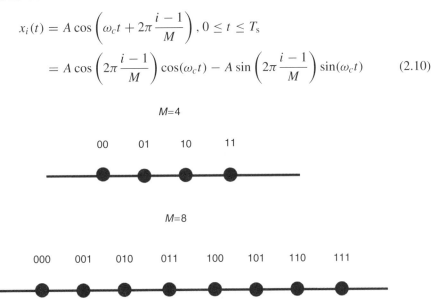

Figure 2.12 Constellation diagram of MASK signals (Goldsmith, 2005).

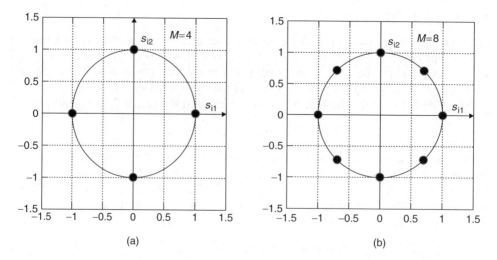

Figure 2.13 Constellation diagram of MPSK; (a) QPSK and (b) 8-PSK.

Each symbol is thus represented by a two-dimensional point (s_{i_1}, s_{i_2}) where $s_{i_1} = A\cos(2\pi\frac{i-1}{M})$ and $s_{i_2} = A\sin(2\pi\frac{i-1}{M})$. Note that the MPSK modulation maps each combination of $k = \log_2 M$ bits into a phase $\theta_i = 2\pi\frac{i-1}{M}$ that defines the angle of each symbol in the signal constellation. These phases define the different states of the transmitted signal and are uniformly distributed in the unit circle, as shown in Figure 2.13.

If the I_i and Q_i are allowed to take the following values $(2i - 1 - L)d, i = 1, 2, \ldots, L = 2^l$, this results in QAM where the transmitted signal in a symbol period is given by

$$x_i(t) = A_i \cos\left(\omega_c t + 2\pi\frac{i-1}{M}\right), 0 \le t \le T_s$$

$$= A_i \cos\left(2\pi\frac{i-1}{M}\right)\cos(\omega_c t) - B_i \sin\left(2\pi\frac{i-1}{M}\right)\sin(\omega_c t) \qquad (2.11)$$

MQAM maps each group of bits $(k = \log_2 M)$ into a symbol represented by an amplitude (A_i) and an angle $\left(\theta_i = 2\pi\frac{i-1}{M}\right)$ and hence, the signal constellation will have a square shape with $M = 2^{2l} = L^2$ constellation points, as shown in Figure 2.14

The constellation diagram gives the location of different states of a linearly modulated signal and hence, it gives an indication of the capability of the receiver for correct detection of the transmitted symbols. The location of these states gives an indication of the probability of symbol errors, which is governed by the distance between adjacent constellation points. For a given power level of the transmitted signal, this distance depends on the mapping rule and on the number of states (M). Hence, a densely packed constellation diagram (large M) (which means high bandwidth efficiency) results in higher error rates. Therefore, the power level of the modulated signal must be increased so that the distance between constellation points increases in order to allow low error rates. Increasing the power levels means a tradeoff between power efficiency and bandwidth efficiency.

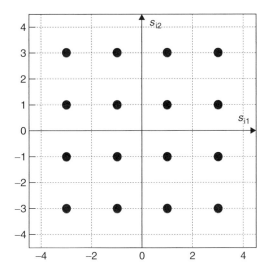

Figure 2.14   Constellation diagram of MQAM ($M = 16$).

Although the constellation diagram is useful in representing the states of a given modulation scheme, it does not give any indication about the transition from one state to another. The transition from one state to another is mainly governed by the bit mapping and also by the pulse shaping involved in the modulation process. Pulse shaping, which will be discussed in the next subsection, is the process of converting the square pulses of digital data into another shape for bandwidth reduction consideration. Pulse shaping results in a time-varying signal envelope where the transition between states on the constellation diagram determine the range of these variations which is characterized by the Peak-to-Average Power Ratio (PAPR) of the modulated signal. The PAPR of a modulated signal defines how much the transmitted signal is affected by nonlinear amplification, where the increase in this ratio causes the input signal to drive the amplifier into a nonlinear region of operation. If the transition between constellation points passes through zero (as in QPSK), the signal will have a high PAPR and this means that the signal will drive an amplifier more into its saturation region. Driving an amplifier into saturation produces a clipping effect to the input signal in the time domain and this is manifested as nonlinear distortion or spectral broadening in the frequency domain that is undesirable (Boccuzzi, 2008; Briffa, 1996).

A solution to the transition of constellation points through zero in QPSK is to limit the transition to 90° instead of 180° by proper bit mapping, which results in OQPSK modulation as shown in Figure 2.15. Other solutions include the introduction of $\pi/4$ shifted QPSK and differential encoded $\pi/4$ shift QPSK (DQPSK) that also aim at avoiding the zero crossing in the constellation diagram of QPSK (Briffa, 1996).

2.2.1.2   Nonlinear Modulation

In nonlinear modulation, information bits are mapped into the frequency of the transmitted carrier. In M-ary FSK modulation, each symbol of $k = \log_2 M$ bits, is mapped

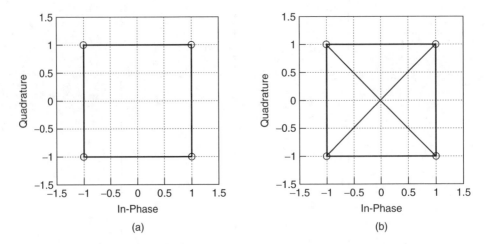

Figure 2.15 Constellation diagrams and state transitions of (a) QPSK and (b) OQPSK.

into a frequency deviation $\alpha_i \Delta\omega$ from a carrier frequency ω_c where $i = 1, 2, ..., M$ is the index of the transmitted symbol and $\alpha_i = (2i - 1 - M)$. Hence, an FSK signal is represented as:

$$x_i(t) = A \cos(\omega_c t + \alpha_i \Delta\omega + \phi_i), 0 \le t \le T_s \qquad (2.12)$$

where ϕ_i is the phase associated with the i-th carrier and $T_s = kT_b$. Figure 2.16 shows the basic concept of FSK modulation.

Note that with the arbitrary choice of the frequency deviations $\alpha_i \Delta\omega$, the transmitted carriers at each symbol period will not be orthogonal (which is required for coherent demodulation) since the basis functions that are in this case represented as $\phi_j(t) = \cos(\omega_j t + \phi_j)$ are not orthogonal ($<\phi_i(t), \phi_j(t)> \neq 0$) over the symbol period T_s. In order to have orthogonal basis functions, a minimum separation between different carrier frequencies of $2\delta\omega = \min_{ij} |\omega_j - \omega_i| = \pi/T_s$ is required. In this case, FSK is called Minimum Shift Keying (MSK).

Note also that the frequency-modulated signal has constant amplitude, hence, nonlinear amplifiers can be used instead of linear amplifiers that have low power efficiency since the signal is less sensitive to amplitude noise and distortion. The main drawback of frequency

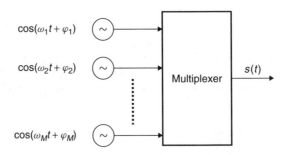

Figure 2.16 A block diagram of an FSK modulator (Goldsmith, 2005).

modulation is its high bandwidth requirements over linear modulation techniques that are usually more bandwidth efficient.

Another issue associated with FSK is the phase discontinuity of the transmitted carriers from one symbol period to another, say the ith and jth symbols where $\phi_i \neq \phi_j$. Phase discontinuities result in broadening of the frequency spectrum of the transmitted signal, which is undesirable. A solution to this problem is to frequency modulate a single carrier with a modulating waveform, as in analogue FM which results in Continuous Phase Frequency Shift Keying (CPFSK) in the case of MSK (Goldsmith, 2005).

2.2.2 Pulse Shaping

The pulse shape of the baseband signal before digital modulation determines the bandwidth of the transmitted passband signal and also determines the amplitude variations of the signal envelope. If the baseband symbols are rectangular pulses of width T_s, then the modulated signal envelope is constant, however, a rectangular pulse exhibit infinite bandwidth by the properties of Fourier transform. The Fourier transform of a rectangular pulse is shown in Figure 2.17, where it is seen that it has infinite side lobes. Theoretically, these side lobes extend to infinity, however, most of the power that spills over to adjacent channels is contained within the first side lobe. To accommodate this power spill-over, a large bandwidth must be reserved for a given frequency channel, which renders the transmission process inefficient.

Pulse-shaping techniques are used in order to band limit the baseband signal to its main frequency channel and hence, to prevent power spill-over to adjacent channels relative to rectangular pulses. However, bandlimiting causes the signal to be infinite in time by the properties of Fourier transform and thus, it introduces intersymbol interference ISI that increases detection errors. Therefore, a pulse-shaping process must be done in a way such that the signal spectrum is confined to a certain bandwidth and at the same time introduces zero ISI. However, these two requirements cannot be achieved in reality and hence, a tradeoff has to be made.

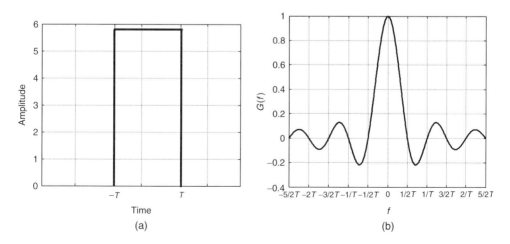

Figure 2.17 Rectangular pulse shaping; (a) impulse response and (b) frequency response.

In mathematical terms, let us assume that the pulse shape is $g(t)$ that also represents the matched filter at the receiver, and given a channel impulse response $h(t)$, the received pulse is $p(t) = g(t) * h(t) * g(t)$. The requirement is to chose $g(t)$ such that the received pulses $p(t)$ have zero ISI given a limited bandwidth of $p(t)$. The Nyquist criterion ensures this requirement when the received pulses have zero ISI only at the ideal sampling times (Briffa, 1996; Goldsmith, 2005; Rappaport, 2000)

$$p(t) = \begin{cases} p_0, & k = 0; \\ 0, & k \neq 0. \end{cases}$$

This can be achieved using Nyquist filters such as the Raise Cosine filter (RC) with an impulse response given by (Rappaport, 2000)

$$p(t) = \frac{\sin \pi t / T_s}{\pi t / T_s} \frac{\cos \alpha \pi t / T_s}{1 - 4\alpha^2 t^2 / T_s^2} \tag{2.13}$$

Its Fourier transform is given by

$$P(f) = \begin{cases} T_s, & 0 \leq |f| \leq (1 - \alpha)/2T_s; \\ \frac{T_s}{2}\left[1 - \sin \frac{\pi T_s}{\alpha}\left(f - \frac{1}{2T_s}\right)\right] & (1 - \alpha)/2T_s \leq |f| \leq (1 + \alpha)/2T_s. \end{cases}$$

where α is called the roll-off factor that controls the spectral roll-off of the pulse and hence its bandwidth. Figures 2.18 shows the time-domain and frequency-domain representations of the RC filter characteristics. RC filters were used in the second-generation mobile standards such as DAMPS and IS-95 systems. Another category of pulse-shaping techniques is the Gaussian pulse-shapes. Gaussian pulse shaping does provide bandwidth bandlimiting, however, it does not satisfy the Nyquist criterion. Hence, a certain level of ISI is introduced and must be tolerated by the receiver if Gaussian pulse shaping filters are to be used. A Gaussian pulse shape is given by (Rappaport, 2000):

$$g(t) = \frac{\sqrt{\pi}}{\alpha} e^{-\pi^2 t^2 / \alpha^2} \tag{2.14}$$

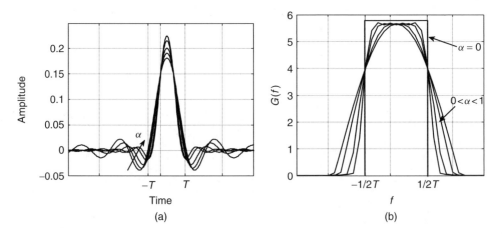

Figure 2.18   RC pulse shaping filter; (a) impulse response and (b) frequency response.

where α is a parameter that controls the bandwidth of the pulse. The Fourier transform of $g(t)$ is given by (Rappaport, 2000):

$$G(f) = e^{-\alpha^2 f^2} \tag{2.15}$$

that also has a Gaussian shape.

Gaussian pulse shaping is used with MSK modulation and in this case the modulation is called Gaussian Minimum Shift Keying (GMSK). GMSK is a well-known modulation format used in GSM standard. Since, Gaussian pulse shaping does not satisfy the Nyquist criterion, high data rates cannot be achieved with GMSK, and hence it only supports voice and low data rate communications.

2.2.3 Orthogonal Frequency Division Multiplexing (OFDM)

Orthogonal Frequency Division Multiplexing (OFDM) is a modulation technique inspired by Multi Carrier Modulation (MCM) where an input bit stream is converted to multiple bit streams with lower data rates and these are modulated on a number of carriers called sub carriers. The benefit of (MCM) is that the transmitted symbols have longer time durations, which makes the modulated signal less affected by channel impairments (specifically multipath effects in mobile communications) that cause ISI. The resulting transmitted signal spectrum will consist of multiple sub channels centered at each sub-carrier and hence, the receiver extracts the data from each sub channel using a bandpass filter. Figure 2.19 shows the spectrum of a MCM signal.

The basic premise of MCM is to transmit a data rate $R = 1/T_b$, where T_b is the bit time, on multiple sub-channels, say N sub channels. In this case each sub channel carries a data rate of R/N, which means that the bit time in each sub-channel is $T_N = NT_b$. If the delay spread of a mobile channel is T_m bandwidth, then by choosing N such that $NT_b \gg T_m$, the system will have almost zero ISI. Thus, a MCM transmission system transmits the same data rate as a single-carrier system, however, ISI caused by multipath in a mobile environment is eliminated.

Although MCM allows higher data rates to be transmitted in a given multipath channel, the need for multiple bandpass filters with high quality factor usually requires guard bands among sub channels which render MCM inefficient with regard to bandwidth occupancy.

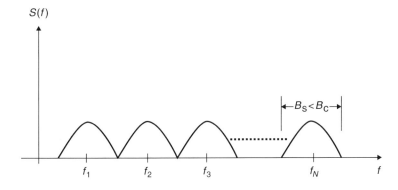

Figure 2.19   Spectrum of multichannel modulated signal (Goldsmith, 2005).

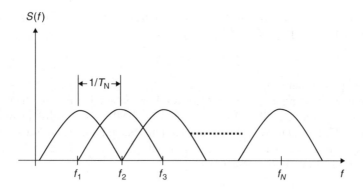

Figure 2.20 Spectrum of an OFDM signal (Goldsmith, 2005).

OFDM is a special form of MCM where the sub carriers are made orthogonal in frequency domain which allows orthogonal detection and hence, eliminates the need for bandpass filters at the receiver. On the other hand, given the orthogonality of the sub carriers, the sub channels are allowed to overlap which result in a better spectral utilization efficiency.

Therefore, given a set of sub carriers $\Phi_i(t) = \cos(\omega_i + 2\pi B_N i + \phi_i)$, $i = 1, 2, \ldots$, an orthogonal set over the symbol interval T_N can be obtained by setting $B_N = 1/T_N$. In this case, the total system bandwidth is N/T_N. Figure 2.20 shows the spectrum of an OFDM signal.

The modulation process in an OFDM system is based on using an Inverse Discrete Fourier Transform (IDFT) that takes multiple input data streams that result from serial/parallel conversion of the original bit stream, and produces orthogonal modulated sub carriers by the properties of DFT. At the receiver, the OFDM signal is processed by DFT, which results in demodulated bit streams and these are then converted to the original bit stream using parallel to serial conversion. Figure 2.21 shows the

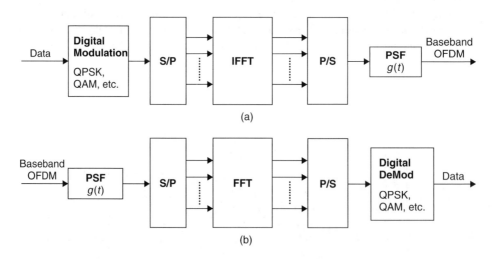

Figure 2.21 OFDM system: (a) transmitter and (b) receiver (Goldsmith, 2005).

modulation/demodulation processes in OFDM systems (Dardari *et al.*, 2000; Goldsmith, 2005; Hasan, 2007). In practice, FFT and IFFT are used instead of DFT and IDFT because these can be implemented on-chip.

To establish the mathematical model of OFDM, consider the OFDM transmitter shown in Figure 2.21, where an input data stream denoted $x(n)$ is first digitally modulated using QAM resulting in a complex symbol stream $X(n), n = 1, 2, \ldots, N - 1$. The resulting symbol stream is broken down into N data streams by an S/P converter where each symbol is to be transmitted over a subcarrier. The output of the S/P converter can be thought of as the Fourier coefficients in a DFT operation. Passing these coefficients into an IDFT results in OFDM symbols of length N of the form:

$$x^m(n) = \frac{1}{\sqrt{N}} \sum_{n=0}^{N-1} X^m(k) e^{j\left(\frac{2\pi nk}{N}\right)}, 0 \le n \le N - 1 \qquad (2.16)$$

Here, m represents the number of OFDM sub-carrier. The symbols $x^m(n)$ belong to an alphabet of M elements, which depend on the modulation format adopted, and have the same probability. Therefore, the complex envelope of the OFDM signal is given by (Dardari *et al.*, 2000)

$$x(t) = I(t) + jQ(t) = \sum_{m=-\infty}^{\infty} \sum_{n=0}^{N-1} x^m(n) g(t - nT - mNT) \qquad (2.17)$$

where T is the channel symbol time and $g(t)$ is a pulse shaping (the waveform used for digital to analogue conversion after IFFT).

The orthogonality of the sub carriers in an OFDM signal can be jeopardized when the signal is passed through a multipath channel. Multipath channels cause transmitted symbols to spread in time which results in ISI. Therefore, a Cyclic Prefix (CP) is usually added to the OFDM signal in order to combat ISI and Inter Channel Interference (ICI). A CP is a copy of the last part of OFDM symbol that is appended to the front of transmitted OFDM symbol. The length of the CP (T_g) is chosen to be longer than the maximum delay spread of the multipath channel. In this case, the ISI is eliminated since the first few samples of the channel output affected by ISI can be discarded without any loss of the original information sequence. However, the addition of CP comes at the cost of a reduction in the data rate since redundant data is transmitted.

An example of a system that uses OFDM is the WLAN standard IEEE 802.11a that occupies 20 MHz of bandwidth in the 5-GHz unlicensed band. This standard uses $N = 64$ sub carrier, of which only 48 sub carriers are used for data transmission, 12 sub carriers are zeroed in order to reduce adjacent channel interference, and 4 used as pilot symbols for channel estimation. The CP consists of 16 samples, so the total number of samples associated with each OFDM symbol, including both data samples and the CP, is 80 (Dardari *et al.*, 2000; Goldsmith, 2005; Hasan, 2007).

2.2.4 Spread Spectrum Modulation

Spread spectrum techniques have become the basis of for most third-generation mobile systems as well as second- and third-generations WLAN standards. Direct-Sequence

Spread Spectrum (DS-SS) is, perhaps, the most popular spread spectrum technique for cellular systems. DS-SS modulation provides ISI resistance capability, narrowband interference rejection as well as bandwidth-sharing capability, which results in a high system capacity.

In DS-SS modulation, spectrum spreading is achieved by multiplying the baseband information signal by a spreading sequence with a much larger bandwidth than the baseband signal. Let the information signal be represented by a binary signal $d(t)$ that takes the values (± 1) with a bit time T and let the spreading sequence be $PN(t)$, then the transmitted signal is (Pickholtz *et al.*, 1991):

$$s(t) = PN(t)Ad(t)(\cos \omega_0 t + \phi_0) \tag{2.18}$$

where A is the amplitude of the BPSK waveform and ϕ_0 is a random phase. The PN spreading sequence waveform is usually pseudo-randomly generated so that each binary chip takes the values of ± 1 and changes every $T_c \ll T$ s with probability $1/2$. Therefore, since $T_c \ll T$, the ratio of the spread bandwidth to the baseband bandwidth is $B_c/B = T/T_c = N$, where N is called the processing gain. Figure 2.22 shows the power spectrum of the signal before and after spreading. At the receiver, despreading produces BPSK signal that is then applied to a standard BPSK detector. Therefore, a DS-SS receiver that has access to the spreading sequence (PN) can retrieve the transmitted data $(d(t))$ by multiplying the received signal by the spreading sequence and then integrating over the symbol period. Thus, if the data symbol in period T is d_1, then:

$$\frac{2}{T} \int_0^T PN(t)s(t)(\cos \omega_0 t + \phi_0)dt = Ad_1 = \pm A \tag{2.19}$$

Figure 2.23 shows the spreading and despreading processes.

DS-SS modulation is considered as a second-level modulation technique in the sense that it follows a standard modulation technique such as BPSK and then a spreading

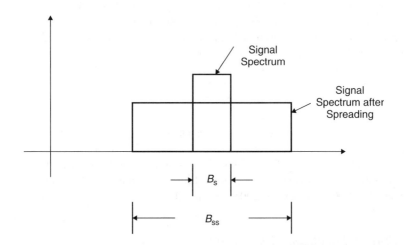

Figure 2.22 Power spectrum of a signal before and after spreading.

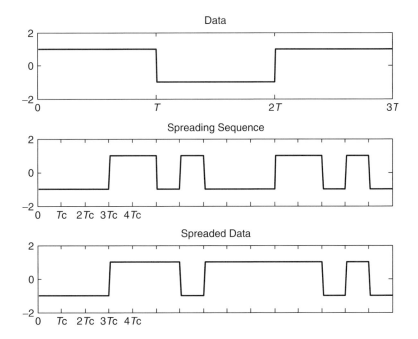

Figure 2.23   Time-domain representation of the spectrum spreading process.

sequence is applied to modulated signal. Note that DS-SS modulation is a form of phase modulation and hence it is usually used with linear modulation formats such as MPSK.

Spreading the spectrum of a baseband signal causes the signal power to spread over a much larger bandwidth than the baseband signal bandwidth. It is this feature that gives DS-SS modulation its essence in the sense that the spread spectrum signal causes little interference to a narrowband user and hence, this characteristic is the basis of the overlay systems that operate spread-spectrum concurrently with existing narrowband systems. On the other hand, the spreading process provides narrowband jammer rejection; where a jammer is rejected by the receiver through the despreading process. The de spreading process has the effect of distributing the interference power over the bandwidth of the spreading code. Hence, the demodulation process effectively acts as a low pass filter where most of the energy of the spread interference is removed and hence, its power is reduced by the processing gain $N = B_c/B$ (Goldsmith, 2005).

ISI caused by multipath effects in mobile wireless systems is also reduced in a DS-SS system based on the same premise. In a multipath channel, replicas of the transmitted signal are received with different delays causing ISI. These multipath components are dealt with as interferers in a DS-SS system where the demodulation process effectively attenuates the multipath component by the processing gain (Goldsmith, 2005; Pickholtz *et al.*, 1991).

DS-SS is in fact used as a multiple access technique when multiple user data is transmitted on the same frequency channel. In this case, DS-SS is called Code Division Multiple

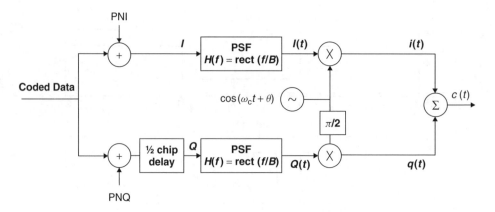

Figure 2.24   A block diagram of an IS95 modulator (Aparin, 2001).

Access (CDMA) where, in addition to the benefits of spreading the spectrum, DS-SS is used to enhance the capacity of a mobile channel by transmitting the data of multiple users (distinguished by their unique spreading codes) on the same frequency channel. CDMA has become an alternative to FDMA and TDMA techniques in mobile systems because of the many benefits it exhibits with regard to system capacity, multipath ISI reduction and narrowband jamming rejection.

2.2.4.1   Signal Model of IS-95 CDMA System

The IS-95 system is 2.5 G mobile phone system based on CDMA technology. According to the IS-95 CDMA reverse-link modulation scheme shown in Figure 2.24. Data is spread using orthogonal Walsh codes and then split into I and Q channels. I and Q data are then spread by the orthogonal PN codes with a chip rate B of 1.2288 Mcps. The Q-channel data is delayed by 1/2 chip resulting in OQPSK spreading. The spread data streams can be represented by impulse trains as (Aparin, 2001)

$$I_{\text{data}}(t) = \sum_{k=-\infty}^{\infty} i_k \delta(Bt + \phi/\pi - k),$$

$$Q_{\text{data}}(t) = \sum_{k=-\infty}^{\infty} q_k \delta(Bt + \phi/\pi - k + \frac{1}{2}) \qquad (2.20)$$

where $\delta(t)$ is the Dirac delta function, ϕ is a random phase uniformly distributed in $(0, 2\pi)$ and i_k and q_k are independent random numbers taking values of $\pm l$ with equal probability. These binary numbers are the results of multiplication of the data symbols with the spreading code chips.

The spread data is then pulse shaped by the IS-95 baseband pulse-shaping filter that can be modeled as an ideal low pass filter with the cutoff frequency $B/2$ and impulse response

$h(t) = B\text{sinc}(Bt)$ where $\text{sinc}(z) = \sin(\pi z)/(\pi z)$. The filtered I and Q spread data are then quadrature modulated, producing a modulated signal of the form (Aparin, 2001):

$$i(t) = \sum_{k=-\infty}^{\infty} i_k \text{sinc}(Bt + \phi/\pi - k) \cos(\omega_c t + \theta),$$

$$q(t) = \sum_{k=-\infty}^{\infty} q_k \text{sinc}(Bt + \phi/\pi - k + 1/2) \sin(\omega_c t + \theta) \tag{2.21}$$

where ω_c is the angular frequency of the carriers and θ is their random phase independent of ϕ and uniformly distributed in $(0, 2\pi)$.

2.2.4.2 Signal Model of Wide band Code Division Multiple Access (WCDMA) Systems

WCDMA is the main modulation/multiple access technique used in most 3G systems. The modulation format used in W-CDMA modulators is the Hybrid PSK (HPSK). The HPSK signal model, assuming that the baseband filters are ideal low-pass filters with a cut off frequency of half the spreading rate, can be described as (Aparin and Larson, 2004):

$$c(t) = i(t) + q(t) \tag{2.22}$$

where

$$i(t) = \sum_{k=-\infty}^{\infty} i_k \text{sinc}(Bt + \phi/\pi - k) \cos(\omega_c t + \theta),$$

$$q(t) = \sum_{k=-\infty}^{\infty} q_k \text{sinc}(Bt + \phi/\pi - k) \sin(\omega_c t + \theta) \tag{2.23}$$

where i_k and q_k are uncorrelated random numbers taking values of $\pm l$ with equal probability, B is the spreading rate, w_c, is the angular frequency of the carriers, θ and ϕ are independent random phases uniformly distributed in $(0, 2\pi)$, and $\text{sinc}(z) = \sin(\pi z)/(\pi z)$.

Due to HPSK mapping, the relationship between the chips can be described as (Aparin and Larson, 2004):

$$i_{2k+1} = i_{2k} r_{2k}$$

$$q_{2k+1} = q_{2k} r_{2k} \tag{2.24}$$

where r_k is a random number independent of i_k and q_k and taking values of ± 1 with equal probability.

2.3 Mobile System Standards

Mobile system standards have evolved through recent decades from the first generation (1G) that only provided mobile voice communications to the latest fourth generation

(4G) that is expected to provide very high speed data transmission enabling seamless multimedia services to users accessing the network through heterogeneous access technologies. The following subsections present the details of major 2G, 3G and 4G mobile system standards.

2.3.1 Second-Generation Mobile Systems

First-generation mobile systems were analogue systems that emerged in the early 1980s and provided only voice communication services. In the 1990s, 1G became obsolete after the penetration of 2G and 2.5G systems that are digital communication systems that use Time Division Multiple Access (TDMA) and provide low data rate communications in addition to voice services. There is a number of systems that belong to the second- and 2.5-generation mobile such as Global System for Mobile (GSM), Enhanced Data rates for Global Evolution (EDGE), Interim Standard (IS)-54 and IS-95.

2.3.1.1 Global System for Mobile (GSM)

GSM was first introduced in 1992 in Europe. Like most 2G systems, GSM is based on TDMA where a frequency channel of 200 kHz bandwidth carries the data of 8 users. Voice data is generated by encoding voice signal samples into 260-bit blocks with duration of 20 ms, which corresponds to a data rate of 13 kbps. Therefore, the maximum data rate is 13 kbps for a full rate speech encoded voice signal. After the addition of error-correction codes, the speech data consists of 456 bits per 20 ms speech frame. Each time frame of length 4.615 ms is divided into 8 time slots where each time slot represents a physical channel of duration 576.92 μs. Each time slot consists of 148 bits, which results in a transmitted data rate of 270.833 kbps. The system uses Frequency Division Duplexing (FDD) where the downlink and uplink use different frequency channels.

The digital GSM signal is modulated using Gaussian Minimum Shift Keying (GMSK). The GMSK modulator uses a Gaussian pulse shaping filter with $\alpha = 0.3$, that is, ($BT = 0.3$ where B is the 3-dB bandwidth of the filter and T is the bit duration). Binary data in a GMSK modulator is represented by shifting the carrier frequency by ± 67.708 kHz, which corresponds to $1/4T = 270.833/4$ (Rappaport, 2000).

Data transmission in GSM is based on two technologies: Circuit Switched Data (CSD) and General Packet Radio Service (GPRS). In CSD a user occupies one time slot during the whole transmission period. This type of data transmission is efficient for real-time applications where data have to be transmitted with constant rate (up to 9.6 kbit/s) during a certain period. High-Speed Circuit Switched Data (HSCSD) is a modified version of CSD where the maximum data rate is 14.4 kbps, which can be achieved using a different type of coding than CSD. The main drawback of CSD is that the user is dedicated a physical channel for data transmission even if he is not transmitting data, which results in inefficient channel utilization.

In GPRS, data is transmitted in packets where the channel is used only when data is to be transmitted, which leads to a better channel utilization than CSD. Therefore, many users can use the same time slot at the same time and a single user can be assigned multiple time slots. Thus, a data rate can theoretically be increased by a factor of 8 (around 100 kbps) (Chabrak, 2006).

The GSM standard is used in about 66% of the world's cell phones, supporting over a billion users in 170 counties (Goldsmith, 2005).

2.3.1.2 Enhanced Data Rates for Global Evolution (EDGE)

EDGE belongs to 2.5-generation mobile systems and is considered as an evolution of GSM. EDGE is designed to support high-rate packet data services through using a high-level modulation format combined with FEC coding. Data modulation is based on 8-PSK instead of GMSK modulation used by GSM. In this way the data rate provided by EDGE is 3 times ($\log_{2}8$) higher than GSM (up to 384 kbps). The evolution from GSM to EDGE with regard to CSD is called in this case Enhanced Circuit Switched Data (ECSD). For packet oriented data transmission, a number of modulation and coding schemes that are either based on GMSK modulation or 8-PSK modulation. GPRS in EDGE systems is called General Packet Radio Service (EGPRS) (Furuskar *et al.*, 1999).

The use of higher-level modulation in EDGE makes the system more sensitive to fading effects. Therefore, EDGE uses adaptive modulation and coding techniques to mitigate this problem. Specifically, EDGE defines 3 different modulation and coding combinations for ECSD (E-TCH/F28.8, E-TCH/F32.0 and ETCH/F43.2) and nine Modulation and Coding Schemes (MCS1 to MCS9) for EGPRS, each optimized to a different value of received SNR. In general, EDGE requires a better transmission quality, which can be achieved by reducing the cell size, and hence increasing the number of the base stations (Chabrak, 2006; Goldsmith, 2005).

2.3.1.3 Interim Standard (IS)-54

The IS-54 standard represents the evolution from analogue 1G systems to digital 2G systems in north America. The IS-54 standard is also called the North American Digital Cellular (NADC) standard. This standard uses the same channel spacing as the first-generation Advanced Mobile Phone System (AMPS) that is 30 kHz, however, each frequency channel supports 6 users using TDMA multiple access scheme. The IS-54 standard was improved over time and evolved into the IS-136 standard uses parity check codes, convolutional codes, interleaving, and equalization. The IS-54 and IS-136 systems provide data rates of 40–60 kbps by assigning multiple time slots to each user and using high-level modulation (Goldsmith, 2005).

2.3.1.4 Interim Standard (IS)-95

The IS-95 standard is considered as the 2G version of CDMA mobile systems that was introduced in the USA, South America and Japan in 1993. IS-95 standard is based on CDMA for multiple access where 64 code channels are transmitted on an RF carrier with a bandwidth of 1.25 MHz. Each user data is spread by a PN code and a Walsh code that provide orthogonal covering resulting in a spread signal that has a chip rate of 1.2288 Mcps. The data rates that can be achieved with IS-95 system is 9.6 kbps that can be increased by 50% to 14.4 kbps by speech coding and interleaving (Rappaport, 2000). The system uses Frequency Division Duplexing (FDD) where the downlink and uplink use different frequency channels.

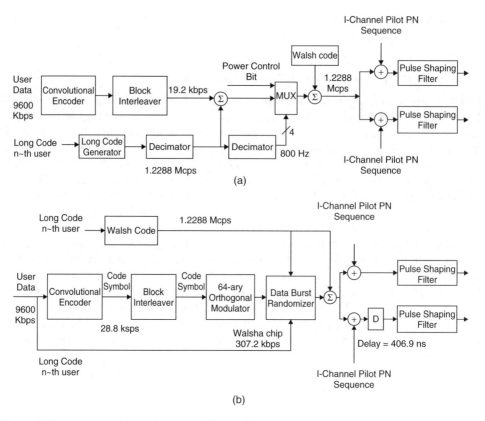

Figure 2.25   IS-95 Modulator: (a) forward link and (b) reverse link (Rappaport, 2000). Reproduced from Wireless Communications: Principles and Practice, Second Edition, Theodore Rappaport, © 2001 Prentice Hall.

Figures 2.25(a) and (b) show block diagrams of the IS-95 modulators for the forward and reverse links. On the forward link, data is first encoded using rate 1/2 convolutional codes, interleaved and then modulated by one of 64 orthogonal Walsh codes. A synchronized scrambling sequence, unique to each cell, is superimposed on the Walsh codes to reduce interference between cells. The spread signal is pulse shaped by an RRC filter (Specifically designed for IS-95 standard) and then modulated using a QPSK modulator. In the reverse link, spreading data is encoded using rate 1/3 convolutional code with interleaving and then spreading is accomplished by modulation by an orthogonal Walsh code and modulation by a nonorthogonal user/base station specific PN code. The spread signal is pulse shaped by the same RRC filter in the forward link and then modulated using a OQPSK modulator.

2.3.2   Third-Generation Mobile Systems

Although 2.5G systems are still being used efficiently throughout the world, some 2.5G systems are being replaced by the new CDMA-based 3G systems that provide higher data

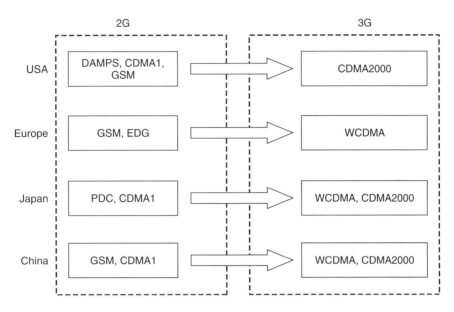

Figure 2.26 Evolution of mobile communication standard from 2G to 3G (Chabrak, 2006).

rates and high system capacity. However, 3G systems are facing standardization problems in addition to their high cost and poor performance. In addition, the development of the 4G systems, which have not been standardized yet, has hindered the deployment of 3G system in many countries in the world. Figure 2.26 illustrates the evolution of mobile phone standards from 2G to 3G (Chabrak, 2006).

The following subsections provide the details of a number of 3G systems that have been adopted in different countries in the world.

2.3.2.1 Universal Mobile Telecommunication System (UMTS)

UMTS represents the European 3G mobile system that is used in EU and in many countries around the world. Like most 3G systems, UMTS uses WCDMA with a channel bandwidth of 5 MHz. Orthogonal codes are used for spreading at a chip rate of 3.84 Mcps. In addition to spreading codes, scrambling codes are also used for identification of base stations. UMTS system uses either Frequency Division Duplexing (FDD) where the uplink and downlink use different frequencies or Time Division Duplexing (TDD) where the uplink and downlink use different time slots in a TDMA frame.

In TDD-WCDMA, each radio channel consists of a frame of 10 ms divided into 15 time slots where a group of time slots can be assigned for the uplink and another group can be assigned to the downlink. In TDD-WCDMA, a combination of CDMA and TDMA can be utilized to maximize the transmitted data rate. The maximum data rate that can be achieved with asymmetrical TDD-WCDMA is 2 Mbps. In FDD-WCDMA, data rate of 12 kbps to 384 kbps can be achieved by varying the length of the spreading codes and the type of channel coding (Viterbi, Turbo, etc.) (Chabrak, 2006; Goldsmith, 2005).

After Time Division Multiplexing (TDM), which results in radio frames of 10 ms length, data is mapped into real and imaginary parts that are then multiplied by a real spreading code. The resulting complex signal from different users and control channels are then applied to complex scrambling. The scrambled data is then applied to pulse shaping and RF modulation. The pulse-shaping filter used is a Root-Raised Cosine (RRC) filter with a roll-off factor $\alpha = 0.22$ (Chabrak, 2006).

Data transmission in UMTS is achieved through High-Speed Downlink Packet Access (HSDPA). HSDPA uses adaptive modulation, coding and spreading code adaptation to maximize the data rate for packet transmission. In the case of good SNR, 16-QAM can be used instead of QPSK and turbo codes with a fixed rate of 1/3 is used with coding adaptation based on different rate matching. The high data rates are achieved by using a low spreading factor that is fixed to 16, which can be extended by multiple channelization codes assignment to one user (Chabrak, 2006).

2.3.2.2 CDMA2000

CDMA2000 represents the 3G version of the 2.5G IS-95 CDMA system where higher data rates are offered. CDMA2000 uses convolutional and turbo codes for channel encoding and adopts a number of modulation formats: QPSK, 8PSK and 16QAM modulation. The main advantage of CDMA2000 over IS-95 is the ability to transmit higher data rates (2.457 Mbps) that can be achieved by reducing the spreading rate (Chabrak, 2006).

There are three CDMA20001x sub standards: CDMA2000 1xRTT, CDMA2000 1xEV-DO, and CDMA2000 1xEV-DV. CDMA2000 1xRTT (Single Carrier Radio Transmission Technology) operates in one pair of 1.25 MHz radio channels (compatible with CDMAOne) and offers low data rates (max: 307.2 kbps) however, it doubles the voice capacity of CDMAOne systems (Chabrak, 2006; Goldsmith, 2005).

In the first substandard, CDMA2000 1xRTT (Single Carrier Radio Transmission Technology) data is modulated using QPSK. Depending on the radio configuration, this substandard can provide a data rate of up to 307.2 kbps. The second substandard, CDMA2000 1xEV-DO (single carrier evolution-data optimized), uses a separate 1.25-MHz dedicated high-speed data channel. This standard uses turbo coding and 16-QAM modulation in order to achieve a data rate of 2.4576 Mbps for downlink and 1843.2 Mbps for uplink (an averaged combined rate of 2.4 Mbps). On the other hand, this standard uses a different frame structure than IS-95 standard where traffic, pilot and control channel are time multiplexed instead of code multiplexing. In this standard, two different data rates can be achieved: 307.2 kbps and 2457.6 kbps. With the first data rate, 2 time slots are used for transmission with QPSK used for modulation, while with the second 1 time slot is used for data transmission with 16-QAM modulation. The third substandard, CDMA2000 1xEV-DV (single carrier evolution data and voice), provides integrated voice and simultaneous packet data services (Chabrak, 2006). This substandard supports data rates of up to 4.8 Mbps as well as legacy voice users within the same radio channel. Another enhancement to CDMA2000 is to aggregate three 1.25 MHz channel into one 3.75 MHz channel. This enhancement is called CDMA2000 3X and provides data rates of up to 8 Mbps.

2.3.3   Fourth-Generation Mobile Systems

The fourth generation mobile systems constitute an integration of data communication services and allow most wired and satellite services to be replaced by a single wireless technology. 4G systems consist of all IP packet switched networks, mobile ultra-broadband (gigabit speed) access and multi carrier transmission. 4G standards are expected to achieve data rates of up to 100 Mbps for high-mobility users and up to 1 Gbps for low-mobility users using scalable bandwidths of up to 40 MHz.

The physical layer of a 4G network will replace CDMA technology used in 3G systems by frequency-domain equalization schemes such as OFDMA and also using multiple antennas (Multiple In Multiple Out (MIMO) systems in order to exploit the frequency-selective channel property without complex equalization. Adaptive modulation and coding including turbo error-correcting codes will be used to minimize the required SNR at the receiver side. 4G systems will also use channel-dependent scheduling to utilize the time-varying channel. Multiple access will be done using frequency-domain statistical multiplexing, for example (OFDMA) or (single-carrier FDMA) (Rumney, 2008). Examples of systems that belong to 4G are LTE, WiMAX and UMB (Formerly EV-DO Rev. C).

2.3.4   Summary

Table 2.1 shows the specifications of different 2G mobile standards. Table 2.2 shows the mobile standards that evolved from 2G standards where data communication is introduced and enhanced data rates are obtained using the same infrastructure of 2G systems. Enhanced data rates can be obtained by either aggregating time slots (in GSM or IS-136) or aggregating Walsh codes (in IS95). Table 2.3 shows a summary of the evolution of different mobile standards from 1G to 4G.

Table 2.1   Specifications of 2G mobile communication standard (Rappaport, 2000)

	GSM	IS-136	IS-95
UL Frequency (MHz)	890–915	824–829	824–829
DL Frequency (MHz)	935–960	869–894	869–894
Carrier Separation (kHz)	200	30	1250
Number of Physical Channels	8	6	64
Modulation	GMSK	DQPSK	QPSK/OQPSK
Speech Rate (kbps)	13	7.95	1.2–9.6
Channel Data rate (kbps)	270.833	48.6	1.2288 Mcps
Code Rate	1/2	1/2	1/2 (DL) and 1/3 (UL)
Multiple Access	TDMA	TDMA	CDMA

Table 2.2 Evolution from 2G to 2.5G standard (Chabrak, 2006)

2G	GSM	IS-136	IS-95
2.5G	HSCSD, GPRS, EDGE	GPRS, EDGE	IS-95B

2.4   Wireless Network Standards

Wireless networks have been designed to provide high-speed data communications for either stationary or mobile users. Wireless Local Area Networks (WLAN) provide high-speed wireless transmission within a local area for stationary users or users moving at a pedestrian speed. Most WLANs operate in unlicensed bands of the spectrum such as the ISM band and the NII bands. On the other hand, Wireless Metropolitan Area Networks (WMAN) provide high-speed data communications in wide areas and for mobile users and they use either licensed or unlicensed frequency bands. In the following, a summary of the major wireless network technologies and the main features of their physical layers are presented.

2.4.1   First-Generation Wireless LANs

The first-generation wireless LANs operate within a 26-MHz spectrum of the 900-MHz ISM band. Most of these networks use DS-SS modulation and multiple access and provide data rates of up to 2 Mbps. The first-generation WLANs lacked standardization and hence they were based on incompatible protocols. Examples of first-generation WLANs are the wireless modems known as "packet radio modems" that were developed by amateur radio operators in the early 1980s (Goldsmith, 2005).

2.4.2   Second-Generation Wireless LANs

The second-generation WLANs (in the US) were based on the *IEEE* 802 standards that operate in the unlicensed ISM and NII bands. The most common WLAN standard was the *IEEE* 802.11b (also known as WiFi) that operates in an 80-MHz bandwidth of the 2.4-GHz ISM band. The standard uses DS-SS technology and provides data rates of around 1.6 Mbps and a range of approximately 150 m. Two other standards were developed to provide higher data rates: the IEEE 802.11a and IEEE 802.11g. The IEEE 802.11a operates in 5GHz NII band with 300 MHz of bandwidth and provides a data rate of up to 70 Mbps. This standard uses multicarrier modulation in order to support these high data rates. The 802.11g standard uses multicarrier modulation and operates in either the 2.4 GHz ISM band or the 5 GHz NII band and provides data rates of up to 54 Mbps (Goldsmith, 2005).

The European WLAN standards are based on the High-Performance Radio LAN (HiperLAN) that are similar to the IEEE standards in the US. HiperLAN operate in an

Table 2.3   Mobile communication technologies from 1G to 4G (Akhtar, 2010)

Feature	1G	2G	2.5G	3G	4G
Start	1970–1984	1980–1991	1985–1999	1990–2002	2000–2006
Data rate	–	14.4 Kbps	14.4 Kbps	2 Mbps	2 Mbps
Bandwidth	25–30 kHz	30–200 kHz	30–200 kHz	5 MHz	5–20 MHz
Standards	AMPS, NMT	GSM, IS136, IS95	GSM/GPRS, EDGE	WCDMA, UMTS, CDMA2000	Unified Standard
Technology	Analogue	Digital	Digital	CDMA/IP	IP
Services	Voice	Voice	Voice/Data	Voice/Data	Voice/Data/ Wearable devices
Multiple Access	FDMA	TDMA	TDMA	CDMA	CDMA
Switching	Circuit	Circuit	Circuit	Circuit/Packet	Packet

unlicensed 5-GHz band similar to the NII band in the USA and provide data rates of up to 20 Mbps at a range of around 50 m. HiperLAN type 2 provides a data rate of up to 54 Mbps and supports access to cellular and IP networks (Goldsmith, 2005).

2.4.3   Third-Generation Wireless Networks (WMANs)

The third-generation wireless networks revolve around the IEEE 802.16 standard that is designed to operate in the 10–66 GHz (for line of site connections) and the 2–11 GHz (for non-line-of-sight connections) licensed bands where several hundred megahertz of bandwidth can be used. This standard provides high speed data communications with data rate up to 120 Mbps using frequency channels of 25 or 28 MHz of bandwidth (Hasan, 2007).

2.4.3.1   IEEE 802.16 Standard

The infrastructure of the IEEE 802.16 network resembles a cellular phone network where a number of Base Stations (BS) that are connected to the core network are deployed in a given area. Each BS provides Point to Multi Point (PMP) service where multiple Subscriber Stations (SS) (which can be a customer premise equipment, mobile phone or a laptop) are served.

There are a number of variants to the original IEEE 802.16 standard that evolved through time. These standards vary in the specifications of the physical layer, multiple access and frequency band as well as user mobility. However, most of these standards use OFDM and OFDMA to achieve high-speed data communications in a mobile environment. Examples of these standards and their specifications are shown in Table 2.4 (Hasan, 2007).

Table 2.4 Specification of IEEE 802.16 standards (Goldsmith, 2005; Hasan, 2007). Reproduced from 'Performance Evaluation of WiMAX/IEEE 802.16 OFDM Physical Layer', MSc Thesis, by permission of © Azizul Hasan

Feature	IEEE 802.16-2001	IEEE 802.16a	IEEE 802.16-2004	IEEE 802.16e
Start	2001	2003	2004	2005
Spectrum	10–66 GHz	2–11 GHz	211 GHz	2–6 GHz
Propagation	LOS	NLOS	NLOS	NLOS
Bit Rate	134 Mbps	75 Mbps	75 Mbps	15 Mbps
Modulation	QPSK, QAM	BPSK, QPSK, QAM	OFDM, BPSK, QPSK, QAM	OFDMA, QPSK, QAM
Mobility	Fixed	Fixed	Fixed	Mobile

Table 2.5 Specification of different IEEE 802.16 physical layers (Hasan, 2007). Reproduced from 'Performance Evaluation of WiMAX/IEEE 802.16 OFDM Physical Layer', MSc Thesis, by permission of © Azizul Hasan

Designation	Band	Duplexing
WirelessMANSC	10–66 GHz	TDD, FDD
WirelessMANSCa	2–11 GHz	TDD, FDD
WirelessMANOFDM	2–11 GHz	TDD, FDD
WirelessMANOFDMA	2–11 GHz	TDD, FDD
WirelessHUMAN	2–11 GHz	TDD

The IEEE 802.16 standard supports multiple physical layer specifications that are assigned unique names. The first version of the standard only supported single carrier modulation, however, most of the standards that followed use OFDM and scalable OFDMA in order to provide mobility. Table 2.5 describes the specifications of the different physical layers supported by IEEE 802.16 standard.

2.4.3.2 WiMAX

WiMAX refers to Worldwide Inter-operability for Microwave Access (WiMAX) which is an alliance of telecommunication manufacturers that promotes interoperability of IEEE 802.16 and HiperMAN wireless networks. WiMAX provides wireless networking for users in a wide area and supports mobility through a cellular based network where hand off and roaming are supported. In fact, two types of access are provided: the first is Fixed Access (based on IEEE 802.16-2004) that supports stationary users and operates in 3.5 GHz and 5.8 GHz frequency bands, and the second is the Mobile Access (based on IEEE 802.16e) for mobile users that operates in the 2.3 GHz, 2.5 GHz, 3.3 GHz and

Table 2.6 Main features of WiMAX physical layer (Ghosh, 2005; Hasan, 2007; IEEE 802.162004, 2004)

	Feature	Benefit
Channel Bandwidth	Integer multiples of 1.25 MHz, 1.5 MHz, 1.75 MHz	Flexible
Error Control Mechanism	Reed–Solomon (RS) code and Convolutional Code (CC) and Turbo codes (optional)	Useful for OFDM links in multipath
Modulation and Coding	Seven combinations of modulation (BPSK, QPSK, 16 QAM and 64QAM) and coding rate	Adaptive: tradeoffs between data rate and robustness given channel conditions
Antenna	Adaptive antenna system	Transmission can be done using directed beams
Transmit Diversity	Space Time Block Codes (STBC) can be implemented in the DL	

3.5 GHz frequency bands. The main features of WiMAX physical layer are summarized in Table 2.6 (Ghosh, 2005; Hasan, 2007; IEEE 802.162004, 2004).

2.5 Nonlinear Distortion in Different Wireless Standards

Wireless communication system performance is usually characterized by system BER. The degradation of system BER is due to a number of factors such as receiver noise, phase noise, nonlinear distortion caused by Tx and Rx nonlinearities, channel impairments, interference and spurious signals and many other factors.

The degradation of system BER can be referred to degradation of SNR at the receiver, which is related to system BER by a relation that depends on the type of modulation, multiple access, etc. SNR is measured as the ratio of the useful signal power to the Additive White Gaussian Noise (AWGN) power. Other system metrics are also used to quantify system performance and can directly be related to the effective SNR such as EVM, ρ, NPR and Nonlinear Noise Figure (NNF).

As discussed in Chapter 1, when nonlinear distortion exists, the effective SNR is defined as the ratio of the useful signal power to the sum of AWGN and nonlinear distortion powers. However, from a designer point of view, it is not easy to design a wireless standard for absolute distortion levels since the level of nonlinear distortion depends on the mechanism by which this distortion degrades system performance. Therefore, wireless standards specify minimum requirements of nonlinear distortion based on the different mechanisms by which that distortion is introduced, such as intermodulation of CW interferers, cross-modulation with a narrowband interferer, etc. These standards also specify standard methods of measurement of different nonlinear distortion parameters. Table 2.7 summarizes the different parameters of nonlinear distortion in different wireless standards and their definition. For the minimum requirements of nonlinear distortion, the reader is referred to the wireless standards such as the GERAN standard (GERAN Standard, 2005), the IS-95 standard (IS-95 Standard, 1993) and the 3GPP2 standard (3GPP2 Standard, 1999).

Table 2.7 Nonlinear distortion parameters in different wireless standards (Chabrak, 2006; 3GPP2 Standard, 1999)

Standard	Minimum Requirements	Definition
GSM (800, GSM 950, PCS and DCS) and EDGE	Intermodulation and Spurious Response Attenuation	Receiver ability to receive a CDMA signal in the presence of 2 interfering CW tones located at frequency offsets 0.8 and 1.6 MHz from carrier where their intermodulation products lie inside the band of the signal.
WCDMA	Intermodulation and Spurious Response Attenuation	Receiver ability to receive a CDMA signal in the presence of 2 interfering CW tones located at frequency offsets 10 and 20 MHz from carrier where their intermodulation products lie inside the band of the signal.
	Receiver Desensitization	Receiver ability to receive a CDMA signal in the presence of a single tone at frequency offset 2.7 or 2.8 MHz from the carrier channel.
	Adjacent Channel Power Ratio	Ratio of the power in the adjacent channel in a 3.84-MHz bandwidth at a 5-MHz offset from carrier to the power in the main channel.
CDMA2000 1xRTT, 1xEV, IS-95	Intermodulation and Spurious Response Attenuation	Receiver ability to receive a CDMA signal in the presence of 2 interfering CW tones located at frequency offsets 0.9 and 1.7 MHz from carrier where their intermodulation products lie inside the band of the signal.
	Receiver Desensitization	Receiver ability to receive a CDMA signal in the presence of a single tone at a frequency offsets 0.9, 1.25 or 2.5 MHz from the carrier channel.
	Adjacent Channel Power Ratio	Ratio of the power in the adjacent channel in a 30-kHz bandwidth at a 1.25-MHz or (2.5-MHz) offset from carrier to the power in the main channel.
	Waveform Quality	Correlation coefficient between transmitted and ideal signal.

2.6 Summary

In this chapter, the basics of wireless communication system design including transmitter/receiver architectures, digital signal processing and signal models have been presented. Common wireless and mobile system standards including cellular and wireless networking standards have also been presented. The concepts discussed in the chapter will be used in the analysis of the performance of communication systems under nonlinear distortion, which will be presented in the following chapters. Signal models and wireless standard specifications are needed to develop distortion metrics that will be used by the different modeling and simulation approaches presented in the following chapters.

3

Modeling of Nonlinear Systems

Behavioral modeling and simulation of nonlinear systems plays an important role in the evaluation of overall communication system performance. On the other hand, behavioral modeling of nonlinear devices is important for the design of linearization techniques used to overcome the effects of nonlinear distortion introduced by nonlinear devices on wireless transmitters. Predistortion, one of the most common and effective approaches, is based on adding a nonlinear circuit that exhibits inverse characteristics of the nonlinear device before the nonlinear device so that the combined output is linear. These inverse characteristics are obtained by the inversion of the nonlinear model that can be developed from I/O measurements.

The traditional way of modeling narrowband nonlinear amplifiers is to use the measured AM–AM and AM–PM characteristics that can be developed from simple single-tone measurements. A polynomial is then fitted to measured data that results in a model that can be used to predict nonlinear distortion in a communication system using computer simulations. However, this type of simple models based on single-tone measurements cannot be used to model the wide band behavior of nonlinear amplifiers that exhibit memory effects because of the dependence of the single-tone characteristics on frequency.

The most common approaches to modeling the wideband behavior (i.e. memory effects) of nonlinear systems are those based on Volterra series analysis. However, the development of higher-order (above third order) Volterra kernels is cumbersome and its implementation in software is inefficient (Ding *et al.*, 2004). Therefore, a wide variety of models have been developed in the literature to overcome the computational complexity of extracting Volterra kernels when the system has finite memory. These models, although based on special cases of Volterra series, provide great simplification in terms of parameter extraction and software implementation over the general Volterra model.

The objective of this chapter is to present a unified understanding of the nonlinear behavior and the applicability of various models to a given system. Major approaches for nonlinear amplifier modeling are reviewed. First, bandpass memoryless nonlinear models are presented and analyzed. Then, analytical nonlinear models with memory based on a Volterra series model and its variants are presented. Special cases of the Volterra model are presented by defining their kernel relationships. Finally, empirical models that are based on

Nonlinear Distortion in Wireless Systems: Modeling and Simulation with MATLAB®, First Edition.
Khaled M. Gharaibeh.
© 2012 John Wiley & Sons, Ltd. Published 2012 by John Wiley & Sons, Ltd.

fitting measured characteristics to preset models are presented. These models are usually used to generate nonlinear characteristics in computer simulations of nonlinear systems.

3.1 Analytical Nonlinear Models

Most analytical models of nonlinearity are based on the Volterra series analysis and the theory developed by Wiener to model a nonlinear system with memory (Schetzen, 1981). In a general Volterra series analysis, a nonlinear model usually consists of linear or nonlinear filters with finite bandwidth that determine the frequency selectivity of the system. In the time domain, these filters are represented by their kernel functions (impulse response) where the support of the kernel represents the memory in the system. A number of variants to the general Volterra model (including models of memoryless systems) can be defined by defining a Volterra kernel relationship, as will be seen next.

3.1.1 General Volterra Series Model

A general Volterra series model of a nonlinear system is described by the following functional expansion of continuous functions (Lunsford, 1993):

$$y(t) = \sum_{n=1}^{\infty} F_n(x(t)) = \sum_{n=0}^{\infty} y_n(t) \tag{3.1}$$

where $F_n(x(t))$ is the Volterra functional and is defined as

$$F_n(x(t)) = \int_{-\infty}^{\infty} \cdots \int_{-\infty}^{\infty} h_n(t; \lambda_1, \ldots, \lambda_n) \prod_{i=1}^{n} x(\lambda_i) d\lambda_i \tag{3.2}$$

where $h_n(\lambda_1, \ldots, \lambda_n)$ is the n-dimensional Volterra kernel that can be symmetric without loss of generality. The support of the Volterra kernel defines the memory of the system since it defines the time interval in which past inputs can influence the current output of the system. Figure 3.1 shows a graphical representation of the Volterra model.

The general Volterra series analysis represents an analytical approach to modeling nonlinearity since it represents nonlinearity in a similar way to which Taylor series does for analytic functions. Hence, a Volterra series can be described as a Taylor series with memory. The convergence of the series can only be guaranteed for a limited range of input amplitudes (or powers). In general, for any continuous nonlinear system to be approximated by a Volterra series expansion, the input signals must form a compact subset of the input function space. In practice, the series approximation holds only on finite time intervals where the input signal is nonzero (Franz and Scholkopf, 2003).

If the nonlinear system is causal and time invariant then it can be expressed as the convolution of the powers of the input signal $x(t)$ with the n-dimensional system kernels (assuming symmetric kernels) as (Lunsford, 1993):

$$\begin{aligned} F_n(x(t)) &= \int_{-\infty}^{\infty} \cdots \int_{-\infty}^{\infty} h_n(\lambda_1, \ldots, \lambda_n) \prod_{i=1}^{n} x(t - \lambda_i) d\lambda_i \\ &= \int_{-\infty}^{\infty} \cdots \int_{-\infty}^{\infty} h_n(t - \lambda_1, \ldots, t - \lambda_n) \prod_{i=1}^{n} x(\lambda_i) d\lambda_i. \end{aligned} \tag{3.3}$$

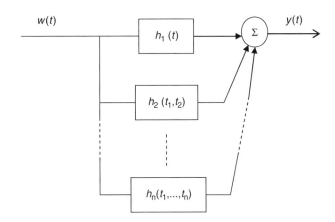

Figure 3.1   A block diagram of the general Volterra model (Gharaibeh, 2004).

It is more convenient to write Equation (3.1) in terms of the n-dimensional Volterra transfer functions since these functions are usually easier to extract than their corresponding time kernels. Therefore, $y_n(t)$ can be expressed as

$$y_n(t) = F_n(x(t)) = \int_{-\infty}^{\infty} \cdots \int_{-\infty}^{\infty} H_n(f_1, \ldots, f_n) \times \prod_{i=1}^{n} X(f_i) e^{j2\pi f_i t} df_i \qquad (3.4)$$

where $H_n(f_1, \ldots, f_n)$ is the n-dimensional Volterra Transfer Function (TF) which results from the n-dimensional Fourier transform of $h_n(\lambda_1, \ldots, \lambda_n)$ (Lunsford, 1993):

$$H_n(f_1, \ldots, f_n) = \int_{-\infty}^{\infty} \cdots \int_{-\infty}^{\infty} h_n(\lambda_1, \ldots, \lambda_n)$$
$$\times e^{-j2\pi(f_1\lambda_1 + \ldots + f_n\lambda_n)} d\lambda_1 \ldots d\lambda_n \qquad (3.5)$$

and consequently

$$h_n(\lambda_1, \ldots, \lambda_n) = \int_{-\infty}^{\infty} \cdots \int_{-\infty}^{\infty} H_n(f_1, \ldots, f_n)$$
$$\times e^{j2\pi(f_1\lambda_1 + \cdots + f_n\lambda_n)} d\lambda_1 \ldots d\lambda_n. \qquad (3.6)$$

Therefore, the frequency-domain description of the nth-order response can be expressed as (Lunsford, 1993)

$$Y_n(f) = \int_{-\infty}^{\infty} \cdots \int_{-\infty}^{\infty} H_n(f_1, \ldots, f_n)$$
$$\times \delta(f - f_1 \ldots - f_n) \prod_{i=1}^{n} X(f_i) e^{j2\pi f_i t} df_i. \qquad (3.7)$$

This form of the Volterra model is most frequently used in modeling electronic systems with sinusoidal or quasi-sinusoidal excitation.

A Volterra series is, in fact, a series representation of a nonlinear analytic function and so it diverges when the nonlinear characteristics are strong. For weakly nonlinear systems only a few terms of the series (usually third order) are required to represent the system with acceptable fidelity. On the other hand, higher order Volterra kernels are hard to develop from measurements and to implement in software (Ding *et al.*, 2004). Therefore, to overcome the computational complexity of extracting higher order Volterra kernels, a number of simplified models based on Voterra series have been developed in the literature. These models, although based on special cases of Volterra series, provide great simplification in terms of parameter extraction and software implementation over the general Volterra model. Some of these models are based on the structural classification of nonlinear systems developed by Korenberg (Korenberg, 1991), which states that any finite memory nonlinear system can be represented by a finite number of parallel Linear-Nonlinear (L-N) or Nonlinear-Linear (N-L) cascades of alternating linear and nonlinear operators. However, these models require that kernel relationships hold for a particular structure to represent the nonlinear system in hand.

In the following we discuss variants of the general Volterra model that make the development of system parameters more tractable by simplifying the kernel relationships. However, kernel relationships must hold for all of these models. This can be verified in some cases through the intuitive knowledge about the system.

3.1.2   Wiener Model

As stated above, Volterra series can be viewed as a Taylor series with memory. This means that the series converges when the error and its derivative approaches zero with increasing number of terms. A less stringent requirement for convergence can be obtained if a given function is approximated by a series of orthogonal functionals where convergence in this case is defined only in terms of the mean squared error. Wiener converted the nonorthogonal Volterra functionals into orthogonal ones using a procedure similar to a Gram–Schmidt orthogonalization with respect to white Gaussian input signals. Therefore, a Wiener model can be described by replacing the Volterra F functionals by the orthogonal Wiener G functionals (Chen *et al.*, 1989) as:

$$y(t) = \sum_{n=1}^{\infty} G_n\{x(t)\} \tag{3.8}$$

where $G_n(x(t))$ represent the Wiener functional which are related to the Volterra functionals by

$$F_n(x(t)) = \sum_{m=0}^{[n/2]} G_{p-2m}\{x(t)\} \tag{3.9}$$

where [..] denotes integer truncation and

$$G_{n-2m}\{x(t)\} = \int_{-\infty}^{\infty} \cdots \int_{-\infty}^{\infty} k_{n-2m}(\sigma_1, \ldots, \sigma_{n-2m}) \prod_{i=1}^{n-2m} x(t - \sigma_i) d\sigma_i \tag{3.10}$$

The kernels k_n are called Wiener kernels and are related to Volterra h_n kernels by:

$$k_{p-2m}(\sigma_1, \ldots, \sigma_{n-2m}) = \frac{(-1)^m n! A^m}{(n-2m)! m! 2^m}$$

$$\times \int_{-\infty}^{\infty} \ldots \int_{-\infty}^{\infty} h_n(\lambda_1, \lambda_1, \ldots, \lambda_m, \lambda_m, \sigma_1, \ldots, \sigma_{n-2m}) d\lambda_1 \ldots d\lambda_m.$$

(3.11)

The zeroth- and first-degree Wiener functionals are equal to Volterra functionals and given by

$$G_0\{x(t)\} = k_0 = h_0$$

$$G_1\{x(t)\} = F_1\{x(t)\} = \int_{-\infty}^{\infty} h_1(\lambda_1 x(t - \lambda_1) d\lambda_1 \qquad (3.12)$$

For example, the second-order Wiener functional can be written using Equation (3.10) as:

$$G_2\{x(t)\} = \int_{-\infty}^{\infty} \int_{-\infty}^{\infty} h_2(\lambda_1, \lambda_2) d\lambda_1 d\lambda_2 - A \int_{-\infty}^{\infty} h_2(\lambda_1, \lambda_1) d\lambda_1 \qquad (3.13)$$

and the third-order Wiener functional as:

$$G_3\{x(t)\} = \int_{-\infty}^{\infty} \int_{-\infty}^{\infty} \int_{-\infty}^{\infty} h_3(\lambda_1, \lambda_2, \lambda_3) d\lambda_1 d\lambda_2 d\lambda_2$$

$$-3A \int_{-\infty}^{\infty} \int_{-\infty}^{\infty} h_3(\lambda_1, \lambda_2, \lambda_2) x(t - \lambda_1) d\lambda_1 d\lambda_2 \qquad (3.14)$$

3.1.3   Single-Frequency Volterra Models

The greatest difficulty in the construction of Volterra model and its parameter extraction from measured data is its multidimensionality in time and frequency. Single-frequency Volterra models represent a simplified version of a Volterra system by simplifying the branches of the general model, see Figure 3.2, into cascades of Linear-Nonlinear (L-N) or Nonlinear-Linear (N-L) subsystems. These configurations simplify parameter estimation and reduce computational complexity since all spectral calculations required involve only a single frequency variable instead of several (Silva *et al.*, 2001). Single-frequency Volterra models take one of the following topologies: filter-nonlinearity model and nonlinearity-filter.

A single-frequency filter-nonlinearity Volterra model is shown in Figure 3.2(a) and is characterized by a Volterra kernel that has the following property:

$$h_n(\lambda_1, \ldots, \lambda_n) = a_n h_n(\lambda_1) h_n(\lambda_2) \ldots h_n(\lambda_n)$$

$$= a_n \prod_{i=1}^{n} h_n(\lambda_i). \qquad (3.15)$$

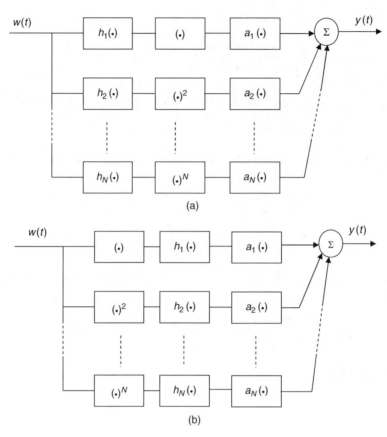

Figure 3.2 Single-frequency Volterra models; (a) filter-nonlinearity Volterra model and (b) nonlinearity-filter model (Gharaibeh, 2004).

This can be contrasted to the general form in Equation (3.5) and it follows that the Volterra TF is

$$H_n(f_1, \ldots, f_n) = a_n \prod_{i=1}^{n} H_n(f_i). \tag{3.16}$$

For the nonlinearity-filter Volterra model shown in Figure 3.2(b), the Volterra kernel has the property:

$$h_n(\lambda_1, \ldots, \lambda_n) = h_n(\lambda_1)\delta(\lambda_1 - \lambda_2) \ldots \delta(\lambda_{n-1} - \lambda_n) \tag{3.17}$$

and it follows that:

$$H_n(f_1, \ldots, f_n) = H(f_1 + \ldots + f_n). \tag{3.18}$$

The primary advantage of the single-frequency Volterra model is the simplicity of its parameter extraction and a significant simplicity of nonlinear system analysis. However, it does not represent the broad variety of systems that the general Volterra model does.

Parameter extraction of the single-frequency Volterra model was discussed in Silva *et al.* (2001) using cross-correlation spectral analysis performed on time-domain measured data. These models are regarded as polyspectral models and they showed good performance for Traveling Wave Tube (TWT) high-power amplifiers. The identification process used measured AM–AM and AM–PM characteristics and the measured small-signal transfer function of the amplifier as a priori information. Then, time-domain measured I/O data is used to extract the remaining blocks of the model using cross-correlation techniques. A detailed analysis of the identification process can be found in (Bendat, 1990; Silva *et al.*, 2001). Model parameters were also extracted using multifrequency excitation (Maqusi, 1985) but only third-order nonlinearity was considered.

3.1.4 The Parallel Cascade Model

The parallel cascade model enables the realization of the Wiener model by a parallel cascade of linear and nonlinear elements, as shown in Figure 3.3 (Chen, 1989; Kenney *et al.*, 2002; Korenberg, 1991). The model enables the formulation of the system equations as Chen (1989).

$$y(t) = \sum_{p=1}^{P} y_p(t)$$

$$y_p(t) = \sum_{r=0}^{N_p} a_{r,p} u_p^r(t) \tag{3.19}$$

where

$$u_p(t) = \int_{-\infty}^{\infty} h_p(\lambda) x(t - \lambda) d\lambda = h_p(t) * x(t). \tag{3.20}$$

Here, p is the number of branches and it represents the memory depth and N_p is the maximum order of the polynomial in branch p.

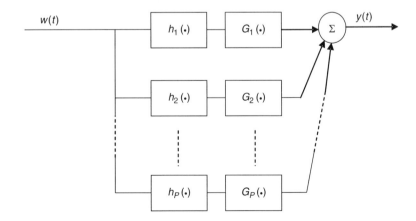

Figure 3.3   Parallel cascade nonlinear model (Gharaibeh, 2004).

Note that the parallel cascade model represents the nonlinear system by a finite number of branches with polynomials of finite order. This model was proven adequate for modeling a finite-memory Volterra model having a finite order. In Korenberg (1991), Korenberg derived an upper bound for the number of cascades required for representing a finite-memory system. Parameter estimation for this model was discussed in Korenberg (1991), Kenney *et al.* (2002) and Shi and Sun (1990) using time-domain measurements. In Kenney *et al.* (2002), system parameters were developed using measured two-tone test with varying frequency separation.

3.1.5 Wiener–Hammerstein Models

A Hammerstein model takes the following forms of the Volterra kernel in the frequency domain (Chen *et al.*, 1989; Korenberg, 1991, Stanley, 2000; Scott and Mulgrew, 1997; Shi and sun, 1990; Westwick and Kearney, 2000):

$$H_n(f_1, \ldots, f_n) = a_n H(f_1) \ldots H(f_n) \tag{3.21}$$

and a Wiener model is represented by the following kernel relationship:

$$H_n(f_1, \ldots, f_n) = a_n H(f_1 + \ldots + f_n). \tag{3.22}$$

The Hammerstein block model is realized as a linear filter that represents the finite memory of the system followed by a memoryless nonlinearity. Similarly, the Wiener block model is realized by a memoryless nonlinearity followed by a linear filter as shown in Figure 3.4(a). A combination of both models results in Wiener–Hammerstein model that consists of a cascade of a linear filter, a memoryless nonlinearity and another linear filter, as shown in Figure 3.4(b). The kernel relationship of this model is defined as:

$$H_n(f_1, \ldots, f_n) = a_n H_1(f_1) \ldots H_1(f_n) H_2(f_1 + \ldots + f_n). \tag{3.23}$$

The above forms of the Volterra kernels provide a great simplification over the general form since they result in fewer parameters being needed for system identification. Parameter extraction of such models are usually based on measuring first- and second-order Volterra kernels (Chen, 1989). These structures were successfully used in modeling a wide variety of nonlinear systems (Chen *et al.*, 1989; Korenberg, 1991; Shi and Sun, 1990).

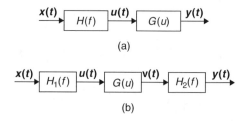

(a)

(b)

Figure 3.4 (a) Wiener model and (b) Wiener–Hammersien model (Gharaibeh, 2004).

3.1.6 Multi-Input Single-Output (MISO) Volterra Model

A MISO Volterra model represents a multilateral system or device and is characterized by (Chen *et al.*, 1989):

$$y(t) = \sum_{n_1=0}^{\infty} \cdots \sum_{n_K=0}^{\infty} \int_{-\infty}^{\infty} \cdots \int_{-\infty}^{\infty} h_{n_1,n_1,\dots,n_K}(\lambda_{11}, \dots, \lambda_{1n_1}, \dots, \lambda_{1n_{K1}}, \dots, \lambda_{Kn_K})$$

$$x_1(t-\lambda_{11}) \dots x_1(t-\lambda_{1n_1}) \dots x_K(t-\lambda_{K1}) \dots \dots x_K(t-\lambda_{Kn_K})$$

$$d\lambda_{11} \dots d\lambda_{1n_1} \dots d\lambda_{1n_1} \dots d\lambda_{K1} \dots d\lambda_{Kn_K}. \tag{3.24}$$

A simplified version of the Volterra model is the *G*-structure (Chen, 1989) that takes the following form of the Volterra kernel in the frequency domain:

$$H_{n_1,n_2,\dots,n_K}(\omega_{11}, \dots, \omega_{1n_1}, \dots, \omega_{1n_K}, \dots, \omega_{Kn_K})$$

$$= a_n \binom{n}{n_1, \dots, n_K} H_{11}(\omega_{11}) \dots H_{11}(\omega_{1n_1}) \dots H_{1K}(\omega_{K1}) \dots H_{1K}(\omega_{Kn_K})$$

$$\times H_2(\omega_{11} + \dots + \omega_{1n_1} + \dots + \omega_{K1} \dots + \omega_{Kn_K}) \tag{3.25}$$

where

$$\binom{n}{n_1, \dots, n_K} = \frac{n!}{n_1! \dots, n_K!}. \tag{3.26}$$

In a similar way to the single input case, this form of the Volterra kernel enables the system model to be realized by the model shown in Figure 3.5. MISO models are usually used to model systems with multiple inputs where each input sees different input impedance such as a mixer. Parameter estimation for this model is similar to its single-input counterpart, however, it requires more extensive computations (Chen *et al.*, 1989).

3.1.7 The Polyspectral Model

Polyspectral models were pioneered by (Silva *et al.*, 2001) and represent the nonlinearity as a parallel cascade of two branches. The first branch consists of a linear filter $H_1(f)$ while the second branch consists of a cascade of a linear filter $H_2(f)$ and a memoryless

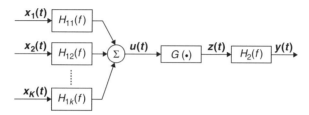

Figure 3.5 MISO nonlinear model (Chen *et al.*, 1989).

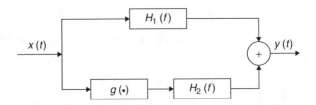

Figure 3.6 Polyspectral model (Silva *et al.*, 2002).

nonlinearity $g(\cdot)$, as shown in Figure 3.6. The kernel relationship of this model can be developed as

$$H_n(f_1, \ldots, f_n) = \begin{cases} a_1 H_2(f), & n = 1; \\ a_n H_2(f_n) H_2(f_1 + \ldots + f_n), & n = 2, \ldots, N. \end{cases} \quad (3.27)$$

Therefore, the polyspectral model is a single-frequency Volterra model, however, it is still analytical and possesses several features with regard to parameter extraction from measured data (Jeruchem *et al.*, 2000).

3.1.8 Generalized Power Series

A Generalized Power Series (GPS) model was pioneered by Steer (Rhyne and Steer, 1987) and is used to determine the steady state frequency-domain description of the output of a nonlinear system. For a multifrequency input of the form (Lunsford, 1993):

$$x(t) = \sum_{k=0}^{K} x_k(t) = \sum_{k=0}^{N} |X_k| \cos(2\pi \xi_k t + \phi_k) = \sum_{k=-N}^{K} X_k e^{\xi_k t} \quad (3.28)$$

the GPS model describes the response of a nonlinear system as (Lunsford, 1993):

$$y(t) = A \sum_{n=0}^{\infty} a_n \left[\sum_{k=0}^{K} b_k x_k(t - \tau_{n,k}) \right]^n \quad (3.29)$$

where $b_{k,i}$ are real coefficients, a_n are complex coefficients and $\tau_{n,k}$ is a time delay that depends on the power series order and the index of the input frequency component.

The output phasor Y_k at a given output frequency ξ_q can then be developed in closed form in terms of the input phasor magnitude, the GPS model coefficients and time delays and the model order. For each output frequency, a table of a detailed analysis of the output of the GPS can be found in Lunsford (1993). This analysis can also be extended to bivariate power series which is used to analyze the response of a nonlinear system to two multifrequency signals.

The GPS reduces to a memoryless polynomial by setting $\tau = 0$ and $b_k = 1$. The inclusion of complex coefficients and frequency-dependent time delays allows a variety of systems to be modeled. The GPS was proved to be equivalent to the single-frequency approximation of the Volterra series as shown in (Lunsford, 1993):

$$H_n(f_{q_1}, \ldots, f_{q_n}) = A a_n \prod_{i=0}^{n} b_{q_i} e^{j 2\pi f_{q_i} \tau_{n,q_i}} \tag{3.30}$$

Parameter extraction can be done by measuring the response to a multifrequency input. In fact, the equivalence of Volterra and GPS series makes the multifrequency excitation suitable to developing the Volterra transfer functions in the frequency domain (Lunsford, 1993).

3.1.9 Memory Polynomials

Memory polynomials represent a special case of the general Volterra model in discrete time and have been extensively used in the literature in a wide variety of applications; especially in the linearization of power amplifiers with memory. To develop the memory polynomial, let us consider a general Volterra model of finite memory that can be expressed in discrete time as (Kokkeler, 2005; Morgan *et al.*, 2006)

$$y(n) = \sum_{k=1}^{N} y_k(n) \tag{3.31}$$

where $y_k(n)$ is the k-dimensional convolution of the input with the kth-order Volterra kernel:

$$y_k(n) = \sum_{m_1=0}^{P-1} \cdots \sum_{m_k=0}^{P-1} h_k(m_1, \ldots, m_k) \prod_{l=1}^{k} x(n - m_l) \tag{3.32}$$

which represents a generalization of a power series representation with a finite memory of length P. By making a change of variables to the diagonal index variables:

$$n_l = m_{l+1} - m_1, l = 1, 2, \ldots, k - 1. \tag{3.33}$$

then Equation (3.32) can be written in the form (Morgan *et al.*, 2006):

$$y_k(n) = \sum_{n_1=-P+1}^{P-1} \cdots \sum_{n_{k-1}=-P+1}^{P-1} g_{n_1, \ldots n_{k-1}}^{k}(n) * \left[x(n) \prod_{l=1}^{k-1} x(n - n_l) \right] \tag{3.34}$$

where $*$ denotes a one-dimensional convolution and

$$g_{n_1, \ldots n_{k-1}}^{k}(n) = h_k(n, n + n_1, \ldots, n + n_{k-1}) \tag{3.35}$$

represents a linear filter that can be determined from the Volterra kernel h_k.

This form of Volterra series can be realized as a parallel cascade where each cascade consists of a linear filter with impulse response defined as in Equation (3:35) and has an input

$$x(n) \prod_{l=1}^{k-1} x(n - n_l) \tag{3.36}$$

A Hammerstein model that is formed by a nonlinearity followed by a linear filter (two-box structure) can be defined by setting

$$g^k_{n_1,...n_{k-1}}(n) = a_{km}g(m) \tag{3.37}$$

thus, the generalized Hammerstein model can be formulated as

$$y(n) = \sum_{k=1}^{N}\sum_{m=0}^{P-1} a_{km}x(n-m)^k \tag{3.38}$$

A narrowband form of the Hammerstein model can be obtained by taking the components of the nonlinear output that lie on the fundamental frequency. Thus, only combinations of the form $x(n)|x(n-m)|^{k-1}$ are considered, and hence the complex envelope of the output can be written in terms of the complex envelope of the input in the form of a *memory polynomial* of the form:

$$\tilde{y}(n) = \sum_{k=1}^{N}\sum_{m=0}^{P-1} a_{km}\tilde{x}(n-m)|\tilde{x}(n-m)|^k \tag{3.39}$$

where the tilde indicates the complex envelope with respect to some carrier frequency f_c. Memory polynomials provide a simplification to the discrete-time general Volterra model and have been used extensively in the literature for the design of predistortion schemes (Morgan *et al.*, 2006; Raich *et al.*, 2005; Zhou, 2000). Note that the number of coefficients in a memory polynomial is of the order of $P + 1$ as compared to the general Volterra model, where the number of unknowns is of the order of $(P + 1)^N$, where N is the nonlinear order.

3.1.10 Memoryless Models

The classification of nonlinearities as memoryless or with memory refer to the dependency of the nonlinear output on either the instantaneous value of the input (as in memoryless) or on its values at past time instances (memory case). These models are usually used to model a bandpass nonlinearity where the input has a finite bandwidth concentrated around a carrier frequency (a modulated carrier) and the output is band limited by a zonal filter to the same bandwidth of the signal. Therefore, a memoryless system is usually referred to as an instantaneous model where it is assumed that the amplitude of the modulated carrier is slowly varying.

3.1.11 Power-Series Model

Perhaps the most commonly used model for modeling distortion in power amplifiers is the memoryless model. This model is characterized by a constant Volterra transfer function over frequency, i.e.

$$H_n(\underline{f}) = H_n(\underline{0}) = a_n \tag{3.40}$$

where the under bar indicate an n-dimensional vector and it follows that the corresponding Volterra kernel is:

$$h_n(\lambda_1, \ldots, \lambda_n) = a_n \delta(\lambda_1) \delta(\lambda_2) \ldots \delta(\lambda_n). \tag{3.41}$$

Therefore, the model reduces to a power series model with real coefficients of the form:

$$y(t) = N(x(t)) = \sum_{n=1}^{N} a_n x^n(t). \tag{3.42}$$

where a_n are the instantaneous coefficients.

The power series model is popular for its simplicity and the fact that distortion can be directly related to its parameters (coefficients), as will be seen in the following chapter. Although the power series model is an analytical model that can be developed from circuit analysis, it can also be developed by fitting a polynomial to measured input/output characteristics. The coefficients a_n can be developed from AM–AM power sweep measurements of single tones using a Vector Network Analyzer (VNA) where a polynomial fitting to the measured data yields a power series with real coefficients. For a quasi-memoryless system, a complex power series model that relates the complex envelope of the output to the complex envelope of the input is obtained from AM–AM and AM–PM measurements, as will be shown in the next chapter.

The output of the memoryless nonlinearity around the carrier frequency f_c (first zonal output) can be expressed as

$$y(t) = g_1 \left[A(t) \right] \cos(2\pi f_c t + \theta(t)) + g_2 \left[A(t) \right] \sin(2\pi f_c t + \theta(t)) \tag{3.43}$$

where $g_1(\cdot)$ and $g_2(\cdot)$ are real functions that can be expressed in terms of N using a Tchebychev transform as in (Jeruchem *et al.*, 2000):

$$g_1(A(t)) = \frac{1}{\pi} \int_0^{2\pi} N[A(t) \cos(\alpha)] \cos(\alpha) d\alpha \tag{3.44}$$

and

$$g_2(A(t)) = \frac{1}{\pi} \int_0^{2\pi} N[A(t) \cos(\alpha)] \sin(\alpha) d\alpha \tag{3.45}$$

where $\alpha = 2\pi f_c t + \theta(t)$. This representation is similar to the quadrature representation of digitally modulated signals where g_1 represents the in-phase component and g_2 represents the quadrature component.

When $g_2 = 0$ the system is called a strictly memoryless system. In the general case when $g_1, g_2 \neq 0$, the system is called a quasi-memoryless system since it introduces not only amplitude distortion but also phase distortions. Therefore, the quasi-memoryless model takes into account short-term memory effects that are manifested as an amplitude-dependent phase shift in the output waveform. However, it has its limitations since it does not take into account long-term memory effects, which makes it inadequate for modeling wideband and multichannel systems.

3.1.12   The Limiter Family of Models

The importance of studying distortion introduced by the family of limiter amplifiers arises from the wide variety of applications of those amplifiers either on the conceptual level or on the practical level. In a multicarrier system, a limiter of clipping characteristics is used before a power amplifier to limit the peak envelope power of the input signals in order to prevent the signal from exceeding the power rating of a power amplifier (Kashyab *et al.*, 1996). On the other hand, a soft-limiter amplifier model is used to model distortion introduced by linearized power amplifiers in CDMA systems (Bennet, 1997). A predistorted power amplifier can be modeled by a soft-limiter model (clipper) that models the perfect cancelation of AM–PM conversion and clipping of AM–AM characteristics by predistortion. A class of memoryless limiter power amplifiers can be modeled by the following parametric form developed in (Jeruchem *et al.*, 2000):

$$y(t) = f(x(t)) = \frac{L\mathrm{sgn}(x(t))}{[1 + (l/|x(t)|)^s]^{1/s}} \tag{3.46}$$

where $\mathrm{sgn}(x)$ is the sign function, L is the asymptotic output level that is used to scale the output, l is the input limit level and s is the knee sharpness parameter. Figure 3.7 shows the input output characteristics of a limiter amplifier for different values of s. In the following, we discuss special cases of Equation (3.46).

3.1.12.1   Hard-Limiter Amplifier Model

A hard-limiter amplifier can be characterized by Equation (3.46) with $l = 0$ and choosing s arbitrarily. The resulting transfer characteristics can be written as:

$$y(t) = f(x(t)) = L\mathrm{sgn}(x(t)) \tag{3.47}$$

Figure 3.8(a) shows the hard-limiter characteristics.

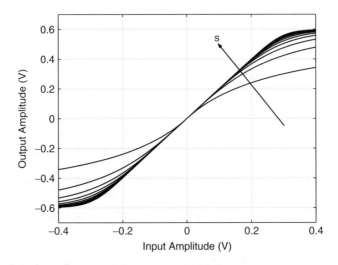

Figure 3.7   Limiter function for different values of smoothing factor (s) (Gharaibeh, 2004).

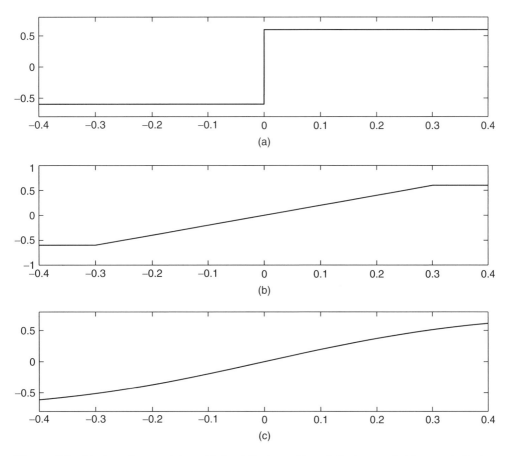

Figure 3.8 Limiter characteristics: (a) hard limiter; (b) soft limiter; and (c) smooth limiter (Gharaibeh, 2004).

3.1.12.2 Soft-Limiter Amplifier Model

A soft limiter or a clipper characteristics shown in Figure 3.8(b), can be obtained by setting $s = \infty$ and therefore the output of the soft limiter can be written as:

$$y(t) = f(x(t)) = \begin{cases} -l & |x(t)| \leq -l \\ x(t) & -l \leq |x(t)| \leq l \\ l & |x(t)| \geq l \end{cases} \tag{3.48}$$

3.1.12.3 Smooth-Limiter Amplifier Model

A smooth limiter is characterized by the error function as:

$$y(t) = f(x(t)) = \frac{1}{\sqrt{2\pi}} \int_0^{x(t)} e^{\frac{-x^2}{2l^2}} dx \tag{3.49}$$

where l is used to set the smoothness of the filter. Figure 3.8(c) shows the smooth-limiter characteristics.

3.2 Empirical Nonlinear Models

Empirical models refer to a class of models where a system is represented by a preset block structure of linear and nonlinear elements. Empirical models of nonlinearity are usually based on the intuitive knowledge about the system where a block structure is assumed and then its parameters are obtained by fitting measured data to the model using Least Squares (LS) techniques. These models were suggested in the literature because of the simplicity of their parameter extraction using either direct Vector Network Analyzer (VNA) measurements (measured AM–AM and AM–PM characteristics) or cross-correlation of time-domain measured data (Silva *et al.*, 2001). In PA modeling, parameter extraction of these models is usually done by fitting power (or/and frequency) swept single-tone measurements to the model. A detailed description of a variety of empirically based models can be found in Jeruchem *et al.* (2000). These include block models, the PSB model, Abuelma'ati model, Saleh Model, Ghorbani model and hyperbolic tangent model. Some of these models have been used in the literature to generate nonlinearity in simulations by providing a model for the nonlinear transfer characteristics. In the following, the details of such models are presented.

3.2.1 The Three-Box Model

The three-box model consists of an input linear filter, a quasi-memoryless nonlinearity, which represents the AM–AM and AM–PM conversion at a given reference frequency followed by another linear filter (Jeruchem *et al.*, 2000), see Figure 3.9. The three-box model, despite its simplicity, captures the behavior of a class of wideband power amplifiers where memory effects are characterized by multiple AM–AM curves at different frequencies that have the same shape as shown in Figure 3.10. If the amplifier circuit exhibits different shapes for the nonlinear characteristics, then the system cannot be approximated by the three-box model and hence, it requires a more sophisticated model. Dissimilar nonlinear characteristics are due to memory effects caused by nonlinear parasitic capacitances of the nonlinear device. The three-box structure was investigated in Muha *et al.* (1999) and Silva *et al.* (2002) and was used to adequately model memory effects of single-input nonlinear amplifiers where its parameters were extracted using time-domain measurements.

The three-box model was used for studying the interaction of the multiple channels by the amplifier nonlinearity in Gharaibeh and Steer (2005), where it was shown that the

Figure 3.9 A three-box model with the middle block being a memoryless nonlinearity (Jeruchem *et al.*, 2000).

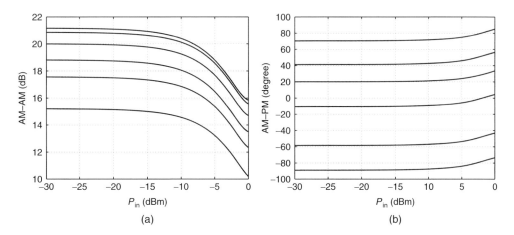

Figure 3.10 AM–AM and AM–PM characteristics at different frequencies (Gharaibeh and Steer, 2005).

model was entirely satisfactory when used to characterize the multichannel response of the amplifier.

3.2.2 The Abuelma'ati Model

The Abuelma'ati Model is used to model frequency-dependent nonlinear characteristics, that is, AM–AM and AM–PM characteristics at different frequencies. Using the representation of a bandpass nonlinearity in Equation (3.43), this model represents a nonlinearity using the following form for the in-phase and quadrature amplitude nonlinearities (Jeruchem *et al.*, 2000):

$$g_1(A, f) = \sum_{n=1}^{N} \gamma_I(n, f) J_1\left(\frac{n\pi A}{D}\right)$$

$$g_2(A, f) = \sum_{n=1}^{N} \gamma_Q(n, f) J_1\left(\frac{n\pi A}{D}\right) \tag{3.50}$$

where J_1 is the first-order Bessel function of the first kind, D is the dynamic range of the input, N is the model order and

$$\gamma_I(n, f) = \gamma_{I,0} G_I(f, n)$$

$$\gamma_Q(n, f) = \gamma_{I,0} G_Q(f, n) \tag{3.51}$$

where $\gamma_{I,0}$ and $\gamma_{Q,0}$ are defined as $\gamma_I(n, f)$ if $G_I = 1$ and $\gamma_I(n, f)$ if $G_Q = 1$ respectively. Note that the functions $G_I(f, n)$ and $G_Q(f, n)$ represent linear filters, while the Bessel functions represent memoryless nonlinearities. This model reduces to a memoryless model when the frequency parameter is fixed.

3.2.3 Saleh Model

In this model, nonlinearity is represented using a model that requires only four parameters. This model was originally used to model TWT amplifier but it has also been used to model solid-state power amplifiers (Jeruchem *et al.*, 2000). In this model, the AM–AM and AM–PM characteristics of the amplifier are given by:

$$F[A, f] = \frac{\alpha_a(f)A}{1 + \beta_a(f)A^2} \tag{3.52}$$

and

$$\Theta(A, f) = \frac{\alpha_\theta(f)A^2}{1 + \beta_\theta(f)A^2} \tag{3.53}$$

where α_a, α_θ, β_a and β_θ represent the parameter, of the model that are, in general, frequency dependent. These parameters describe the hardness of the nonlinear characteristics and can be adjusted to give the desired nonlinear characteristics. If A is very large, $F(A)$ is proportional to $1/A$ and $\Theta(A)$ approaches a constant.

The Saleh model reduces to the memoryless case when the model parameters are frequency independent. In this case, the complex envelope of the output of the amplifier is given by (Saleh, 1981):

$$\tilde{y}(t) = \frac{\alpha_a|\tilde{x}(t)|}{1 + \beta_a|\tilde{x}(t)|^2} \exp\left\{ j\left(\theta(t) + \frac{\alpha_\theta|\tilde{x}(t)|^2}{1 + \beta_\theta|\tilde{x}(t)|^2} \right) \right\} \tag{3.54}$$

3.2.4 Rapp Model

The Rapp model models a memoryless nonlinearity with no AM–PM conversion. The Rapp model is given by (Rapp, 1991):

$$\tilde{y}(t) = \frac{G\tilde{x}(t)}{\left(1 + |\frac{\tilde{x}(t)}{V_{\text{sat}}}|^{2p} \right)^{1/2p}} \tag{3.55}$$

where G represents the linear gain of the amplifier, V_{sat} represents the saturation input voltage and p is a parameter that describes the hardness of the nonlinear characteristics. The Rapp model converges to a hard-limiter model when p approaches infinity.

3.3 Parameter Extraction of Nonlinear Models from Measured Data

Parameter extraction of nonlinear models parameters from measured data is a classical problem in engineering. There is a huge number of parameter extraction techniques of nonlinear models presented in the literature (Bendat, 1990; Chen *et al.*, 1989; Kenney *et al.*, 2002; Korenberg, 1991; Silva *et al.*, 2002, 2001; Shi and Sun, 1990). However, the objective of this section is to discuss approaches based on simple measurements that will be used in the simulation of the performance of wireless communication system in later chapters of this book. Therefore, common approaches for parameter extraction of nonlinear models from measured data will be presented.

3.3.1 Polynomial Models

3.3.1.1 Memoryless Polynomial Models

A memoryless power amplifier can be characterized by its AM–AM and AM–PM conversions as

$$y(n) = F(|x(n)|)e^{j[\Theta(|x(n)|)+\theta(n)]} \tag{3.56}$$

where F and Θ represent the AM–AM and AM–PM conversions and $x(n)$ and $y(n)$ represent the complex envelopes of the input and output signals with respect to a certain fundamental frequency.

The envelope version of the power-series model in Equation (3:42) will be developed in the next chapter and can be written as

$$y(n) = \sum_{k=1}^{N} b_k |x(n)|^{k-1} x(n) \tag{3.57}$$

where b_k are the envelope coefficients.

The coefficients of this model (b_n) can be developed from measured input/output data as follows, let x_i, where $i \leq M$, be the data points obtained by sweeping the input power of the a single tone signal in M power steps and $y(i)$ be the measured output of the amplifier at the ith power step. The envelope coefficients can be obtained by polynomial fitting to the measured input/output characteristics using LS optimization. Therefore, given a measured input vector X of length M (samples), $X = [X_1\ X_2\ \ldots\ X_M]^T$ where $X_i = x(i)$ and a measured output complex vector Y of length M, $Y = [Y_1\ Y_2\ \ldots\ Y_M]$, where $Y_i = y(i)$, we form a matrix equation as

$$\begin{pmatrix} Y_1 \\ Y_2 \\ \vdots \\ Y_M \end{pmatrix} = \begin{pmatrix} X_1 & X_1|X_1|^2 & X_1|X_1|^4 & X_1|X_1|^6 & \ldots & X_1|X_1|^{N-1} \\ X_2 & X_2|X_2|^2 & X_2|X_2|^4 & X_2|X_2|^6 & \ldots & X_2|X_2|^{N-1} \\ \vdots & \vdots & \vdots & \vdots & \vdots & \vdots \\ X_M & X_M|X_M|^2 & X_M|X_M|^4 & X_M|X_M|^6 & \ldots & X_M|X_M|^{N-1} \end{pmatrix} \begin{pmatrix} b_1 \\ b_3 \\ \vdots \\ b_{\frac{N-1}{2}} \end{pmatrix}$$

or in matrix form as:

$$\mathbf{Y} = \Phi_x \mathbf{b}. \tag{3.58}$$

Therefore, the problem leads to a linear system optimization and the coefficients can be obtained using the method of least squares. The least squares problem reduces to solving a system of linear equations of the form

$$\mathbf{b} = \Phi_x{}^+ \mathbf{Y} \tag{3.59}$$

where $\Phi_x{}^+ = (\Phi_x{}^H \Phi_x)^{-1} \Phi_x{}^H$ is the Moore–Penrose pseudo-inverse of Φ_x and H indicates the Hermitian transpose.

3.3.1.2 Memory Polynomial Model

To develop a memory polynomial model of a nonlinear system with memory, we consider the Weiner–Hammerstein model that is composed of a linear filter in cascade with a memoryless nonlinearity. This model can be represented by a memory polynomial model as (Morgan *et al.*, 2006)

$$y(n) = \sum_{k=1}^{N} \sum_{m=0}^{P-1} b_{km} x(n-m) |x(n-m)|^{k-1}$$

$$= \sum_{k=1}^{N} \sum_{m=0}^{P-1} b_{km} u^k(n-m) \tag{3.60}$$

where $u^k(n) = x(n){-}x(n){-}^{k-1}$ and P is the memory depth.

The memory polynomial coefficients can be obtained using LS in a similar way to memoryless polynomials, however, the number of coefficients is equal to NP. Therefore, given measured or simulated input/output data, the estimation problem is formulated as in Equation (3.58) with

$$\Phi_x = \begin{pmatrix} X_{11} & \dots & X_{1P} & X_{11}|X_{11}|^2 & \dots & X_{1P}|X_{1P}|^2 & \dots & X_{11}|X_{11}|^{N-1} & \dots & X_{1P}|X_{1P}|^{N-1} \\ X_{21} & \dots & X_{2P} & X_{21}|X_{21}|^2 & \dots & X_{2P}|X_{2P}|^2 & \dots & X_{21}|X_{21}|^{N-1} & \dots & X_{2P}|X_{2P}|^{N-1} \\ \vdots & \vdots & \vdots & \vdots & & \vdots & & & & \\ X_{M1} & \dots & X_{MP} & X_{M1}|X_{M1}|^2 & \dots & X_{MP}|X_{MP}|^2 & \dots & X_{M1}|X_{M1}|^{N-1} & \dots & X_{MP}|X_{MP}|^{N-1} \end{pmatrix}$$

and

$$\mathbf{b} = \begin{pmatrix} b_{11} \\ b_{12} \\ \vdots \\ b_{1P} \\ b_{31} \\ b_{32} \\ \vdots \\ b_{3P} \\ \vdots \\ b_{\frac{N-1}{2}1} \\ b_{\frac{N-1}{2}2} \\ \vdots \\ b_{\frac{N-1}{2}P} \end{pmatrix}$$

where the coefficients vector can be estimated from measured data using LS as in Equation (3.59).

3.3.1.3 Model Accuracy and Stability

The extraction of the model coefficients of both memory and memoryless power-series models requires the inversion of the matrix $\Phi_x{}^H \Phi_x$ in Equation (3.59), which usually experiences numerical instability problems as the model order is increased (Raich *et al.*, 2004). The stability of the LS solution can be understood as the effect of the change in Y on the change in the estimated model parameters a. This effect is usually characterized by the condition number of the regressor matrix Φ_x. The condition number of the matrix is defined as (Gharaibeh, 2009):

$$\text{cond}(\Phi_x) = \|\Phi_x\| \|\Phi_x^+\| \tag{3.61}$$

where $\| \cdot \|$ represents the Euclidian norm. The condition number characterizes the rate at which the solution a, will change with respect to a change in Y. Thus, if the condition number is high, even a small error in Y may cause a large error in a. On the other hand, if the condition number is low then the error in a will not be much higher than the error in Y. Therefore, numerical errors involved in matrix inversion are high whenever the condition number of the matrix is higher than 1 (Raich *et al.*, 2004).

The condition number of the matrix is mainly dependent on the structure of the regressor matrix Φ_x (which is model dependent) and on the input vector X. The high condition number of a matrix results from the high correlation between its columns. In a polynomial model the high correlation between the columns of the matrix Φ_x results from their monomial basis functions $x, x^2, \ldots$ that are highly correlated, and also from the correlation of the data sample $x(t)$ at different time instants (Raich *et al.*, 2004). Therefore, the condition number associated with the parameter estimation of the polynomial model increases as the order of the polynomial is increased. This makes polynomials unsuitable for modeling hard nonlinearities because these require high orders to achieve good modeling accuracy (Gharaibeh, 2009).

The accuracy of the power-series model can be assessed using the concept of the Normalized Mean Squared Error (NMSE) that is defined as (Raich *et al.*, 2004)

$$\text{NMSE} = 10 \log_{10} \left[\frac{\sum_{i=1}^{M} |y_i - \hat{y}_i|^2}{\sum_{i=1}^{M} |y_i|^2} \right] \tag{3.62}$$

where $y_i = y(i)$ is the measured PA output and $\hat{y}_i = \hat{y}(i)$ is the modeled PA output. Note that the NMSE decreases as the model order is increased, however, increasing the nonlinear order results in increasing the instability of the model, as discussed above. Therefore, a compromise between the model accuracy and model stability has to be made when a model is to be chosen.

3.3.2 Three-Box Model

Parameter extraction of the three-box model has been extensively investigated (Bendat, 1990; Jeruchem *et al.*, 2000; Muha *et al.*, 1999; Silva *et al.*, 2002, 2000). A simplified intuitive approach was presented in (Jeruchem *et al.*, 2000) where the transfer functions of the linear filters are obtained by measuring the gain characteristics at saturation $H_{\text{sat}}(f)$ (for example at the 1-dB compression point), and measuring the small signal linear frequency response $H_{\text{ss}}(f)$. The transfer functions of the linear filters, $H_1(f)$

and $H_2(f)$, are then computed from the measured filter transfer functions as follows (Jeruchem *et al.*, 2000):

$$H_1(f) = \frac{H_{ss}(f)}{G_{ss}|H_{sat}(f)|} \tag{3.63}$$

and

$$H_2(f) = \frac{|H_{sat}(f)|}{|G_{sat}|} \tag{3.64}$$

where $H_{ss}(f)$ is the measured small signal-transfer function, $H_{sat}(f)$ is the measured large signal transfer function, $G_{ss} = |H_{ss}(f_{ref})|$ and $G_{sat} = |H_{sat}(f_{ref})|$ are the small-signal gain and the 1-dB compression gain at the reference frequency. The normalization of the filter responses is included in the memoryless nonlinearity.

The static nonlinearity, the middle block in Figure 3.9, is extracted as the measured single-tone AM–AM and AM–PM characteristics taken at a reference carrier frequency f_{ref}. The required input-output characteristics of the nonlinear block are the memoryless nonlinear characteristics given by Equation (3.42). The reference AM–AM and AM–PM characteristics are usually measured at the middle frequency of the band of interest. The choice of the reference is arbitrary when the phase-frequency response is linear.

3.3.3 Volterra Series

The attractiveness of Volterra series representation comes from the fact that its kernels can be related to circuit parameters. For black-box models, Volterra kernels can be developed from input-output measured data using cross-correlation techniques (Bendat, 1990; Korenberg, 1991, Schetzen, 1981). Boyd and Chua (1993) showed that the I/O measurements alone are sufficient to classify these structures because of the unique I/O mappings. Therefore, depending on the measured data, a certain structure can be accepted or rejected based on satisfying the kernel relationships constituting a sufficient and necessary condition for a system to have a given structure. Parameter estimation of these structures is usually based on measuring the first-and second-order Volterra or Weiner kernels that are used to define the linear filters incorporated in a given structure. Lee and Schetzen (1981) developed an approach for measuring Volterra kernels leading to the complete identification process of a nonlinear system. On the other hand, frequency-domain methods based on the multifrequency excitation were developed in Chang (1990), Lunsford (1993), Rhyne and Steer (1987) and Rhyne (1988) to extract the Volterra transfer functions. However, these methods become inefficient and even computationally prohibitive for high-order kernels.

Since the parameter extraction of Volterra-based models is not within the scope of this book, the readers are referred to the following references for more details: Boyd and Chua (1993), Bendat (1990), Chang (1990), Korenberg (1991), Lunsford (1993), Rhyne and steer (1987), Rhyne (1988) and Schetzen (1981).

3.4 Summary

In this chapter, the most common models of nonlinearity and their parameter extraction have been reviewed. Two different types of models have been presented: the first are analytical models based on Volterra series analysis and its variants and the second are empirical models that can be developed from intuition about the nonlinear system supported by measurements. Volterra series provides an analytical approach to arbitrary system level modeling and can be related to circuit parameters. The choice of the model is a tradeoff between the level of needed accuracy and the simplicity of the model with respect to parameter extraction and computer simulations. Volterra-based models as well as other nonlinear models will be revisited in the following chapters with the objective of analyzing distortion introduced by nonlinear amplifiers.

4

Nonlinear Transformation of Deterministic Signals

Nonlinear distortion can be defined as the signal components, other than the original signal, which are produced by nonlinear transformation of an input signal. The simplest form of nonlinear distortion can be understood when a discrete tone is input to a saturation device. Saturation characteristics are the basic form of nonlinearity where the output power of a saturation device increases linearly on increasing the input power up to a certain saturation point and then becomes constant beyond that point. In this case, when the input power is close to the saturation power of the device, nonlinear distortion is manifested as the production of a large number of harmonics at frequencies commensurate with the input tone frequency. This means that the output power of the nonlinearity is distributed between the harmonics and the fundamental tone, and hence the output power of the fundamental tone will be reduced over the case when the same single tone is input to a linear device, a phenomena called gain compression. This view of nonlinear distortion implies that such a phenomena needs to be observed in the frequency domain, however, its manifestation in the time domain can also be observed, where nonlinear distortion appears as a transformation of the signal into another signal with regard to amplitude and frequency.

More complex input signals (such as multiple tones) can also be used to view and understand nonlinear distortion. In this case, nonlinear distortion takes more sophisticated manifestations in both time and frequency domains. However, the generation of extra signal components and the gain compression of the fundamental signal components are still the main manifestation of the nonlinear phenomena. With multiple tones, and as will be seen next, the nonlinear transformation produces signal components at frequencies that depend on the mixing properties of the nonlinear characteristics. This makes analyzing the nonlinear output a more challenging task as it becomes hard to distinguish the effective signal and the effective distortion components. In these cases, a mathematical process needs to be performed in order to develop the analysis of the nonlinear output into signal and effective nonlinear distortion.

Nonlinear Distortion in Wireless Systems: Modeling and Simulation with MATLAB®, First Edition.
Khaled M. Gharaibeh.
© 2012 John Wiley & Sons, Ltd. Published 2012 by John Wiley & Sons, Ltd.

In circuit- and system-level simulations, multisine signals have been the main tool used for predicting nonlinear distortion in circuits and systems. The rationale for using discrete tones is that they require lower computational complexity than that required by direct use of the actual communication signals. Discrete-tone representations of signals have direct application in harmonic balance simulation (Rizzoli *et al.*, 1999) and in measurement of nonlinear microwave circuits (Vandermot *et al.*, 2006). Moreover, discrete tone analysis leads to an analytic evaluation of distortion with simple expressions for nonlinear system figures of merit such as Intermodulation Ratio (IMR), SNR, Adjacent-Channel Interference (ACI), etc.

This chapter presents analysis of nonlinear transformation of deterministic signals using different models of nonlinearity. The analysis is done at the complex envelope level where the complex envelope of the output signal is determined using single and multiple input signals. The analysis enables distortion in nonlinear amplifiers to be predicted for any deterministic signal including single-tone and multisine signals.

4.1   Complex Baseband Analysis and Simulations

The very high carrier frequency of RF systems make computer simulations of communication signals inefficient since the sampling frequency of RF modulated signals is proportional to the carrier frequency. This means that a high number of samples need to be processed by a simulator, which results in long simulation time and the need for a large memory size.

Complex baseband analysis of nonlinear RF systems provides a tool for the efficient computer simulations of such systems. In complex baseband simulations, only the information signal (the complex envelope of the modulated signal), and not the carrier, is processed by a simulator. Therefore, with complex baseband simulations, the carrier frequency of a modulated signal is translated to zero, which results in a much lower sampling rate, and hence, more efficient simulations. Furthermore, in system simulations, the transfer functions of the system must also be transformed into the complex baseband by the same process. In this case, the resulting system transfer function is called the complex baseband equivalent of the system.

Complex baseband analysis is based on deriving the complex envelope of a bandpass signal (or the baseband equivalent of system impulse response) which has a non-negligible energy around some carrier frequency f_c. Bandpass signals result from modulating a carrier by a baseband signal, or from filtering a signal with a bandpass filter. Bandpass signals are commonly used to model transmitted and received signals in communication systems. The complex envelope of a bandpass signal represents the baseband equivalent of a bandpass signal that consists of the information signal only while excluding the carrier. Bandpass signals are real signals since they are generated by circuitry that can only generate real sinusoids. Bandpass signals could also be transmitted through bandpass channels where the channel just introduces an amplitude and phase change at each frequency of the real transmitted signal.

With nonlinear systems, complex baseband analysis is needed to study the effect of nonlinearity on the performance of communication systems in a specified frequency band of interest. Therefore, out-of-band signal components that result from nonlinear behavior outside the band of interest (for example harmonics and intermodulation products) are lost. If these frequency bands are to be included in the simulations, then the complex baseband analysis needs to be extended to deriving the baseband nonlinear model, and hence the nonlinear output. Therefore, with envelope simulations, the nonlinear transfer characteristics are converted to envelope transfer characteristics that relate the complex envelope of the output to the complex envelope of the input with respect to a certain carrier frequency. However, nonlinear system analysis and simulations are not always based on this form since we are also interested in estimating harmonics that might be generated by nonlinearity.

4.1.1 Complex Envelope of Modulated Signals

A bandpass signal is a real signal with frequency spectrum concentrated around some carrier frequency f_c in a finite bandwidth, say $2B$. Bandpass signals result either from modulation of a baseband signal or from bandpass filtering of a signal with a bandpass filter. Bandpass signals are real signals that are used to model transmitted signal in communication systems where the transmitted signal is generated as a sinusoid with varying amplitude, phase or frequency.

In computer simulations, bandpass signals require high sampling rates since the sampling frequency must be multiple of the highest frequency in the signal by the Nyquist theorem. This means high memory storage requirements and long simulation times since simulation of each period of the modulating signal requires considering a large number of periods of the carrier signal. Because the carrier signal does not carry any information, it is usually useful to simulate the "complex envelope" of the modulated signal since this contains all the information in the modulated signal. The complex envelope of a modulated signal is the baseband equivalent signal of a modulated signal that contains all the information and at the same time has a maximum frequency that is much lower than the carrier frequency. This means that much lower sampling frequency is required in simulations and hence, much shorter simulation time than the real modulated signal. A simple illustration of this concept is shown in Figure 4.1 where it is shown that when the modulated signal spectrum is shifted to baseband, the sampling frequency required by computer simulations becomes much lower than the case with a real modulated signal.

To establish the mathematical formulation of the complex envelope of a bandpass signal, consider a modulated signal $x(t)$ at carrier frequency f_c in the following form:

$$x(t) = A(t) \cos(2\pi f_c t + \Phi(t)) \qquad (4.1)$$

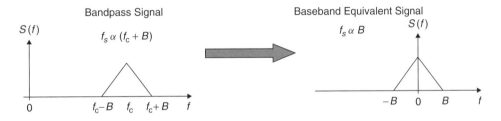

Figure 4.1 Concept of envelope simulations.

where $A(t)$ represents the amplitude modulation and $\Phi(t)$ represents the phase modulation. Using trigonometric identities, the modulated signal can be written as

$$x(t) = i(t)\cos(2\pi f_c t) - q(t)\sin(2\pi f_c t) \tag{4.2}$$

where $i(t)$ and $q(t)$ are real lowpass (baseband) signals of bandwidth $B << f_c$. The signal $i(t)$ is called the *in-phase* component and the signal $q(t)$ is called the *quadrature* component.

Define the signal $\tilde{x}(t) = i(t) + jq(t)$, where $i(t) = \Re\{\tilde{x}(t)\}$ and $\tilde{q}(t) = \Im\{\tilde{x}(t)\}$, then $\tilde{x}(t)$ is called the complex envelope of $x(t)$, which is a complex signal with bandwidth B. The modulated signal can then be written as

$$x(t) = \Re(\tilde{x}(t))\cos(2\pi f_c t) - \Im(\tilde{x}(t))\sin(2\pi f_c t) = \Re(\tilde{x}(t))e^{j2\pi f_c t} \tag{4.3}$$

The complex envelope of the modulated signal $\tilde{x}(t)$ can also be written as

$$\tilde{x}(t) = A(t)e^{j\Phi(t)} \tag{4.4}$$

where

$$A(t) = \sqrt{i^2(t) + q^2(t)} \tag{4.5}$$

and

$$\Phi(t) = \tan^{-1}\frac{i(t)}{q(t)} \tag{4.6}$$

and hence the modulated signal can be written as

$$x(t) = \Re\{A(t)e^{j\Phi(t)}e^{j2\pi f_c t}\} = A(t)\cos(2\pi f_c t + \Phi(t)) \tag{4.7}$$

Another equivalent form can be developed from Equation (4.3) and Equation (4.4)

$$x(t) = \frac{1}{2}\tilde{x}(t)e^{j2\pi f_c t} + \frac{1}{2}\tilde{x}(t)e^{-j2\pi f_c t} \tag{4.8}$$

In the frequency domain, the spectrum of the modulated signal in terms of the spectrum of the complex envelope can be developed using properties of the Fourier transform. It can be shown, by taking the Fourier transform of Equation (4.8), that

$$X(f) = \frac{1}{2}\left[\tilde{X}(f - f_c) + \tilde{X}(-f - f_c)\right] \tag{4.9}$$

where $\tilde{X}(f)$ is the Fourier transform of $\tilde{x}(t)$ with a bandwidth of B. Note that since $x(t)$ is real, its Fourier transform is symmetric around $f = 0$, however, $\tilde{X}(f)$ is not necessarily symmetric around $f = 0$ since it is complex in general. If $\tilde{x}(t)$ is real, that is, if $q(t) = 0$, then $\tilde{X}(f)$ is only conjugate symmetric about $f = 0$. Note that, in this case where $x(t)$ is symmetric about $f = 0$, the modulated signal $x(t)$ will be symmetric about carrier frequency f_c (Goldsmith, 2005).

The process of deriving the complex envelope of a bandpass signal consists of three main sub processes as shown in Figure 4.1. First, a quadrature representation of the real modulated signal is performed. Then, a Hilbert transformed version of the real modulated signal is added to the real modulated signal in order to remove the negative frequency part of the modulated signal. The Hilbert Transformation (HT) can be viewed as a $90°$ phase shifting of the real bandpass signal that results in an analytical signal and is defined as

$$\hat{x}(t) = \mathrm{HT}[x(t)] = \frac{1}{\pi} \int_{-\infty}^{\infty} \frac{x(\tau)}{t - \tau} d\tau \qquad (4.10)$$

In the frequency domain, HT can be defined using Fourier transform of Equation (4.10):

$$\hat{X}(f) = -j\mathrm{sgn}(f)X(f) \qquad (4.11)$$

In a final step, the analytical signal is down converted to baseband resulting in the complex envelope of the modulated signal.

Therefore, given a bandpass signal $x(t)$, its complex envelope $\tilde{x}(t)$ can be found as follows (See Figure 4.2):

- For a bandpass signal that has a frequency response as shown in Figure 4.2(a), find the Hilbert transform of the signal; $\hat{x}(t)$ as:

$$\hat{x}(t) = \mathrm{HT}\left[i(t)\cos(2\pi f_c t) - q(t)\sin(2\pi f_c t)\right] = i(t)\sin(2\pi f_c t) + q(t)\cos(2\pi f_c t) \qquad (4.12)$$

- Add $\hat{x}(t)$ to the bandpass signal as an imaginary component:

$$y(t) = x(t) + j\hat{x}(t) \qquad (4.13)$$

which yields

$$y(t) = (i(t) + jq(t))\, e^{j2\pi f_c t} \qquad (4.14)$$

This results in removing the negative frequency part of the signal spectrum as show in Figure 4.2(b).
- The complex envelope of $x(t)$ is found by down conversion of $y(t)$ to the baseband (multiplying $y(t)$ by $e^{-j2\pi f_c t}$):

$$\tilde{x}(t) = y(t)e^{-j2\pi f_c t} = (i(t) + jq(t))\, e^{j2\pi f_c t} e^{-j2\pi f_c t} = i(t) + jq(t) \qquad (4.15)$$

This will shift the positive frequency part of the spectrum of $x(t)$ to the baseband and yields the spectrum of the complex envelope of the bandpass signal as shown in Figure 4.2(c).

Note that the complex envelope of a bandpass signal consists of the amplitude- and phase-modulation information of the modulated signal. Because the resulting baseband signal is complex valued, it has double the bandwidth of the original bandpass signal.

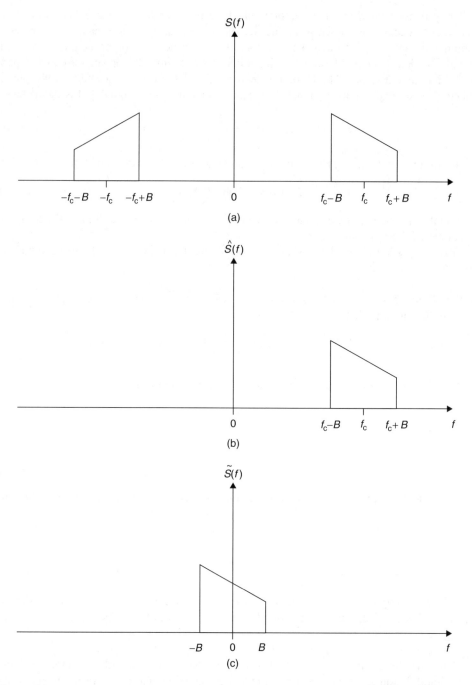

Figure 4.2 Complex envelope; (a) Hilbert transform (b) multiplication by $e^{j2\pi f_c t}$ and (c) complex envelope.

4.1.2 Baseband Equivalent of Linear System Impulse Response

Bandpass linear systems such as linear filters, linear bandpass amplifiers, etc. are usually used in RF front ends of wireless systems and are designed to have a finite bandwidth around a center RF frequency. For linear bandpass systems (filters), the system impulse response, denoted by $h(t)$, is real with frequency response $H(f)$ centered at some carrier frequency f_c and has a bandwidth $B << f_c$. The low-pass equivalent of the system impulse response can be developed in a similar way to the complex envelope of bandpass signals. Therefore, in a similar manner to Equation (4.7) we can write:

$$h(t) = \Re\{\tilde{h}(t)e^{j2\pi f_c t}\} \tag{4.16}$$

where $\tilde{h}(t)$ is called the baseband equivalent of the system impulse response. Another form is similar to Equation (4.8):

$$h(t) = \frac{1}{2}\tilde{h}(t)e^{j2\pi f_c t} + \frac{1}{2}\tilde{h}(t)e^{-j2\pi f_c t} \tag{4.17}$$

In the frequency domain, the system frequency response can be found from the Fourier transform of Equation (4.17):

$$H(f) = \frac{1}{2}\tilde{H}(f - f_c) + \frac{1}{2}\tilde{H}^*(-f - f_c) \tag{4.18}$$

where $\tilde{H}(f)$ is the Fourier transform of $\tilde{h}(t)$ and is called the baseband equivalent of the frequency response of the system. Theretofore, $H(f)$ consists of two components: $\tilde{H}(f)$ shifted up by f_c, and $\tilde{H}(f)$ shifted down by f_c. Similar to bandpass signals, the frequency response of bandpass systems possess the property that if $H(f)$ is conjugate symmetric about the carrier frequency f_c within the bandwidth B then $\tilde{h}(t)$ will be real and its frequency response $\tilde{H}(f)$ is conjugate symmetric about zero. However, in many applications, $H(f)$ is not conjugate symmetric about f_c and hence, $\tilde{h}(t)$ is complex with in-phase component $\tilde{h}_I(t) = \Re\{\tilde{h}(t)\}$ and quadrature component $\tilde{h}_q(t) = \Im\{\tilde{h}(t)\}$. Note that if $\tilde{h}(t)$ is complex then $\tilde{H}(f)$ is not conjugate symmetric about zero (Goldsmith, 2005).

To study the baseband equivalent of the response of a bandpass linear system to a bandpass input signal, let $y(t)$ denote the output of the system then

$$y(t) = x(t) * h(t) = \int_{-\infty}^{\infty} h(\tau)x(t - \tau)d\tau \tag{4.19}$$

where $x(t)$ and $h(t)$ are both real. Given that $x(t)$ and $h(t)$ are bandpass, the output of the system is also bandpass and hence, it has a complex envelope representation of the form

$$y(t) = \Re\left\{\tilde{y}(t)e^{j2\pi f_c t}\right\} \tag{4.20}$$

In the frequency domain, the spectrum of the output can be written using the Fourier transform of Equation (4.19) as

$$Y(f) = H(f)X(f) = \frac{1}{2}\left[\tilde{H}(f - f_c) + \tilde{H}^*(-f - f_c)\right]\left[\tilde{X}(f - f_c) + \tilde{X}^*(-f - f_c)\right] \tag{4.21}$$

For bandpass signals and systems, the bandwidth of $\tilde{x}(t)$ and $\tilde{h}(t)$ is much less than the carrier frequency f_c. Hence,

$$\tilde{H}(f - f_c)\tilde{X}^*(-f - f_c) = 0 \tag{4.22}$$

and

$$\tilde{H}^*(-f - f_c)\tilde{X}(f - f_c) = 0. \tag{4.23}$$

Substituting Equation (4.22) and Equation (4.23) into Equation (4.21), we have

$$Y(f) = \frac{1}{2}\left(\tilde{Y}(f - f_c) + \tilde{Y}^*(-f - f_c)\right) \tag{4.24}$$

where

$$\tilde{Y}(f - f_c) = \tilde{X}(f - f_c)\tilde{H}(f - f_c)$$
$$\tilde{Y}^*(-f - f_c) = \tilde{X}^*(-f - f_c)\tilde{H}^*(-f - f_c). \tag{4.25}$$

or, equivalently, that

$$\tilde{Y}(f) = \tilde{H}(f)\tilde{X}(f) \tag{4.26}$$

Taking the inverse Fourier transform yields:

$$\tilde{y}(t) = \tilde{x}(t) * \tilde{h}(t). \tag{4.27}$$

Thus, the complex envelope of the system output can be obtained by taking the convolution of $\tilde{h}(t)$ and $\tilde{x}(t)$. The received signal is, therefore, given by

$$y(t) = \Re\{\left(\tilde{x}(t)\tilde{h}(t)\right)e^{j2\pi f_c t}\} \tag{4.28}$$

Note that the complex envelope of the system output is complex with nonzero in-phase and quadrature components if either $\tilde{x}(t)$ or $\tilde{h}(t)$ is complex. In general, if $\tilde{x}(t) = i(t) + jq(t)$ and $\tilde{h}(t) = h_I(t) + jh_Q(t)$ then

$$\tilde{y}(t) = [i(t) + jq(t)][h_I(t) + jh_Q(t)]$$
$$= [i(t)h_I(t) + q(t)h_Q(t)] + j[i(t)h_Q(t) + q(t)h_Q(t)]. \tag{4.29}$$

which is always a complex quantity.

Note that the sampling frequency required for simulation of the system is commensurate with the bandwidth of its baseband equivalent and not its center frequency. This means that great savings in computation time and hence, CPU usage can be achieved while maintaining all the details of the system.

4.2   Complex Baseband Analysis of Memoryless Nonlinear Systems

A general bandpass nonlinearity consists of a memoryless nonlinearity followed by a bandpass filter. The bandpass filter works as a zonal filter and extracts the output of the nonlinearity at a given center frequency f_c and a given bandwidth. To establish the model for a bandpass nonlinearity, consider a bandpass input signal of the form:

$$x(t) = A(t) \cos(2\pi f_c t + \theta(t)) \tag{4.30}$$

where $A(t)$ represents the amplitude modulation and $\theta(t)$ represents the phase modulation. A general bandpass memoryless system is defined as:

$$y(t) = N[x(t)] \tag{4.31}$$

where N is an arbitrary memoryless nonlinear operator.

The output of the memoryless nonlinearity around the carrier frequency f_c (first zonal output) can be expressed as shown in Chapter 3 as

$$y(t) = g_1 \left[A(t)) \cos(2\pi f_c t + \theta(t) \right] + g_2 \left[A(t) \right] \sin(2\pi f_c t + \theta(t)) \tag{4.32}$$

where $g_1(\cdot)$ and $g_2(\cdot)$ are as defined in the previous chapter

$$g_1(A(t)) = \frac{1}{\pi} \int_0^{2\pi} N[A(t) \cos(\alpha)] \cos(\alpha) d\alpha \tag{4.33}$$

and

$$g_2(A(t)) = \frac{1}{\pi} \int_0^{2\pi} N[A(t) \cos(\alpha)] \sin(\alpha) d\alpha \tag{4.34}$$

and $\alpha = 2\pi f_c t + \theta(t)$.

The complex envelope of the nonlinear output with respect to f_c can be expressed in terms of the complex envelope of the input signal as

$$\tilde{y}(t) = \left[g_1(|\tilde{x}(t)|) - j g_2(|\tilde{x}(t)|) \right] e^{j\theta(t)} = F(|\tilde{x}(t)|) e^{j[\Theta(|\tilde{x}(t)|) + \theta(t)]} \tag{4.35}$$

where $\tilde{x}(t) = A(t) e^{j\theta(t)}$ is the complex envelope of $x(t)$ with respect to f_c,

$$F(|\tilde{x}(t)|) = \sqrt{g_1^2(|\tilde{x}(t)|) + g_2^2(|\tilde{x}(t)|)} \tag{4.36}$$

and

$$\Theta(|\tilde{x}(t)|) = \tan^{-1} \left(\frac{g_2(|\tilde{x}(t)|)}{g_1(|\tilde{x}(t)|)} \right) \tag{4.37}$$

Hence, the bandpass output in Equation (4.32) can be written in the form:

$$y(t) = F(|\tilde{x}(t)|) \cos(2\pi f_c t + \Theta(|\tilde{x}(t)|) + \theta(t)) \tag{4.38}$$

The function F describes the input/output amplitude conversion (AM–AM conversion), whereas the function Θ describes the influence of the input amplitude on the output phase (AM-PM conversion). Note that the envelope version of the memoryless model represents the nonlinear relationship between the complex envelopes of the input and the output waveforms denoted by $\tilde{x}(t)$ and $\tilde{y}(t)$. Note also that when both functions (g_1 and g_2) are nonzero, the envelope nonlinear transfer characteristics are complex, which implies both amplitude and phase variations to the input signal. If $g_2 = 0$, then these characteristics reduce to real transfer characteristics that imply only amplitude variations (no AM-PM conversion).

Another commonly used form for the output of a memoryless nonlinearity in a complex envelope form is

$$\tilde{y}(t) = \tilde{x}(t)G\left(\tilde{x}(t)\right) \tag{4.39}$$

where G is a complex gain function defined as

$$G\left(\tilde{x}(t)\right) = \frac{F(|\tilde{x}(t)|)}{|\tilde{x}(t)|}e^{j\Theta(|\tilde{x}(t)|)} \tag{4.40}$$

The gain function G is usually simpler to calculate and measure than the functions F and Θ.

4.2.1 Power-Series Model

For a strictly memoryless system, the power-series model is expressed as

$$y(t) = N(x(t)) = \sum_{n=1}^{N} a_n x^n(t). \tag{4.41}$$

where a_n are the instantaneous coefficients. This implies that $g_2 = 0$ and hence, the complex envelope of the output signal with respect to the fundamental frequency f_c can be expressed as an envelope power-series model as (Gharaibeh, 2004)

$$\tilde{y}(t) = F(|\tilde{x}(t)|)e^{j\Theta(t)} = \sum_{\substack{n=1 \\ n:\text{odd}}}^{N} b_n|\tilde{x}(t)|^{n-1}\tilde{x}(t) \tag{4.42}$$

where the real coefficients (b_n) represent the envelope coefficients that are directly related to the coefficients a_n by:

$$b_n = \frac{1}{2^{n-1}}\binom{n}{\frac{n+1}{2}}a_n \tag{4.43}$$

In the general case when $g_2 \neq 0$, the envelope power series can be expressed in a similar form to Equation (4.40) but with complex coefficients b_n. Note that in this case, the model is an envelope power-series model that is not necessarily directly related to an instantaneous power-series model as in Equation (4.41) (Raich et al., 2004). The complex gain function in Equation (4.40) can be written as (Zhou et al., 2005)

$$G\left(\tilde{x}(t)\right) = \sum_{\substack{n=1 \\ n:\text{odd}}}^{N} b_n|\tilde{x}(t)|^{n-1} \tag{4.44}$$

4.2.2 Limiter Model

A bandpass limiter can be modeled as a limiter followed by a bandpass (zonal) filter according to the general model of a bandpass nonlinearity. Hence, with a bandpass input $x(t)$, the output is given by the first term on the right side of Equation (4.35). The phase is unaffected, but the output envelope, which is the modulus of the complex envelope, is given by

$$|\tilde{y}(t)| = g_1(|\tilde{x}(t)|) \tag{4.45}$$

where $|\tilde{x}(t)|$ is the input envelope (the same as $A(t)$). The function g_1 is calculated from the Chebyshev transforms in Equation (3.33) and using Equation (3.46) for the definition of the nonlinear operator N.

Hard-limiter characteristics are obtained by setting $l = 0$ and arbitrarily setting the parameter s. For convenience s could be set to unity and it can be shown that the transfer characteristics in complex envelope form are (Jeruchem *et al.*, 2000):

$$\tilde{y}(t) = \frac{4L}{\pi} e^{j\phi(t)} \tag{4.46}$$

For soft limiters, we let $s \to \infty$ and hence, the transfer characteristics in the complex envelope form can be obtained using Equation (3.44) as (Jeruchem *et al.*, 2000):

$$\tilde{y}(t) = \begin{cases} g_1(|\tilde{x}(t)|)e^{j\phi(t)}, & |\tilde{x}(t)| \geq l; \\ \frac{L}{l}e^{j\phi(t)}, & |\tilde{x}(t)| < l. \end{cases} \tag{4.47}$$

where

$$g_1(|\tilde{x}(t)|) = \frac{2L|\tilde{x}(t)|}{\pi l}\left[\sin^{-1}\left(\frac{l}{|\tilde{x}(t)|}\right) + \frac{l}{|\tilde{x}(t)|}\sqrt{1 - \frac{l^2}{|\tilde{x}(t)|^2}}\right] \tag{4.48}$$

Figure 4.3 shows the baseband and the passband limiter characteristics for both hard- and soft-limiter models.

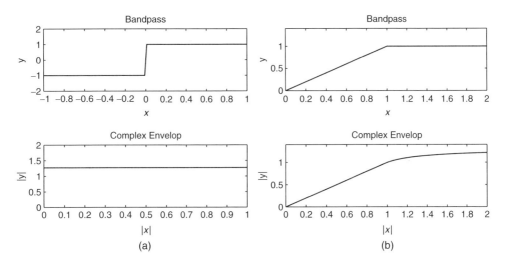

Figure 4.3 Limiter characteristics: (a) hard limiter and (b) soft limiter (Jeruchem *et al.*, 2000).

4.3 Complex Baseband Analysis of Nonlinear Systems with Memory

4.3.1 Volterra Series

Volterra series represents the output of the system as a convolution of powers of the input with the Volterra kernels. In this section we develop a complex envelope analysis of the output waveform for the Volterra series model described in Chapter 3. We start the analysis by considering an input that consists of a single modulated carrier centered at frequency ξ and applied to a nonlinear amplifier. The input signal can be written in terms of its complex envelope as

$$
\begin{aligned}
x(t) &= Re\{\tilde{x}(t)e^{j2\pi\xi t}\} \\
&= \frac{1}{2}\tilde{x}(t)e^{j2\pi\xi t} + \frac{1}{2}\tilde{x}^*(t)e^{-j2\pi\xi t}.
\end{aligned}
\tag{4.49}
$$

The output of the Volterra model can be expressed by substituting Equation (4.49) into Equation (3.2):

$$
\begin{aligned}
y_n(t) = &\int_{-\infty}^{\infty} \dots \int_{-\infty}^{\infty} h_n(\lambda_1, \dots, \lambda_n) \\
&\times \prod_{i=1}^{n} \frac{1}{2}\left(\tilde{x}(t-\lambda_i)e^{j2\pi\xi(t-\lambda_i)} + \tilde{x}^*(t-\lambda_i)e^{-j2\pi\xi(t-\lambda_i)}\right) d\lambda_i.
\end{aligned}
\tag{4.50}
$$

The component of $y_n(t)$ centered at the fundamental frequency can be expressed in complex envelope form as (Jeruchem *et al.*, 2000)

$$
\begin{aligned}
\tilde{y}_n(t) = &\frac{1}{2^{\frac{n+1}{2}}}\binom{n}{\frac{n-1}{2}}\int_{-\infty}^{\infty} \dots \int_{-\infty}^{\infty} \tilde{h}_n(\tau_1, \dots, \lambda_n) \\
&\prod_{i=1}^{\frac{n+1}{2}}\tilde{x}(t-\tau_i) \prod_{i=\frac{n+3}{2}}^{n} \tilde{x}^*(t-\lambda_i)d\lambda_1 \dots d\lambda_n
\end{aligned}
\tag{4.51}
$$

where $\frac{1}{2^{\frac{n+1}{2}}}\binom{n}{\frac{n-1}{2}}\tilde{h}_n(\lambda_1, \dots, \lambda_n)$ is the low pass-equivalent of the Volterra kernel $h_n(\lambda_1, \dots, \lambda_n)$. Hence, the output of the Volterra model in complex envelope form can be written as:

$$
\begin{aligned}
\tilde{y}(t) = &\sum_{n=0}^{N}\frac{1}{2^{\frac{n+1}{2}}}\binom{n}{\frac{n-1}{2}}\int_{-\infty}^{\infty} \dots \int_{-\infty}^{\infty} \tilde{h}_n(\tau_1, \dots, \lambda_n) \\
&\prod_{i=1}^{\frac{n+1}{2}}\tilde{x}(t-\lambda_i) \prod_{i=\frac{n+3}{2}}^{n} \tilde{x}^*(t-\lambda_i)d\lambda_1 \dots d\lambda_n
\end{aligned}
\tag{4.52}
$$

4.3.2 Single-Frequency Volterra Models

For the single-frequency filter-nonlinearity Volterra model, we use the kernel relationship Equation (3.15). It follows that the output $y_n(t)$ can be written as

$$y_n(t) = \int_{-\infty}^{\infty} \cdots \int_{-\infty}^{\infty} a_n \prod_{i=1}^{n} h_n(\lambda_i) x(t - \lambda_i) d\lambda_i$$

$$= a_n u_n^n(t) \tag{4.53}$$

where

$$u_n(t) = \int_{-\infty}^{\infty} h_n(\lambda) x(t - \lambda) d\lambda \tag{4.54}$$

represents the filtered input. Therefore, the nth-order output consists of powers of the filtered input waveforms by the linear filter of the nth branch. Applying the same analysis of the memoryless nonlinearity, the complex envelope of the output signal in the vicinity of a frequency ξ can be written as

$$\tilde{y}(t) = \sum_{\substack{n=1 \\ n:\text{odd}}}^{N} \frac{1}{2^{n-1}} \binom{n}{\frac{n+1}{2}} a_n |\tilde{u}_n(t)|^{n-1} \tilde{u}_n(t) \tag{4.55}$$

where

$$\tilde{u}_n(t) = \int_{-\infty}^{\infty} \tilde{h}_n(\lambda) \tilde{x}(t - \lambda) d\lambda \tag{4.56}$$

and $\tilde{h}_n(t)$ is the baseband equivalent of the filter of the nth branch with respect to frequency ξ.

For the nonlinearity filter model, the kernel relationship Equation (3.17) enables the output of the system to be written as

$$y_n(t) = a_n \int_{-\infty}^{\infty} h_n(\lambda) x^n(t - \lambda) d\lambda. \tag{4.57}$$

Using Equation (4.52), the complex envelope of the nth-order output can be written as

$$\tilde{y}_n(t) = \frac{1}{2^{n-1}} \binom{n}{\frac{n+1}{2}} a_n \int_{-\infty}^{\infty} \tilde{h}_n(\lambda) |\tilde{x}_n(t - \lambda)|^{n-1} \tilde{x}(t - \lambda) \tag{4.58}$$

where $\tilde{h}_n$ is the baseband equivalent of the filter $h_n(t)$ with respect to the frequency ξ. Hence, the complex envelope of the output waveform $y(t)$ is

$$\tilde{y}(t) = \sum_{n=1}^{N} \tilde{y}_n(t) \tag{4.59}$$

4.3.3 Wiener-Hammerstein Model

As shown in Chapter 3, the Wiener-Hammerstein model consists of three major blocks: an input Linear Time Invariant (LTI) filter with frequency response $H_1(f)$, a static non-linearity G and an output LTI filter with a frequency response $H_2(f)$. In our present work the static nonlinearity, is described by the power series

$$z(t) = \sum_{n=0}^{N} a_n u^n(t). \tag{4.60}$$

Now consider an input signal $x(t)$ applied to the nonlinear amplifier, then the signal $u(t)$ at the output of the input filter can be written as a convolution of the input signal with the input filter impulse response as

$$u(t) = \int_{-\infty}^{\infty} h_1(\lambda)x(t - \lambda)d\lambda \tag{4.61}$$

where $h_1(t)$ is the impulse response of the input filter and

$$u(t) = \int_{-\infty}^{\infty} h_1(\lambda)x(t - \lambda)d\lambda. \tag{4.62}$$

Finally, the total output of the model is

$$y(t) = \int_{-\infty}^{\infty} h_2(\lambda)z(t - \lambda)d\lambda \tag{4.63}$$

where $h_2(t)$ is the impulse response of the output filter. In order to estimate distortion (particularly cross-modulation) introduced by the interaction of multiple signals, a closed form expression for the output waveform at a particular frequency in complex envelope form is derived. The input to the nonlinear block, $u(t)$, can be written in complex form as (Jeruchem et al., 2000)

$$\tilde{u}(t) = \int_{-\infty}^{\infty} \tilde{h}_1(\lambda)\tilde{x}(t - \lambda)d\lambda \tag{4.64}$$

where $\tilde{h}_1(t)$ is the baseband equivalent impulse response of the input filter with respect to frequency ξ and $\tilde{u}(t)$ is the complex envelope of the input signal $x(t)$. Following the analysis presented in Section 4.3 for Volterra series, the complex envelope of the nth-order output of the nonlinear block becomes

$$\tilde{z}(t) = \sum_{\substack{n=1 \\ n:\text{odd}}}^{N} \frac{1}{2^{n-1}} \binom{n}{\frac{n+1}{2}} a_n |\tilde{u}_n(t)|^{n-1} \tilde{u}_n(t) \tag{4.65}$$

Thus, the complex envelope of the output signal $y(t)$ can now be written in complex envelope form as

$$\tilde{y}(t) = \sum_{n=1}^{N} \int_{-\infty}^{\infty} \tilde{h}_2(\lambda)\tilde{z}(t - \lambda)d\lambda \tag{4.66}$$

where $\tilde{h}_2(t)$ is the baseband equivalent impulse response of the output filter with respect to frequency ξ.

4.4 Complex Envelope Analysis with Multiple Bandpass Signals

Multichannel wideband nonlinear amplifiers have been introduced to replace multiple single channel narrowband amplifiers in order to achieve better spectrum and hardware utilization. However, using multichannel amplifiers results in interaction among input signals by nonlinearity complicates modeling of nonlinear distortion (Gharaibeh and Steer, 2005). With multichannel systems, the nonlinear behavior results in components at all the intermodulation frequencies of the input frequencies that consist of contributions from all the input signals. Specifically, cross-modulation distortion represents the contribution of other signals in the output spectrum of the desired signal.

The analysis of nonlinear multichannel systems presented in this section enables cross-modulation distortion to be quantified. The complex envelope of the output at a particular frequency (which results from the intermodulation of input channels) will be derived, as shown in Figure 4.4. The figure illustrates the concept of finding the output complex envelope of the output, for example the complex envelope of the signal centered at f_1, for an input that consists of the sum of band limited signals centered at different frequencies. The complex envelope of the output at f_1 consists of the response of the nonlinear system to the input signal at f_1 plus all the interactions of all input signals that lie on f_1. In this case, envelope simulations offer an advantage over time-domain simulations as the sampling frequency is commensurate with the bandwidth of individual channels and not with the frequency separation of input channels. This makes the simulation of multiple channels more practical and more numerically efficient.

4.4.1 Volterra Series

In order to develop a theoretical complex envelope analysis for the interaction of multiple signals by the nonlinear behavior, we use the analysis of general Volterra series addressed in Graham and Ehrmen (1973), Lunsford (1993) and Rhyne (1988) to develop the output waveform for the models described in Chapter 3. We start the analysis by considering an input that consists of the sum of K modulated carriers applied to a nonlinear amplifier so that:

$$x(t) = \sum_{k=1}^{K} x_k(t). \tag{4.67}$$

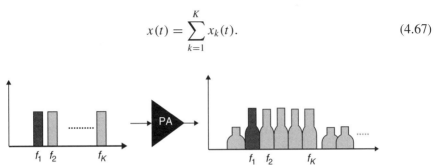

Figure 4.4 Principle of complex envelope analysis of nonlinear systems.

We will assume that the signals $x_k(t)$ are modulated RF carriers with center frequencies ξ_k and have equal bandwidths B_k. Now let $\tilde{x}_k(t)$ be the complex envelope of $x_k(t)$, then the $x_k(t)$ can be written in complex envelope form as

$$x_k(t) = Re\{\tilde{x}_k(t)e^{j2\pi\xi_k t}\}$$

$$= \frac{1}{2}\tilde{x}_k(t)e^{j2\pi\xi_k t} + \frac{1}{2}\tilde{x}_k^*(t)e^{-j2\pi\xi_k t}. \tag{4.68}$$

Taking the Fourier transform of Equation (4.67) yields

$$X_k(f) = \frac{1}{2}\tilde{X}_k(f - \xi_k) + \frac{1}{2}\tilde{X}_k^*(-f - \xi_k). \tag{4.69}$$

Then, $x(t)$ and its Fourier transform can be written as

$$x(t) = \frac{1}{2}\sum_{k=-K}^{K}\tilde{x}_k(t)e^{j2\pi\xi_k t} \tag{4.70}$$

and

$$X(f) = \frac{1}{2}\sum_{k=-K}^{K}\tilde{X}_k(f - \xi_k) \tag{4.71}$$

respectively, where $\tilde{x}_{-k} = \tilde{x}_k^*$ and $\xi_{-k} = -\xi_k^*$. The output of the Volterra system for a multiple input can be expressed by substituting Equation (4.71) into Equation (3.4):

$$y_n(t) = \int_{-\infty}^{\infty}\cdots\int_{-\infty}^{\infty} H_n(f_1,\ldots,f_n)$$

$$\times \prod_{i=1}^{n}\frac{1}{2}\sum_{k=-K}^{K}\tilde{X}_k(f_i - \xi_k)e^{j2\pi f_i t}df_i. \tag{4.72}$$

Now since

$$\prod_{i=1}^{n}\frac{1}{2}\sum_{k=-K}^{K}\tilde{X}_k(f_i - \xi_k) = \frac{1}{2^n}\sum_{k_1=-K}^{K}\cdots\sum_{k_n=-K}^{K}\prod_{i=1}^{n}\tilde{X}_{k_i}(f_i - \xi_{k_i}) \tag{4.73}$$

the output $y_n(t)$ can be written as (Lunsford, 1993)

$$y_n(t) = \frac{1}{2^n}\sum_{k_1=-K}^{K}\cdots\sum_{k_n=-K}^{K}\Lambda(t;\underline{\xi}) \tag{4.74}$$

where $\underline{\xi} = (\xi_{k_1},\ldots,\xi_{k_n})$ is an n-dimensional vector and represents the intermodulation frequency vector where each k_i ranges from $-K$ to K and $\tilde{X}_{k_i}$ is the complex envelope of $\tilde{X}_k$ with respect to the frequency ξ_{k_i}. The function Λ represents the following integral (Graham and Ehrmen, 1973):

$$\Lambda(t;\underline{\xi}) = \int_{-\infty}^{\infty}\cdots\int_{-\infty}^{\infty} H_n(f_1,\ldots,f_n)\prod_{i=1}^{n}\tilde{X}_{k_i}(f_i - \xi_{k_i})e^{j2\pi f_i t}df_i. \tag{4.75}$$

The integral Λ represents the component of the output waveform centered at the sum frequency $\xi = (\xi_{k_1} + \cdots + \xi_{k_n})$ (Graham and Ehrmen, 1973). In the frequency vector $\underline{\xi}$, each distinct frequency ξ_{k_i} occurs n_k times where $\sum_{k=-K}^{K} n_k = n$, therefore, there are $\binom{n}{n_{-K}, \ldots, n_K}$ such integrals. The component of $y_n(t)$ centered at ξ due to the intermodulation of input components centered at the frequency vector $(\xi_{k_1}, \ldots, \xi_{k_n})$ can be expressed in complex envelope form as (Graham and Ehrmen, 1973)

$$\tilde{y}_{n,\xi}(t) = \frac{1}{2^{n-1}} \sum_{n_{-K} + \cdots + n_K = n} \binom{n}{n_{-K}, \ldots, n_K}$$

$$\times \int_{-\infty}^{\infty} \cdots \int_{-\infty}^{\infty} \tilde{H}_{n,\underline{\xi}}(\underline{f}) \prod_{i=1}^{n} \tilde{X}_{k_i}(f_i) e^{j2\pi f_i t} df_i \tag{4.76}$$

where $\tilde{H}_{n,\underline{\xi}}(\underline{f})$ is the baseband equivalent of $H_n(\underline{f})$ with respect to the frequency vector $\underline{\xi}$. The total output of the system at a certain frequency ξ can then be expressed as

$$\tilde{y}_\xi(t) = \sum_{n=0}^{\infty} \frac{1}{2^{n-1}} \sum_{n_{-K} + \cdots + n_K = n} \binom{n}{n_{-K}, \ldots, n_K}$$

$$\times \int_{-\infty}^{\infty} \cdots \int_{-\infty}^{\infty} \tilde{H}_{n,\underline{\xi}}(\underline{f}) \prod_{i=1}^{n} \tilde{X}_{k_i}(f_i) e^{j2\pi f_i t} df_i. \tag{4.77}$$

This formulation establishes the basis for the analysis that follows for the models discussed in Chapter 2. In the following the output of each of these models will be derived for multiple input excitation.

4.4.2 Single-Frequency Volterra Models

For the single-frequency filter-nonlinearity Volterra model, we use the kernel relationship in Equation (3.15) and then it follows that the output $y_n(t)$ can be written as

$$y_n(t) = \int_{-\infty}^{\infty} \cdots \int_{-\infty}^{\infty} a_n \prod_{i=1}^{n} h_n(\lambda_i) x(t - \lambda_i) d\lambda_i$$

$$= \int_{-\infty}^{\infty} \cdots \int_{-\infty}^{\infty} a_n \prod_{i=1}^{n} H_n(f_i) X(f_i) e^{j2\pi f_i t} df_i$$

$$= \int_{-\infty}^{\infty} \cdots \int_{-\infty}^{\infty} a_n \prod_{i=1}^{n} U_n(f_i) e^{j2\pi f_i t} df_i$$

$$= a_n u_n^n(t) \tag{4.78}$$

where

$$u_n(t) = \int_{-\infty}^{\infty} H_n(f) X(f) e^{j2\pi ft} e^{j2\pi ft} df$$

$$= \int_{-\infty}^{\infty} U_n(f) e^{j2\pi ft} df \tag{4.79}$$

represents the filtered input. Therefore, the nth-order output consists of powers of the filtered input waveforms by the linear filter of the nth-branch. Applying the same analysis of the memoryless nonlinearity, the complex envelope of the output signal in the vicinity of a frequency ξ due to the intermodulation of input components centered at a frequency vector $\underline{\xi}$ can be written as

$$\tilde{y}_{n,\xi}(t) = \frac{a_n}{2^{n-1}} \sum_{n_{-K}+\cdots+n_K=n} \binom{n}{n_{-K}, \ldots, n_K} \prod_{i=1}^{n} \tilde{u}_{n,k_i}(t) \tag{4.80}$$

where

$$\tilde{u}_{n,k}(t) = \int_{-\infty}^{\infty} \tilde{h}_{n;\xi_k}(\lambda)\tilde{x}_k(t-\lambda)d\lambda \tag{4.81}$$

and $\tilde{h}_{n;\xi_k}(t)$ is the baseband equivalent of the filter of the nth-branch with respect to frequency ξ_k. Using Equation (4.77), the complex envelope in Equation (4.78) reduces to a power-series model with filtered inputs $u(t)$:

$$\tilde{y}_\xi(t) = \sum_{n=1}^{N} \frac{a_n}{2^{n-1}} \sum_{n_{-K}+\cdots+n_K=n} \binom{n}{n_{-K}, \ldots, n_K} \prod_{k=-K}^{K} \tilde{u}_{n,k}^{n_k}(t) \tag{4.82}$$

For the nonlinearity-filter model, the kernel relationship in Equation (3.17) enables the output of the system to be written as

$$y_n(t) = \int_{-\infty}^{\infty} h_n(\lambda)x^n(t-\lambda)d\lambda. \tag{4.83}$$

Now, let $x^n(t) = v_n(t)$, then the complex envelope of the output $v_n(t)$ with respect to frequency ξ is

$$\tilde{v}_{n,\xi}(t) = \frac{1}{2^{n-1}} \sum_{n_{-K}+\cdots+n_K=n} \binom{n}{n_{-K}, \ldots, n_K} \prod_{k=-K}^{K} \tilde{x}_k^{n_k}(t) \tag{4.84}$$

and hence, the complex envelope of the output waveform $y(t)$ is

$$\tilde{y}_\xi(t) = \sum_{n=1}^{N} \frac{1}{2^{n-1}} \sum_{n_{-K}+\cdots+n_K=n} \binom{n}{n_{-K}, \ldots, n_K} \int_{-\infty}^{\infty} \tilde{h}_{n,\xi}(\lambda) \prod_{k=-K}^{K} \tilde{x}_k^{n_k}(t-\lambda)d\lambda \tag{4.85}$$

where $\tilde{h}_{n,\xi}$ is the baseband equivalent of the filter $h_n(t)$ with respect to the frequency ξ.

4.4.3 Wiener-Hammerstein Model

For the Wiener-Hammerstein model, the static nonlinearity, is described by a power series as

$$z(t) = \sum_{n=0}^{N} a_n u^n(t). \tag{4.86}$$

Now consider a multichannel input signal $x(t)$ defined in Equation (4.67) applied to the nonlinear amplifier then the signal $u(t)$ at the output of the input filter can be written as a convolution of the input signal with the input filter impulse response as

$$u(t) = \int_{-\infty}^{\infty} h_1(\lambda)x(t - \lambda)d\lambda$$

$$= \sum_{k=1}^{K} u_k(t) \tag{4.87}$$

where $h_1(t)$ is the impulse response of the input filter and

$$u_k(t) = \int_{-\infty}^{\infty} h_1(\lambda)x_k(t - \lambda)d\lambda. \tag{4.88}$$

The output of the nonlinear block is then

$$z(t) = \sum_{n=0}^{N} z_n(t) \tag{4.89}$$

where

$$z_n(t) = a_n \left(\sum_{k=1}^{K} u_k(t) \right)^n \tag{4.90}$$

is the n-th order nonlinear response. Finally, the total output of the model is

$$y(t) = \int_{-\infty}^{\infty} h_2(\lambda)z(t - \lambda)d\lambda$$

$$= \sum_{n=1}^{N} \int_{-\infty}^{\infty} h_2(\lambda)z_n(t - \lambda)d\lambda \tag{4.91}$$

where $h_2(t)$ is the impulse response of the output filter. In order to estimate distortion (particularly cross-modulation) introduced by the interaction of multiple signals, a closed-form expression for the output waveform at a particular frequency in complex envelope form is derived. The input to the nonlinear block, $u_k(t)$, can be written in complex form as (Jeruchem *et al.*, 2000)

$$\tilde{u}_k(t) = \int_{-\infty}^{\infty} \tilde{h}_{1;\xi_k}(\lambda)\tilde{x}_k(t - \lambda)d\lambda \tag{4.92}$$

where $\tilde{h}_{1;\xi_k}(t)$ is the baseband equivalent impulse response of the input filter with respect to frequency ξ_k and $\tilde{x}_k(t)$ is the complex envelope of the input signal $x_k(t)$. Hence, the total input to the nonlinear block can be written as a complex conjugate pair as

$$u(t) = \sum_{k=-K}^{K} \frac{1}{2}\tilde{u}_k(t)e^{j2\pi\xi_k t} \tag{4.93}$$

where the minus sign notation here indicates complex conjugation so that $\tilde{u}_{-k} = \tilde{u}_k^*$ and $\xi_{-k} = -\xi_k$. Following the analysis in Section 4.3.1 for Volterra series, the nth-order output of the nonlinear block becomes

$$
z_n(t) = \frac{a_n}{2^n} \left(\sum_{k=-K}^{K} \tilde{u}_k(t) e^{j2\pi\xi_k t} \right)^n
$$

$$
= \frac{a_n}{2^n} \sum_{k_1=-K}^{K} \cdots \sum_{k_n=-K}^{K} \prod_{i=1}^{n} \tilde{u}_{k_i}(t) e^{j2\pi\xi_{k_i} t}
$$

$$
= \frac{a_n}{2^n} \sum_{k_1=-K}^{K} \cdots \sum_{k_n=-K}^{K} A(t) e^{j2\pi(\xi_{k_1}+\cdots+\xi_{k_n})t} \tag{4.94}
$$

where $A(t) = \prod_{i=1}^{n} \tilde{u}_{k_i}(t)$. The output $y(t)$ consists of components centered at all the intermodulation frequencies that result from the permutations of the input carrier frequencies represented by the frequency vector $(\xi_{k_1}, \ldots, \xi_{k_n})$. Therefore, the component of $z_n(t)$ centered at frequency $\xi = \xi_{k_1} + \cdots + \xi_{k_n}$ due to the intermodulation of input components (centered at the frequency vector $(\xi_{k_1}, \ldots, \xi_{k_n})$) can be expressed as (Graham and Ehrmen, 1973)

$$
z_{n,\xi}(t) = \sum_{n_{-K}\xi_k + \ldots n_K \xi_k = \xi} \frac{a_n}{2^n} \binom{n}{n_{-K}, \ldots, n_K} \left(A(t) e^{j2\pi\xi t} + A^*(t) e^{-j2\pi\xi t} \right) \tag{4.95}
$$

where

$$
\binom{n}{n_{-K}, \ldots, n_K} = \frac{n!}{n_K! \ldots n_{-K}!} \tag{4.96}
$$

is the multinomial coefficient. Thus the complex envelope of the output signal at the output of the nonlinear block and in the vicinity of a frequency ξ is

$$
\tilde{z}_{n,\xi}(t) = \sum_{\xi_{k_1}+\cdots+\xi_{k_n}=\xi} \frac{a_n}{2^{n-1}} \binom{n}{n_{-K}, \ldots, n_K} A(t)
$$

$$
= \sum_{n_{-K}+\cdots+n_K=n} \frac{a_n}{2^{n-1}} \binom{n}{n_{-K}, \ldots, n_K} \prod_{k=-K}^{K} \tilde{u}_k^{n_k}(t) \tag{4.97}
$$

where $\xi = \sum_{k=-K}^{K} n_k \xi_k = \sum_{i=1}^{n} \xi_{k_i}$ and $\sum_{k=-K}^{K} n_k = n$. The complex envelope of the output $y(t)$ can now be written in complex envelope form as

$$
\tilde{y}_\xi(t) = \sum_{n=1}^{N} \int_{-\infty}^{\infty} \tilde{h}_{2;\xi}(\lambda) \tilde{z}_{n,\xi}(t-\lambda) d\lambda \tag{4.98}
$$

where $\tilde{h}_{2;\xi}(t)$ is the baseband equivalent impulse response of the output filter with respect to frequency ξ.

4.4.4 Multi-Input Single-Output Nonlinear Model

Using Equation (3.24) and the frequency-domain kernel relationship in Equation (3.25), the output $z(t)$ of the memoryless block in the MISO system can be written as

$$
z(t) = \sum_{n_1=0}^{\infty} \cdots \sum_{n_K=0}^{\infty} a_n \binom{n}{n_1, \ldots, n_K} \int_{-\infty}^{\infty} \cdots \int_{-\infty}^{\infty}
$$

$$
h_{11}(\lambda_{11})x_1(t - \lambda_{11}) \ldots h_{11}(\lambda_{1n_1})x_1(t - \lambda_{1n_1}) \ldots
$$

$$
h_{1K}(\lambda_{K1})x_K(t - \lambda_{K1}) \ldots h_{1K}(\lambda_{Kn_K})x_K(t - \lambda_{Kn_K})
$$

$$
d\lambda_{11} \ldots d\lambda_{1n_1} \ldots d\lambda_{1n_1} \ldots d\lambda_{K1} \ldots d\lambda_{Kn_K}
$$

$$
= \sum_{n=0}^{\infty} \sum_{n_1+\cdots+n_K=n} a_n \binom{n}{n_1, \ldots, n_K} \int_{-\infty}^{\infty} \cdots \int_{-\infty}^{\infty} h_{11}(\lambda_{11})x_1(t - \lambda_{11}) \ldots
$$

$$
h_{11}(\lambda_{1n_1})x_1(t - \lambda_{1n_1}) \ldots h_{1K}(\lambda_{K1})x_K(t - \lambda_{K1}) \ldots h_{1K}(\lambda_{Kn_K})x_K(t - \lambda_{Kn_K})
$$

$$
d\lambda_{11} \ldots d\lambda_{1n_1} \ldots d\lambda_{1n_1} \ldots d\lambda_{K1} \ldots d\lambda_{Kn_K}. \tag{4.99}
$$

Note that because of the kernel relationship Equation (3.25), the integrand now is separable, which means that the multiple integral in Equation (4.99) can be written as the product of individual integrals and hence the output of the memoryless block can be written as

$$
z(t) = \sum_{n=0}^{\infty} a_n \left(\sum_{k=1}^{K} \int_{-\infty}^{\infty} h_{1k}(\lambda)x_k(t - \lambda)d\lambda \right)^n
$$

$$
= \sum_{n=0}^{\infty} z_n(t) \tag{4.100}
$$

where

$$
z_n(t) = a_n \left(\sum_{k=1}^{K} u_k(t) \right)^n \tag{4.101}
$$

and

$$
u_k(t) = \int_{-\infty}^{\infty} h_{1k}(\lambda)x_k(t - \lambda)d\lambda. \tag{4.102}
$$

The complex envelope of the output $z(t)$ with respect to the frequency ξ can be developed in a similar way to the memoryless nonlinearity. Therefore:

$$
\tilde{z}_{n;\xi}(t) = \frac{a_n}{2^{n-1}} \sum_{n_{-K}+\cdots+n_K} \binom{n}{n_{-K}, \ldots, n_K} \prod_{k=-K}^{K} \tilde{u}_k^{n_k}(t) \tag{4.103}
$$

where

$$
\tilde{u}_k(t) = \int_{-\infty}^{\infty} \tilde{h}_{1k;\xi}(\lambda)\tilde{x}_k(t - \lambda)d\lambda \tag{4.104}
$$

and hence, the complex envelope of the total output waveform $y(t)$ is

$$\tilde{y}_\xi(t) = \sum_{n=1}^{N} \frac{1}{2^{n-1}} \sum_{n_{-K}+\cdots+n_K=n} \binom{n}{n_{-K},\ldots,n_K} \int_{-\infty}^{\infty} \tilde{h}_{2,\xi}(\lambda) \prod_{k=-K}^{K} \tilde{u}_k^{n_k}(t-\lambda)d\lambda$$

(4.105)

where $\tilde{h}_{2,\xi}(t)$ is the baseband equivalent of the output filter $h_2(t)$ with respect to frequency ξ.

4.4.5 Memoryless Nonlinearity-Power-Series Model

As discussed in Chapter 3, a memoryless system is characterized by a constant Volterra transfer function over frequency. Therefore, using the kernel relationship in Equation (3.40) and substituting in Equation (4.77) yields

$$\tilde{y}_{n,\xi}(t) = \frac{1}{2^{n-1}} \sum_{n_{-K}+\cdots+n_K=n} \binom{n}{n_{-K},\ldots,n_K} \int_{-\infty}^{\infty} \cdots \int_{-\infty}^{\infty} \tilde{a}_n \prod_{i=1}^{n} \tilde{X}_{k_i}(f_i)e^{j2\pi f_i t}df_i$$

$$= \frac{a_n}{2^{n-1}} \binom{n}{n_{-K},\ldots,n_K} \prod_{i=1}^{n} \tilde{x}_{k_i}(t).$$

(4.106)

Now since

$$\prod_{i=1}^{n} \tilde{x}_{k_i}(t) = \prod_{k=-K}^{K} \tilde{x}_k^{n_k}(t)$$

(4.107)

the complex envelope of the output $y(t)$ reduces to

$$\tilde{y}_\xi(t) = \sum_{n=1}^{N} \frac{a_n}{2^{n-1}} \sum_{n_{-K}+\cdots+n_K=n} \binom{n}{n_{-K},\ldots,n_K} \prod_{k=-K}^{K} \tilde{x}_k^{n_k}(t)$$

(4.108)

where $\sum_{k=-K}^{K} n_k = n$.

The above development applies for an arbitrary number of channels (K). In the following subsections, two particular cases will be considered: the case of a single-channel input $(K=1)$ and the case of a two-channel input $(K=2)$. The single-channel input is used as a verification tool where the nonlinear response reduces to the single-channel analysis presented in the previous section. The two-channel case is an important special case since it represents the phenomena of intermodulation and cross-modulation in wireless systems.

4.4.5.1 Single Channel

Consider the case where the input consists of a single channel, $K=1$, with carrier frequency ξ_1. Then, the output complex envelope at the fundamental frequency, ξ_1, is, from Equation (4.107),

$$\tilde{y}_{n,\xi_1}(t) = \sum_{n_{-1}+n_1=n} \frac{a_n}{2^{n-1}} \binom{n}{n_{-1},n_1} \tilde{x}_{-1}^{n_{-1}} \tilde{x}_1^{n_1}(t)$$

(4.109)

Table 4.1 Contributing vectors of intermodulation at a given frequency (Gharaibeh and Steer, 2005)

No. of Channels	Frequency	Contributing Vectors $n = 3$	Contributing Vectors $n = 5$
Single Channel	f_1	$(\xi_1, \xi_1, -\xi_1)$	$(\xi_1, \xi_1, \xi_1, -\xi_1, -\xi_1)$
Two Channels	f_1	$(\xi_1, \xi_1, -\xi_1),$	$(\xi_1, \xi_1, \xi_1, -\xi_1, -\xi_1),$
		$(\xi_1, \xi_2, -\xi_2)$	$(\xi_1, \xi_1, -\xi_1, \xi_2, -\xi_2),$
			$(\xi_1, \xi_2, \xi_2, -\xi_2, -\xi_2)$
Two Channels Intermod	$2f_1 - f_2$	$(\xi_1, \xi_1, -\xi_2)$	$(\xi_1, \xi_1, \xi_1, -\xi_1, -\xi_2),$
			$(\xi_1, \xi_1, -\xi_2, -\xi_2, \xi_2)$
Two Channels Intermod	$2f_2 - f_1$	$(-\xi_1, \xi_2, \xi_2)$	$(-\xi_1, -\xi_2, \xi_2, \xi_2, \xi_2),$
			$(\xi_1, -\xi_1, -\xi_1, \xi_2, \xi_2)$

The contributing frequency vectors at frequency ξ_1, are as shown in Table 4.1 for $n = 3$ and $n = 5$. Therefore, $n_{-1} = (n-1)/2$ and $n_1 = (n+1)/2$, and so, Equation (4.109) becomes

$$\tilde{y}_{\xi_1}(t) = \sum_{n=1}^{N} b_n \tilde{x}_{-1}^{\frac{n-1}{2}} \tilde{x}_1^{\frac{n+1}{2}} \tag{4.110}$$

where

$$b_n = \frac{a_n}{2^{n-1}} \binom{n}{\frac{n-1}{2}, \frac{n+1}{2}}. \tag{4.111}$$

The output complex envelope at any of the harmonics can also be derived in compact form as in Equation (4.110) using the appropriate frequency vectors.

This result means that in a single-channel system, the complex envelope of the nth-order output signal $\tilde{y}_{\xi_1}(t)$ centered at the carrier frequency ξ_1, is obtained from the complex envelope of the input signal given the nonlinear behavioral model (described by the power-series coefficients and filter impulse responses after applying Equation (4.110)). The output complex envelope at any of the harmonics can also be derived in compact form as in Equation (4.110) using the appropriate frequency vectors.

4.4.5.2 Two Channels

Following a similar approach to that used above, the complex envelope of the output waveform at the first carrier (ξ_1) is

$$\tilde{y}_{n,\xi_1}(t) = \sum_{n_{-2}+n_{-1}+n_1+n_2=n} \frac{a_n}{2^{n-1}} \binom{n}{n_{-2}, n_{-1}, n_1, n_2} \tilde{x}_{-2}^{n-2} \tilde{x}_{-1}^{n-1} \tilde{x}_1^{n_1} \tilde{x}_2^{n_2} \tag{4.112}$$

The contributing vectors at frequency ξ_1 are all permutations of the vectors shown in Table 4.1 for $n = 3$ and $n = 5$. By induction, $n_{-2} = l, n_{-1} = ((n-1)/2) - l, n_1 =$

$((n + 1)/2) - l, n_2 = l$, and hence, Equation (4.112) can be written as the sum of the contributions defined by each of the above frequency vectors as

$$\tilde{y}_{\xi_1}(t) = \sum_{n=1}^{N} \sum_{l=0}^{\frac{n-1}{2}} b_{n,l} \tilde{x}_{-2}^{l} \tilde{x}_{-1}^{\frac{n-1}{2}-l} \tilde{x}_{1}^{\frac{n+1}{2}-l} \tilde{x}_{2}^{l} \tag{4.113}$$

where

$$b_{n,l} = \frac{a_n}{2^{n-1}} \left(l, \frac{n-1}{2} - l, \frac{n+1}{2} - l, l \right). \tag{4.114}$$

This expression captures gain compression that describes the effect on the output $\tilde{y}_{\xi_1}$ due to the level of the input $\tilde{x}_1$ centered at the same frequency ξ_1; and gain saturation that describes the effect on the output $\tilde{y}_{\xi_1}$ due to the level of the input $\tilde{x}_2$ in the other channel centered at ξ_2. If the second channel is modulated (with a finite bandwidth), gain saturation manifests itself as cross-modulation.

Intermodulation products can be derived in the same way by describing the nonlinear components by their frequency vectors. The lower intermodulation component IM3$_L$ is centered at frequency $(2\xi_1 - \xi_2)$ and the contributing vectors at frequency $2\xi_1 - \xi_2$ are all the permutations of the vector in Table 4.1. By induction, $n_{-2} = l + 1, n_{-1} = ((n - 3)/2) - l, n_1 = ((n + 1)/2) - l, n_2 = l$. Thus, the output complex envelope centered at frequency $(2\xi_1 - \xi_2)$ is

$$\tilde{y}(t)_{2\xi_1 - \xi_2} = \sum_{n=3}^{N} \sum_{l=0}^{\frac{n-3}{2}} b_{n,l} \tilde{x}_{-2}^{l+1} \tilde{x}_{-1}^{\frac{n-3}{2}-l} \tilde{x}_{1}^{\frac{n+1}{2}-l} \tilde{x}_{2}^{l} \tag{4.115}$$

For the upper intermodulation component, IM3$_U$, centered at frequency $(2\xi_1 - \xi_2)$, and the contributing vectors at frequency $2\xi_2 - \xi_1$ are all the permutations of the vector in Table 4.1. The output complex envelope IM3$_U$ is

$$\tilde{y}(t)_{2\xi_2 - \xi_1} = \sum_{n=3}^{N} \sum_{l=0}^{\frac{n-3}{2}} b_{n,l} \tilde{x}_{-2}^{\frac{n-3}{2}-l} \tilde{x}_{-1}^{l+1} \tilde{x}_{1}^{l} \tilde{x}_{2}^{\frac{n+1}{2}-l} \tag{4.116}$$

where

$$b_{n,l} = \frac{a_n}{2^{n-1}} \left(\frac{n-3}{2} - l, l + 1, l, \frac{n+1}{2} - l \right). \tag{4.117}$$

4.5   Examples–Response of Power-Series Model to Multiple Signals

In this section, examples of the response of a multichannel nonlinear system at the complex envelope level are given. These examples consider deterministic input signals such as discrete tones and finite bandwidth deterministic signals.

4.5.1 Single Tone

Consider the case where the input consists of a single tone signal, $K = 1$, with carrier frequency ξ_1:

$$x(t) = A_1 \cos(2\pi \xi_1 t + \phi_1) \qquad (4.118)$$

then $\tilde{x}_1 = Ae^{j\phi_1}$, $\tilde{x}_{-1} = Ae^{-j\phi_1}$ and the output complex envelope at the fundamental frequency, ξ_1, is, from Equation (4.110):

$$\tilde{y}_{\xi_1}(t) = \sum_{n=1}^{N} b_n (Ae^{-j\phi_1})^{\frac{n-1}{2}} (Ae^{j\phi_1})^{\frac{n+1}{2}} = \sum_{n=1}^{N} b_n A^n e^{j\phi_1} \qquad (4.119)$$

where

$$b_n = \frac{a_n}{2^{n-1}} \left(\frac{n}{\frac{n-1}{2}, \frac{n+1}{2}} \right). \qquad (4.120)$$

The expression in Equation (4.119) represents the gain-compression characteristics of nonlinearity which is used to model the AM–AM and AM–PM characteristics of an amplifier.

4.5.2 Two-Tone Signal

For a two-tone input ($K = 2$) we have:

$$x_1(t) = A_1 \cos(2\pi \xi_1 t + \phi_1)$$
$$x_2(t) = A_2 \cos(2\pi \xi_2 t + \phi_2) \qquad (4.121)$$

where A_i are the amplitudes and ϕ_i are the phase of the input tones. Then $\tilde{x}_1 = A_1 e^{j\phi_1}$, $\tilde{x}_{-1} = A_1 e^{-j\phi_1}$, $\tilde{x}_2 = A_2 e^{j\phi_1}$ and $\tilde{x}_{-2} = A_2 e^{-j\phi_1}$. Therefore, using Equation (4.113)

$$\tilde{y}_{\xi_1}(t) = \sum_{n=1}^{N} \sum_{l=0}^{\frac{n-1}{2}} b_{n,l} A_1^{n-2l} A_2^{2l} e^{j\phi_1} \qquad (4.122)$$

and for the second tone:

$$\tilde{y}_{\xi_2}(t) = \sum_{n=1}^{N} \sum_{l=0}^{\frac{n-1}{2}} b_{n,l} A_2^{n-2l} A_1^{2l} e^{j\phi_2} \qquad (4.123)$$

where

$$b_{n,l} = \frac{a_n}{2^{n-1}} \left(\frac{n}{l, \frac{n-1}{2} - l, \frac{n+1}{2} - l, l} \right). \qquad (4.124)$$

These expressions capture gain compression and gain saturation of discrete tones in conventional two-tone test. The intermodulation products can be derived using

Equation (4.115) and Equation (4.116), thus, for the lower intermodulation component, IM3$_L$, the output complex envelope centered at frequency $(2\xi_1 - \xi_2)$ is

$$\tilde{y}(t)_{2\xi_1-\xi_2} = \sum_{n=3}^{N} \sum_{l=0}^{\frac{n-3}{2}} b_{n,l} A_1^{n-2l-1} A_2^{2l+1} e^{j(2\phi_1-\phi_2)} \tag{4.125}$$

and for the upper intermodulation component, IM3$_U$, the output complex envelope centered at frequency $(2\xi_2 - \xi_1)$ is

$$\tilde{y}(t)_{2\xi_2-\xi_1} = \sum_{n=3}^{N} \sum_{l=0}^{\frac{n-3}{2}} b_{n,l} A_1^{2l+1} A_2^{n-2l-1} e^{j(2\phi_2-\phi_1)} \tag{4.126}$$

where

$$b_{n,l} = \frac{a_n}{2^{n-1}} \left(\frac{n}{\frac{n-3}{2} - l, l+1, l, \frac{n+1}{2} - l} \right). \tag{4.127}$$

These expressions represent the output sinusoids that appear at the intermodulation frequencies due to the mixing properties of nonlinearity. IMR can be defined as the ratio of the power of the intermodulation component to the total power of the fundamental tones at the output of nonlinearity.

4.5.3   Single-Bandpass Signal

Consider the case where the input consists of a single narrowband signal, $K = 1$, with carrier frequency ξ_1.

$$x(t) = A\,\text{sinc}(Bt)\cos(2\pi\xi_1 t + \theta) \tag{4.128}$$

where $\text{sinc}(x) = \frac{\sin(\pi x)}{\pi x}$. In this case, the complex envelope of the signal is $\tilde{x}_1 = \tilde{x}_{-1} = A\,\text{sinc}(Bt)$. Then, the output complex envelope at the fundamental frequency, ξ_1, is, from Equation (4.110):

$$\tilde{y}_{\xi_1}(t) = \sum_{n=1}^{N} b_n \tilde{x}_{-1}^{\frac{n-1}{2}} \tilde{x}_1^{\frac{n+1}{2}} = \sum_{n=1}^{N} b_n A^n \text{sinc}^n(Bt) \tag{4.129}$$

Again, this expression represents gain compression effects on the bandpass signal that results in reduction of the signal power but retains the signal spectrum (no frequency distortion).

4.5.4   Two-Bandpass Signals

Consider a two-channel input $(K = 2)$ of the form:

$$x(t) = A_1\text{sinc}(2\pi B_1 t)\cos(2\pi\xi_1 t) + A_2\text{sinc}(2\pi B_2 t)\cos(2\pi\xi_2 t) \tag{4.130}$$

an expression for gain compression and gain saturation will be derived by deriving the output envelope response centered at one of the carriers.

The two channels are described by their complex envelopes $\tilde{x}_1(t) = A_1\text{sinc}(B_1t)$ and $\tilde{x}_2(t) = A_2\text{sinc}(B_2t)$ (at the output of the input filter) with respect to the frequencies ξ_1 and ξ_2, respectively. Following Equation (4.113), the above equation can be written as the sum of the contributions defined by each of the above frequency vectors as

$$\tilde{y}_{\xi_1}(t) = \sum_{n=1}^{N} \sum_{l=0}^{\frac{n-1}{2}} b_{n,l} A_1^{n-2l} A_2^{2l} \text{sinc}^{n-2l}(2\pi B_1t)\text{sinc}^{2l}(2\pi B_2t) \tag{4.131}$$

where

$$b_{n,l} = \frac{a_n}{2^{n-1}} \left(l, \frac{n-1}{2} - l, \frac{n+1}{2} - l, l \right). \tag{4.132}$$

The expression in Equation (4.131) describes cross-modulation introduced by the mixing property of nonlinearity in multichannel systems where part of the power of the second signal is transferred to the first signal. This results in changing (distorting) the signal spectrum of the first signal by co-channel interference within the signal bandwidth.

The intermodulation products can be derived in the same way as in Equation (4.115) and Equation (4.116). Thus, the output complex envelope centered at frequency $(2\xi_1 - \xi_2)$ is

$$\tilde{y}(t)_{2\xi_1-\xi_2} = \sum_{n=3}^{N} \sum_{l=0}^{\frac{n-3}{2}} b_{n,l} A_1^{n-2l-1} A_2^{2l+1} \text{sinc}^{n-2l-1}(2\pi B_1t)\text{sinc}^{2l+1}(2\pi B_2t) \tag{4.133}$$

and for the upper intermodulation component, IM3_U, the output complex envelope centered at frequency $(2\xi_2 - \xi_1)$ is

$$\tilde{y}(t)_{2\xi_2-\xi_1} = \sum_{n=3}^{N} \sum_{l=0}^{\frac{n-3}{2}} b_{n,l} A_1^{2l+1} A_2^{n-2l-1} \text{sinc}^{2l+1}(2\pi B_1t)\text{sinc}^{n-2l-1}(2\pi B_2t) \tag{4.134}$$

where

$$b_{n,l} = \frac{a_n}{2^{n-1}} \left(\frac{n-3}{2} - l, l+1, l, \frac{n+1}{2} - l \right). \tag{4.135}$$

Intermodulation components represent extra signal components that appear at the intermodulation frequencies when two finite bandwidth signals are amplified by a nonlinear multichannel amplifier. These signals have finite bandwidth spectra centered at the intermodulation frequencies and consist of the interaction of the two signal spectra at the intermodulation frequencies. The bandwidth of the intermodulation signals depends on the bandwidth of the input signals and also on the order of nonlinearity.

4.5.5 Single Tone and a Bandpass Signal

Now, consider the sum of a single tone and a bandpass signal of the form:

$$x(t) = A_1 \cos(2\pi \xi_1 t + \phi) + A_2 \text{sinc}(2\pi B_2 t) \cos(2\pi \xi_2 t) \tag{4.136}$$

The two channels are described by their complex envelopes $\tilde{x}_1(t) = A_1 e^{j\phi}$ and $\tilde{x}_2(t) = A_2 \text{sinc}(B_2 t)$ with respect to the frequencies ξ_1 and ξ_2, respectively. Following a similar approach to that used above, the complex envelope of the output waveform at the first carrier (ξ_1) is

$$\tilde{y}_{\xi_1}(t) = \sum_{n=1}^{N} \sum_{l=0}^{\frac{n-1}{2}} b_{n,l} \tilde{x}_{-2}^l \tilde{x}_{-1}^{\frac{n-1}{2}-l} \tilde{x}_1^{\frac{n+1}{2}-l} \tilde{x}_2^l$$

$$= \sum_{n=1}^{N} \sum_{l=0}^{\frac{n-1}{2}} b_{n,l} A_1^{n-2l} A_2^{2l} e^{j\phi} \text{sinc}^{2l}(2\pi B_2 t) \tag{4.137}$$

Note that the output at the frequency of the unmodulated carrier is not a pure sinusoid but rather a bandpass signal that result from the cross modulation between the sinusoid and the input bandpass signal. This phenomena is known in wireless systems as cross-modulation that results in desensitization of wireless receivers.

4.5.6 Multisines

A more general case is when the input is a multisine signal $x(t)$ consisting of the sum of K tones of the form:

$$x(t) = \sum_{k=1}^{K} x_k(t). \tag{4.138}$$

Here, the signal $x_k(t)$ is the kth tone with frequency ξ_k and phase ϕ_k:

$$x_k(t) = A_k \cos(2\pi \xi_k t + \phi_k) \tag{4.139}$$

The complex envelope of this signal is equivalent to the phasor form of sinusoids:

$$\tilde{x}_k(t) = A_k e^{j\phi_k} \tag{4.140}$$

The response of the nonlinear system to a multisine signal consists of all intermodulation products of the input tones. Now using Equation (4.108) and applying the multisine signal model in Equation (4.138), the component of $y_n(t)$ centered at frequency $\xi = \xi_{k_1} + \ldots + \xi_{k_n}$ due to the intermodulation of input components (centered at the frequency vector $(\xi_{k_1}, \ldots, \xi_{k_n})$) can be expressed as (Graham and Ehrmen, 1973)

$$\tilde{y}_{n,\xi}(t) = \sum_{\xi_{k_1} + \cdots + \xi_{k_n} = \xi} \frac{a_n}{2^{n-1}} \binom{n}{n_{-K}, \ldots, n_K} \prod_{i=1}^{n} A_{k_i} e^{j\phi_{k_i}}$$

$$\tag{4.141}$$

where $\xi = \sum_{k=-K}^{K} n_k \xi_k = \sum_{i=1}^{n} \xi_{k_i}$, $\sum_{k=-K}^{K} n_k = n$, and

$$\binom{n}{n_{-K}, \ldots, n_K} = \frac{n!}{n_K! \ldots n_{-K}!} \tag{4.142}$$

is the multinomial coefficient.

Therefore, by the proper choice of the contributing vectors, the output can be calculated at a given frequency. This process can be done using computer simulations and using methods like the Arithmetic Operator Method (AOM) presented in Hart *et al.* (2003).

4.5.7 *Multisine Analysis Using the Generalized Power-Series Model*

Using the Generalized Power-Series (GPS) model in Equation (3.29), the output phasor at a particular frequency is given by (Lunsford, 1993)

$$Y_\xi = \sum_{n=0}^{\infty} \sum_{\underbrace{n_1, \ldots, n_K}_{|n_1|+\cdots+|n_K|=n}} \Re\{\varepsilon_n T\}_\xi \tag{4.143}$$

where $\xi = \sum_{k=1}^{K} n_k \xi_k$ and

$$T = \sum_{\alpha=0}^{\infty} \sum_{\underbrace{s_1, \ldots, s_K}_{s_1+\cdots+s_K=\alpha}} (n+2\alpha)! a_{n+2\alpha} \Phi \tag{4.144}$$

where

$$\Phi = \prod_{k=1}^{K} \frac{(b_k)^{n_k+2s_k} |X_k|^{|n_k|+s_k} |X_k^*|^{s_k} \Gamma_k^{|n_k|+s_k} \Gamma_k^{s_k}}{s_k!(|n_k|+s_k)!}. \tag{4.145}$$

and $\Gamma_{k,n} = e^{-j2\pi\xi_k\tau_{k,n}}$.

4.6 Summary

The development in the above sections establishes the relationship between the complex envelope of the input and the output of a nonlinear transformation. A generalized approach has been used where different nonlinear models can be dealt with as special cases. The analysis establishes the basis for dealing with nonlinear systems by simulations where the objective is to estimate distortion in different wireless systems. The above analysis will be used in the following chapters where methods for estimating distortion of deterministic and random signals will be presented.

5

Nonlinear Transformation of Random Signals

Digital communication signals exhibit a random nature where the signal envelope at any point in time does not have a deterministic value that can be predicted from past values, hence, communication signals are considered as random processes. The spectrum of random signals cannot be characterized by the direct voltage Fourier transform because a voltage Fourier transform may not exist for most sample functions of the process. Even if it exists for a realization of the process, it may not exist for other realizations and hence, the Fourier transform of a single realization does not represent the whole process.

Although random signals do not have defined amplitude, frequency or phase values, they do have some known features such as the Probability Density Functions (PDF) of amplitude, phase or frequency values that can be developed from experience gained by observing the signal over a long period of time. This rationale leads to the concept of ergodicity that states that the statistics of a random signal can be obtained from a single realization of the signal. This concept is central to the simulation of communication signals where a long sequence of a signal is used to develop the signal statistical properties and its power spectrum.

Systems response to a random signal is also an important concept that needs to be understood when dealing with system simulations as transformations of random signals usually entail transformation of their statistics. In linear systems, the effect of the linear transformation is manifested as a transformation to the shape of the signal spectrum. Nonlinear systems, however, produce more complex effects on the signal spectrum as well as on signal statistics such as the probability distribution of signal parameters.

This chapter discusses nonlinear system response to single and multiple random input signals that represent real-world communication signals. We present the underlying theory of the probabilistic model that will be used to characterize communication signals and their nonlinear distortion. The basic concepts introduced in this chapter provide the tools needed for understanding nonlinear distortion in real-world communication systems, which will be discussed in the next chapters.

Nonlinear Distortion in Wireless Systems: Modeling and Simulation with MATLAB®, First Edition.
Khaled M. Gharaibeh.
© 2012 John Wiley & Sons, Ltd. Published 2012 by John Wiley & Sons, Ltd.

5.1   Preliminaries

In order to model distortion in a digital communication system, the PSD at the output of a nonlinearity must be estimated. The PSD of a random signal can be developed from the Fourier transform of the autocorrelation function of the signal by the Wiener–Khinchin theorem (Papoulis, 1994). The autocorrelation function of real-world communication signals can be developed from their time autocorrelation functions assuming that those signals are ergodic. The autocorrelation function of the complex envelope of a random signal is defined as

$$R_{\tilde{x}\tilde{x}}(\tau) = E[\tilde{x}(t)\tilde{x}^*(t + \tau)] \tag{5.1}$$

where E is the expected value or the statistical average. The autocorrelation function is a measure of the correlation of the signal with time-shifted replicas of itself. The Power Spectral Density (PSD) defines the frequency contents of the process and can be found from the Fourier transform of the autocorrelation function by Weiner–Khinchin theorem (Papoulis, 1994) as:

$$S_{\tilde{x}\tilde{x}}(f) = \int_{-\infty}^{\infty} R_{\tilde{x}\tilde{x}}(\tau)e^{-j\omega\tau d\tau}. \tag{5.2}$$

In general, the autocorrelation function of the input signal $x(t)$ centered at frequency f_0 is related to the autocorrelation function of its complex envelope by:

$$
\begin{aligned}
R_{xx}(\tau) &= E[x(t + \tau)x(t)] \\
&= \Re\left\{E[\tilde{x}(t + \tau)\tilde{x}^*(t)]e^{j2\pi f_0\tau}\right\} \\
&= \Re\left\{\frac{1}{2}R_{\tilde{x}\tilde{x}}(\tau)e^{j2\pi f_0\tau}\right\}
\end{aligned}
\tag{5.3}
$$

Hence, the PSD of a bandpass process centered at f_0 can be related to the PSD of its complex envelope using the properties of Fourier transform (Peebles, 1987):

$$S_{xx}(f) = \frac{1}{4}\left[S_{\tilde{x}\tilde{x}}(f - f_0) + S_{\tilde{x}\tilde{x}}(f + f_0)\right] \tag{5.4}$$

5.2   Linear Systems with Stochastic Inputs

Linear transformation of random signals produces a new random signal whose statistics are related to the statistics of the input process by the impulse response function of the linear system. In a similar way to deterministic signals, the complex envelope of the output process $\tilde{y}(t)$ can be evaluated using the convolution operation of the complex envelope of the input random process $\tilde{x}(t)$ and the baseband equivalent of the impulse response of the linear system $h(t)$:

$$\tilde{y}(t) = \tilde{x}(t) * \tilde{h}(t) = \int_{-\infty}^{\infty} \tilde{x}(\lambda)\tilde{h}(t - \lambda)d\lambda \tag{5.5}$$

where $*$ indicates the convolution operator. However, this convolution operation cannot be performed since the signal is random and cannot be given by a deterministic function of time. Instead, the statistics of the output signal can be found given the statistic of the input signal. For example, the mean of the output $\tilde{y}(t)$ can be evaluated as:

$$\eta_{\tilde{y}}(t) = E\left[\int_{-\infty}^{\infty} \tilde{x}(\lambda)\tilde{h}(t - \lambda)d\lambda\right]$$

$$= \int_{-\infty}^{\infty} \eta_{\tilde{x}}(\lambda)h(t - \lambda)d\lambda$$

$$= \eta_{\tilde{x}}(t) * \tilde{h}(t). \tag{5.6}$$

If $\tilde{x}(t)$ is Wide Sense Stationary (WSS), then $\tilde{y}(t)$ is WSS and hence (Papoulis, 1994):

$$\eta_{\tilde{y}}(t) = \eta_{\tilde{y}} = \eta_{\tilde{x}} \int_{-\infty}^{\infty} \tilde{h}(\lambda)d\lambda \tag{5.7}$$

The autocorrelation function of the output processes can be related to the autocorrelation of the input process by (Papoulis, 1994):

$$R_{\tilde{y}\tilde{y}}(\tau) = R_{\tilde{x}\tilde{x}}(\tau) * \tilde{h}(t) * \tilde{h}^*(-\tau) \tag{5.8}$$

The cross-correlation function between the input and output processes is an important concept in system identification. This function is related to the input autocorrelation function by (Papoulis, 1994):

$$R_{\tilde{x}\tilde{y}}(\tau) = R_{\tilde{x}\tilde{x}}(\tau) * \tilde{h}(t) \tag{5.9}$$

Hence, the PSD of the output process and can be related to the PSD of the input process using the properties of Fourier transform as:

$$S_{\tilde{y}\tilde{y}}(f) = S_{\tilde{x}\tilde{x}}(f)\tilde{H}(f)\tilde{H}^*(f) = S_{\tilde{x}\tilde{x}}(f)|\tilde{H}(f)|^2 \tag{5.10}$$

Therefore, the spectrum of the output process is a shaped version of the input process by the squared magnitude of the frequency response of the linear system. This is an important result in system identification where the square magnitude of the frequency response of the linear system can be found by comparing the input and output PSDs. An interesting case is when the input PSD is flat over all frequencies (white spectrum). This case will be discussed in the next subsection.

5.2.1 White Noise

For the special case where the input process is white noise (flat spectrum over all frequencies), the autocorrelation of the input process is given by $R_{\tilde{x}\tilde{x}}(\tau) = q\delta(\tau)$. Using Equation (5.8), the autocorrelation of the output of the linear system is given by

$$R_{\tilde{y}\tilde{y}}(\tau) = qh(\tau) * \tilde{h}^*(-\tau) \tag{5.11}$$

and the output spectrum is found by the Fourier transform of Equation (5.11):

$$S_{\tilde{y}\tilde{y}}(f) = q|\tilde{H}(f)|^2 \tag{5.12}$$

This is an important result in system identification where the transfer function of a linear system can be estimated by exciting the linear system with white noise and observing the output spectrum that represents the squared magnitude of the system transfer function.

5.2.2 Gaussian Processes

Given that the convolution operation is a linear operation, it can be proven that the response of a linear system to a Gaussian input process is also a Gaussian process but with a different mean and autocorrelation functions by Equation (5.7) and Equation (5.8). A detailed proof of this case can be found in Papoulis (1994).

5.3 Response of a Nonlinear System to a Random Input Signal

In a similar fashion to linear systems, the response of a nonlinear system to a random input is also a random signal. However, nonlinear systems change the PDF and produce new spectral components in the signal spectrum. The response of a nonlinear system to a random input is evaluated by finding the autocorrelation function and the PSD of the output signal. Specifically, the PSD of the system output is used to estimate nonlinear distortion through the identification of the spectral components introduced by nonlinearity.

In the following subsections, the autocorrelation function and the PSD of the output of different nonlinear models are found using the complex envelope analysis presented in Chapter 4. Closed-form expressions for the autocorrelation function and the PSD will be developed for single- and multiple-input cases. Special cases where the input is assumed to have a Gaussian distribution will also be analyzed. Detailed analysis of Volterra series model with random inputs can be found in Rudko and Wiener (1978).

5.3.1 Power-Series Model

As shown in the previous chapter, the complex envelope at the output of a memoryless nonlinearity is given by an envelope power-series model as in Equation (4.42). The autocorrelation function of the output of the power-series model can be found using Equation (4.42) as

$$R_{\tilde{y}\tilde{y}}(\tau) = E[\tilde{y}(t)\tilde{y}^*(t+\tau)]$$

$$= \sum_{n=1}^{N} \sum_{m=1}^{N} b_n b_m^* R_{\tilde{x}_n \tilde{x}_m}(\tau) \tag{5.13}$$

where

$$R_{\tilde{x}_n \tilde{x}_m}(\tau) = E\left[\tilde{x}^{\frac{(n+1)}{2}}(t)\tilde{x}^{*\frac{(n-1)}{2}}(t)\tilde{x}^{\frac{(m-1)}{2}}(t+\tau)\tilde{x}^{*\frac{(m+1)}{2}}(t+\tau)\right]. \tag{5.14}$$

is the $(n+m)$th-order autocorrelation function of the input signal. The PSD of the output of the model is obtained from the Fourier transform of the output autocorrelation function using Equation (5.2):

$$S_{\tilde{y}\tilde{y}}(f) = \sum_{n=1}^{N} \sum_{m=1}^{N} b_n b_m^* S_{\tilde{x};nm}(f) \tag{5.15}$$

where $S_{nm}(f)$ is the $(n+m)$th-order PSD of the input signal which can be found from the Fourier transform of the $(n+m)$th-order autocorrelation function in Equation (5.14):

$$S_{\tilde{x};nm}(f) = \int_{-\infty}^{\infty} R_{\tilde{x}_n \tilde{x}_m}(\tau) e^{-j\omega\tau} d\tau. \tag{5.16}$$

Therefore, the output PSD is a sum of the Fourier transform of each component of the autocorrelation function weighted by the appropriate power-series coefficient. This formulation of the PSD of the output of the power-series model enable the partition of the output spectrum into the following spectral components:

- The components $S_{\tilde{x};nm}(f)$, where $n, m \neq 1$ correspond to intermodulation (spectral regrowth).
- The components $S_{\tilde{x};nm}(f)$, where $n, m = 1$ correspond to linear output terms.
- The components $S_{\tilde{x};nm}(f)$, where $n = 1$ or $m = 1$ but $n \neq m$ correspond to gain compression.

Note that this formulation is useful for the identification of distortion components responsible of the degradation of communication system performance. Figure 5.1 shows the output spectrum of the nonlinear model for a WCDMA input signal partitioned into linear (with gain compression) and intermodulation components (spectral regrowth).

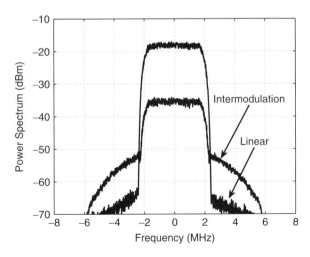

Figure 5.1 Output power spectral density of a nonlinear PA partitioned into linear (with gain compression) and intermodulation components.

5.3.2 Wiener–Hammerstein Models

Referring to the Wiener–Hammerstein model in Figure 3.4(b), and considering only the output component centered at the fundamental frequency ξ (first zonal output), then by using Equation (4.65), the autocorrelation function of the output of the second block can be found as (Gharaibeh, 2004)

$$R_{\tilde{z}\tilde{z}}(\tau) = E[\tilde{z}(t)\tilde{z}^*(t+\tau)]$$

$$= \sum_{n=1}^{N}\sum_{m=1}^{N} b_n b_m^* R_{\tilde{u}_n \tilde{u}_m}(\tau) \tag{5.17}$$

where

$$R_{\tilde{u}_n \tilde{u}_m}(\tau) = E\left[\tilde{u}^{\frac{(n+1)}{2}}(t)\tilde{u}^{*\frac{(n-1)}{2}}(t)\tilde{u}^{\frac{(m-1)}{2}}(t+\tau)\tilde{u}^{*\frac{(m+1)}{2}}(t+\tau) \right]. \tag{5.18}$$

The PSD of the output signal $\tilde{z}(t)$ at the fundamental frequency ξ is obtained from the Fourier transform of the autocorrelation function as (Gharaibeh, 2004)

$$S_{\tilde{z}\tilde{z}}(f) = \sum_{n=1}^{N}\sum_{m=1}^{N} b_n b_m^* S_{\tilde{u};nm}(f) \tag{5.19}$$

where $S_{\tilde{u}}(f)$ is the PSD of the output of the first block, which is a linear filter. If this filter response is constant within the signal bandwidth, then $\tilde{H}_1(f) = \tilde{H}_1(0)$, where $\tilde{H}_1(f)$ is the baseband equivalent of the transfer functions of the input filter with respect to frequency ξ. Hence, Equation (5.19) can be written as

$$S_{\tilde{u};nm}(f) = |\tilde{H}_{1;\xi}(0)|^{n+m} S_{\tilde{x};nm}(f) \tag{5.20}$$

where $S_{\tilde{x};nm}(f)$ is the $(n+m)$th-order PSD of the input signal as defined in Equation (5.16).

The PSD at the output of the output filter can be found using the relationship between the PSDs at the input and the output of a linear system as in Equation (5.10):

$$S_{\tilde{y}\tilde{y}}(f) = |\tilde{H}_{2;\xi}(f)|^2 S_{\tilde{z}\tilde{z}}(f)$$

$$= |\tilde{H}_2(f)|^2 |\tilde{H}_1(0)|^{n+m} S_{\tilde{x}\tilde{x}}(f) \tag{5.21}$$

where $\tilde{H}_2(f)$ is the baseband equivalent of the transfer function of the output filter with respect to frequency ξ. Note that the output spectrum is weighted by powers of the impulse responses of the input and output filter. The derivations here assume that the input filter response in constant within the signal bandwidth. This is an acceptable assumption since the amplifier is expected to have more significant variations of the impulse response at high drive levels and this is included by the response of the output filter (Gharaibeh and Steer, 2005).

5.4 Response of Nonlinear Systems to Gaussian Inputs

The formulation of the autocorrelation function of the output of nonlinearity is greatly simplified if the input signal is assumed to have Gaussian probability distribution. In most CDMA-based wireless systems, the Gaussian assumption is an acceptable approximation of the forward link signal since these signals consists of the sum of multiple user data, which means that its probability distribution, by the central limit theorem, approaches a Gaussian distribution. However, and as shown in Aparin (2001), the Gaussian assumption is not an acceptable assumption for reverse-link signals.

A basic theorem for the calculation of the autocorrelation function of the output of a memoryless nonlinearity is Price's theorem (Price, 1958). Using this theorem, the output autocorrelation of a memoryless nonlinearity can be written as:

$$R_{yy}(\tau) = E[y(t)y(t+\tau)]$$

$$= E[f(x(t))f(x(t+\tau))]. \qquad (5.22)$$

where $y(t) = f(x(t))$. Now, let $x(t) = x_1$ and $x(t+\tau) = x_2$ then, by using Price's theorem (Price, 1958), the output autocorrelation function for a zero mean Gaussian input process can be computed using the following relation:

$$\frac{\partial^k R_{yy}(\tau)}{\partial R_{xx}^k(\tau)} = E[f^{(k)}(x_1)f^{(k)}(x_2)]$$

$$= \int_{-\infty}^{\infty} \int_{-\infty}^{\infty} \frac{f^{(k)}(x_1)f^{(k)}(x_2)}{2\pi\sqrt{1 - R_{xx}^2(\tau)}}$$

$$\times \exp\left\{ \frac{-1}{2(1 - R_{xx}^2(\tau))} \left[\frac{x_1^2}{\sigma_{x_1}^2} + \frac{x_2^2}{\sigma_{X_2}^2} - \frac{2R_{xx}(\tau)}{\sigma_{x_1}^2\sigma_{x_2}^2} \right] \right\} dx_1 dx_2 \qquad (5.23)$$

where $f^{(k)}(x)$ denotes the kth derivative of $f(x)$ with respect to x and $\sigma_{x_1}^2, \sigma_{x_2}^2$ are the variances of the random variables x_1, x_2, respectively. In the following subsections, this theorem will be used to derive the autocorrelation function of the output of various limiter models.

Another important theorem in this context is the Bussgang theorem (Bussgang, 1952), which states that if an input to a memoryless nonlinearity is a zero-mean complex Gaussian process, then the output of the nonlinearity can be written as:

$$y(t) = \alpha(t)x(t) + d(t) \qquad (5.24)$$

where $E[x(t)d(t)] = 0$ and $\alpha(t)$ is a gain function defined by:

$$\alpha(t) = \frac{E[y(t)x^*(t+\tau)]}{E[x(t)x^*(t+\tau)]} \qquad (5.25)$$

Note that $\alpha(t)$ depends on the input power $(\sigma_x^2(t))$ of the signal and on the nonlinear model.

The autocorrelation function at the output of the nonlinearity can then be written as (Rugini *et al.*, 2002):

$$R_{yy}(t, t+\tau) = \alpha(t)\alpha^*(t+\tau)R_{xx}(t, t+\tau) + R_{dd}(t, t+\tau) \qquad (5.26)$$

If the input signal is assumed to be a WSS process (which is a realistic assumption in most communications signals), then the linear gain in Equation (5.25) is a constant $\alpha(t) = \alpha_0$, and hence the autocorrelation in Equation (5.26) can be written as:

$$R_{yy}(\tau) = |\alpha_0|^2 R_{xx}(\tau) + R_{dd}(\tau) \tag{5.27}$$

which means that the output autocorrelation function is directly related to the input auto-correlation function by a scaling factor $|\alpha_0|$. Furthermore, this formulation means that nonlinear distortion is expressed as an additive component $R_{dd}(\tau)$, which greatly simplifies the analysis of nonlinear distortion and enables system performance to be estimated analytically in a similar manner to AWGN.

On the other hand, if a nonlinearity can be modeled by monomials of the form x^n (e.g. as in a power-series model), the response of a memoryless nonlinearity to a Gaussian input can be analyzed using the properties of the moments of complex Gaussian processes (Papoulis, 1994). Therefore, if the random process $\tilde{x}(t)$ is a zero-mean Gaussian then the following property of zero-mean Gaussian random variables applies (Gard et al., 2005; McGee, 1971):

$$E[x_1 x_2 \ldots x_s x_1^* x_2^* \ldots x_t^*]$$
$$= \begin{cases} \sum_\pi E[x_{\pi(1)} x_1^*] E[x_{\pi(2)} x_2^*] \ldots E[x_{\pi(s)} x_s^*] & s = t \\ 0 & \text{otherwise.} \end{cases} \tag{5.28}$$

The summation is over all the permutations π of the set of integers $\{1, 2, \ldots, s\}$. Equation (5.28) means that higher-order autocorrelation functions such as the expression in Equation (5.14) can be expressed in terms of the second-order statistic $R_{xx}(\tau)$ that again greatly simplifies the analysis of nonlinear distortion.

In the following subsections, we use the properties of the moments of complex Gaussian random variables to obtain a simplified formulation for the autocorrelation function at the output of a nonlinear model.

5.4.1   Limiter Model

In order to formulate the output autocorrelation function of a limiter model for a Gaussian input, we use Price's theorem as discussed in Baer (1982), Jones (1963), Kirlin (1977), Max (1970), McGuffin (1992), Roberts et al. (1979) and Sevy (1966), for the special cases of hard, soft and smooth limiters.

5.4.1.1   Hard-Limiter Model

A hard limiter is characterized by the model Equation (3.47). For the hard-limiter models, we take the case where $k = 1$, then $f^{(k)}(x)$ is a first-order delta function of area $2L$ and centered at $x = 0$. Substituting into Equation (5.23) and integrating, we get (Price, 1958):

$$\frac{\partial R_{yy}(\tau)}{\partial R_{xx}(\tau)} = \frac{2L^2}{\pi \sqrt{1 - R_x^2(\tau)}} \tag{5.29}$$

The output autocorrelation function can be obtained by integrating Equation (5.29) and applying the condition that $R_{yy}(\tau) = 0$ when $R_{xx}(\tau) = 0$ and hence:

$$R_{yy}(\tau) = \frac{2L^2}{\pi\sqrt{1 - R_x^2(\tau)}} dR_x$$

$$= \frac{2L^2}{\pi} \sin^{-1}\left[\frac{R_{xx}(\tau)}{R_{xx}(0)}\right]. \tag{5.30}$$

This form is called Van Vleck's result (Price, 1958) that relates the input autocorrelation to the output autocorrelation by the inverse sine function.

5.4.1.2 Soft-Limiter Model

For a soft limiter or a clipper, the output autocorrelation can be obtained using Equation (3.48) and setting $k = 2$ in Equation (5.23):

$$\frac{\partial^2 R_{yy}(\tau)}{\partial R_x^2(\tau)} = \frac{\exp\left[\frac{l^2}{1+\rho_x(\tau)}\right] - \exp\left[\frac{l^2}{1-\rho_x(\tau)}\right]}{\pi\sqrt{1 - \rho_x^2(\tau)}} \tag{5.31}$$

where $\rho_{x(\tau)} = R_{x(\tau)}/R_{xx}(0)$ and the output autocorrelation function can be found by integrating Equation (5.31) where we get (Gross and Veeneman, 1994):

$$R_{yy}(\tau) = R_{xx}(0)\left[\text{erf}^2\left(\frac{l}{\sqrt{2}\sigma_x}\right)\frac{R_{xx}(\tau)}{R_{xx}(0)} + \sum_{n=2}^{\infty} C_n\left[\frac{R_{xx}(\tau)}{R_{xx}(0)}\right]^{n+1}\right] \tag{5.32}$$

where erf indicates the error function and C_n is defined as

$$C_n = \frac{4H_{n-1}^2\left(\frac{l}{\sqrt{2}\sigma_x}\right)}{\pi 2^n(n+1)!} e^{\frac{-l^2}{\sigma_x^2}} \tag{5.33}$$

where H_n is a Hermite polynomial of order n.

5.4.1.3 Smooth-Limiter Model

A smooth limiter is characterized by the error function as in Equation (3.49). The output autocorrelation function of a smooth limiter can be found using Price's theorem in a simple compact form as (Price, 1958):

$$\frac{\partial R_{yy}(\tau)}{\partial R_{xx}(\tau)} = \frac{L^2}{2\pi\sqrt{(1 + l^{-2})^2 - l^{-4}R_x^2(\tau)}} \tag{5.34}$$

Now, by integrating Equation (5.34), we get the output autocorrelation function as:

$$R_{yy}(\tau) = \frac{(lL)^2}{2\pi} \sin^{-1}\left[\frac{R_{xx}(\tau)}{(1 + l^2)R_{xx}(0)}\right]. \tag{5.35}$$

Figure 5.2 shows the output PSD of three kinds of limiter amplifier using the calculated autocorrelation functions for an input with box shaped spectral density (Narrowband Gaussian Noise (NBGN)). The statistical properties of NBGN can be found in Appendix B.

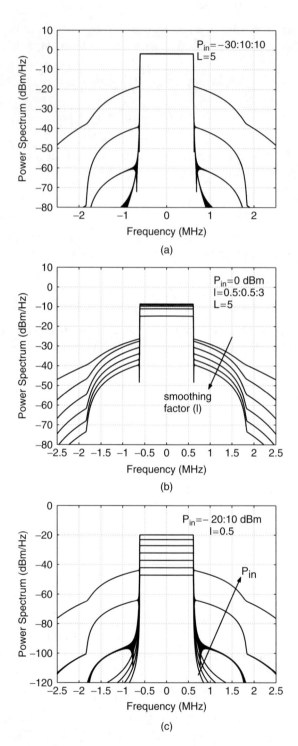

Figure 5.2 Output power spectral densities: (a) hard limiter; (b) smooth limiter; and (c) soft limiter (Gharaibeh, 2004).

5.4.2 Memoryless Power-Series Model

The autocorrelation function at the output of a power-series model can be found using Equation (5.13) and applying the properties of the moments of Gaussian random variables in Equation (5.28). Therefore, the output autocorrelation function for third-order nonlinearity can be evaluated as

$$R_{\tilde{y}\tilde{y}}(\tau) = |b_1|^2 E[\tilde{x}_1 \tilde{x}_2^*] + b_1 b_3^* E[\tilde{x}_1 \tilde{x}_2^{*^2} \tilde{x}_2] + b_1^* b_3 E[\tilde{x}_1^2 \tilde{x}_1^* \tilde{x}_2^*]$$

$$+ |b_3|^2 E[\tilde{x}_1^2 \tilde{x}_1^* \tilde{x}_2^{*^2} \tilde{x}_2] \qquad (5.36)$$

For illustration purposes, consider a third-order nonlinearity ($n = 3$) then, the autocorrelation function at the output of the nonlinearity ($n = 3$) reduces to

$$R_{\tilde{y}\tilde{y}}(\tau) = R_{\tilde{x}\tilde{x}}(\tau) \left[|b_1|^2 + 4Re[(b_1 b_3^*)] R_{\tilde{x}\tilde{x}}(0) + 4|b_3|^2 R_{\tilde{x}\tilde{x}}^2(0) \right]$$

$$+ 2|b_3|^2 R_{\tilde{x}\tilde{x}}^3(\tau) \qquad (5.37)$$

This result means that the output autocorrelation function can be written as a power series in the input autocorrelation function $R_{\tilde{x}\tilde{x}}(\tau)$) which provides a great simplification over the generalized autocorrelation analysis.

The Gaussian assumption leads to inaccurate results when modeling distortion in the some wireless communication signals. For example, and as shown in (Aparin, 2001), higher-order moments of a reverse link IS-95 CDMA signals are different from those obtained from the properties of Gaussian moments although the second-order moments (the input autocorrelation function) are the same. The Gaussian assumption was shown to be fairly acceptable for the forward link since the transmitted signal consist of a large number of Walsh-coded data sequences that by the central limit theorem, approaches the Gaussian distribution. However the accuracy of such an assumption decreases if the composite signal is lightly loaded.

Figure 5.3 shows the output spectrum computed using signal realizations of a forward link WCDMA signal compared to the spectrum generated using the NBGN assumption. It is clear that the Gaussian assumption overestimated nonlinear distortion for the case of a lightly loaded signal (1 DPCH) while it gives a good estimate for distortion for a heavily loaded signal (16 DPCH) and this is evident from the level of adjacent channel power in both cases.

5.5 Response of Nonlinear Systems to Multiple Random Signals

The previous sections provided analysis of the autocorrelation function and the PSD at the output of the nonlinear system for a random input. The analysis was limited to a single-channel input. In this section, multichannel analysis developed in Chapter 4 is used to develop the autocorrelation function and the output spectrum for a nonlinear system with two random input signals.

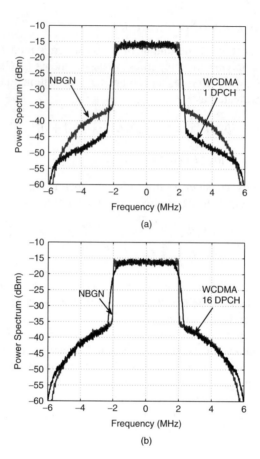

Figure 5.3 Simulated output spectrum using signal realization (solid) and NBGN model (dashed); (a) forward link and (b) reverse link (Gharaibeh, 2004).

5.5.1 Power-Series Model

For the case of two-channel input, the output autocorrelation function can be derived using Equation (4.110) where the autocorrelation function of the output signal centered at the first carrier frequency ξ_1 can be written as (Gharaibeh and Steer, 2005)

$$R_{yy;\xi_1}(\tau) = \sum_{n=1}^{N}\sum_{m=1}^{N}\sum_{l=0}^{\frac{n-1}{2}}\sum_{k=0}^{\frac{m-1}{2}} b_{n,l}b_{m,k}^{*} R_{x_n x_m x_l x_k;\xi_1}(\tau) \tag{5.38}$$

where

$$R_{x_n x_m x_l x_k;\xi_1}(\tau) = E[\tilde{x}_1^{*\frac{(n+1)}{2}-l}(t)\tilde{x}_1^{*\frac{(n-1)}{2}-l}(t)\tilde{x}_1^{\frac{(m-1)}{2}-k}(t+\tau)$$

$$\times \tilde{x}_1^{*\frac{(m+1)}{2}-k}(t+\tau)\tilde{x}_2^{l}(t)\tilde{x}_2^{*l}(t)\tilde{x}_2^{k}(t+\tau)\tilde{x}_2^{*k}(t+\tau)] \tag{5.39}$$

The above expression reduces to the single-channel formulation when $\tilde{x}_2(t) = 0$ and setting k and l to 0. The PSD of the output signal can be computed from the Fourier transform of the autocorrelation function as

$$S_{yy;\xi_1}(f) = \sum_{n=1}^{N}\sum_{m=1}^{N}\sum_{l=0}^{\frac{n-1}{2}}\sum_{k=0}^{\frac{m-1}{2}} b_{n,l}b_{m,k}^{*} S_{\tilde{x};nmlk;\xi_1}(\tau) \tag{5.40}$$

where $S_{\tilde{x};nmlk;\xi_1}(f)$ is the $(nmlk)$th-order PSD that can be found from the Fourier transform of the $nmlk$ autocorrelation function in Equation (5.39).

The output autocorrelation function of the spurious intermodulation components can be developed in the same way. Therefore, for the upper and lower intermodulation components, using Equation (4.117), the output autocorrelation function at the intermodulation frequency (ξ_{IM3}) can be written as

$$R_{\tilde{y}\tilde{y};\xi_{IM3}}(\tau) = \sum_{n=3}^{N}\sum_{m=3}^{N}\sum_{l=0}^{\frac{n-3}{2}}\sum_{k=0}^{\frac{m-3}{2}} b_{n,l}b_{m,k}^{*} R_{\tilde{x}_n\tilde{x}_m\tilde{x}_l\tilde{x}_k}(\tau). \tag{5.41}$$

where for the upper intermodulation component (at $\xi_{IM3U} = 2\xi_2 - \xi_1$) we have

$$R_{\tilde{x}_n\tilde{x}_m\tilde{x}_l\tilde{x}_k;\xi_{IM3X}}(\tau) = E\left[\tilde{x}_1^l(t)\tilde{x}_1^{*l+1}(t)\tilde{x}_1^{k+1}(t+\tau)\tilde{x}_1^{*k}(t+\tau)\tilde{x}_2^{\frac{(n+1)}{2}-l}(t) \right.$$
$$\left. \times \tilde{x}_2^{*\frac{(n-3)}{2}-l}(t)\tilde{x}_2^{\frac{(m-3)}{2}-k}(t+\tau)\tilde{x}_2^{*\frac{(m+1)}{2}-k}(t+\tau) \right] \tag{5.42}$$

and for the lower intermodulation component (at $\xi_{IM3L} = 2\xi_1 - \xi_2$):

$$R_{\tilde{x}_n\tilde{x}_m\tilde{x}_l\tilde{x}_k;\xi_{IM3L}}(\tau) = E\left[\tilde{x}_1^{\frac{(n+1)}{2}-l}(t)\tilde{x}_1^{*\frac{(n-3)}{2}-l}(t)\tilde{x}_1^{\frac{(m-3)}{2}-k}(t+\tau) \right.$$
$$\left. \times \tilde{x}_1^{*\frac{(m+1)}{2}-k}(t+\tau)\tilde{x}_2^l(t)\tilde{x}_2^{*l+1}(t)\tilde{x}_2^{k+1}(t+\tau)\tilde{x}_2^{*k}(t+\tau) \right]$$
$$\tag{5.43}$$

Equation (5.40) consists of $[(N+1)/2]^2 \times [(N+3)/4]^2$ terms. These terms can be divided into three groups (Gharaibeh and Steer, 2005):

- the linear output with gain compression ($l = 0, k = 0$ and $n = 1$ or $m = 1$);
- intermodulation distortion ($l = 0, k = 0$, $n > 1$ and $m > 1$);
- cross-modulation distortion caused by the presence of the second signal ($l > 0, k > 0$).

The distinction between intermodulation and cross-modulation terms is clear in this formulation since the cross-modulation is described by the cross terms (the terms that consist of the product of both signal envelops). This enables different distortion terms to be identified and estimated. Note also that the output spectrum can be further partitioned into a number of spectral components based on the frequency band of interest as follows (Gharaibeh and Steer, 2005):

- the linear signal output;
- in-band gain compression;

- in-band spectral regrowth from intermodulation;
- in-band spectral regrowth from cross-modulation;
- out-of-band spectral regrowth from intermodulation;
- out-of-band spectral regrowth from cross-modulation;
- spurious components centered at the intermodulation frequencies of the two carriers.

The designation of "in-band" components corresponds to the in-band components that appear within each signal bandwidth around its carrier frequency while designation of out-of-band components corresponds to the spectral components which appear in the vicinity of each signal bandwidth and are responsible for adjacent channel interference. However, not all the in-band components contribute to the effective nonlinear distortion responsible of degrading system performance, as will be seen in the next chapter.

The above development provides accurate characterization of intermodulation as well as cross-modulation distortions. Note that the formulation of the autocorrelation functions in Equation (5.39) and Equation (5.41) provides an insight into the effect of the self and joint statistics of the two signals on the level of distortion. This means that this distortion can be controlled by changing the statistics of the two signals. For example, in CDMA systems, this can be done by the proper choice of the number of users in the two channels.

Figure 5.4 shows simulated output spectra of the response of a nonlinear amplifier to the sum of two IS-95-CDMA signals centered at frequencies f_1 and f_2 with the level of the second channel at f_2 held constant and the power of the first channel at f_1 varied from 0 to 15 dBm above the power of the second channel. The figure shows how ACPR at the second carrier increases as the power level of the first carrier is swept.

Figure 5.5 shows the output spectrum at f_1 partitioned into linear, intermodulation and cross-modulation components assuming that the two CDMA channels have the same input power. The output spectrum shows the increase in spectral regrowth and in-band distortion over single-channel excitation due to cross-modulation.

5.5.2   Wiener–Hammerstein Model

The autocorrelation function of the output of the second block ($z(t)$) considering only the output component centered at frequency ξ_1 can be written as (Gharaibeh and Steer, 2005)

$$R_{\tilde{z}\tilde{z};\xi_1}(\tau) = E[\tilde{z}_{\xi_1}(t)\tilde{z}_{\xi_1}(t+\tau)]. \tag{5.44}$$

Now, using Equation (4.97), the autocorrelation function in Equation (5.44) becomes

$$R_{\tilde{z}\tilde{z};\xi_1}(\tau) = \sum_{n=1}^{N}\sum_{m=1}^{N}\sum_{l=0}^{\frac{n-1}{2}}\sum_{k=0}^{\frac{m-1}{2}} b_{n,l}b_{m,k}^{*} R_{\tilde{u}_n\tilde{u}_m\tilde{u}_l\tilde{u}_k;\xi_1}(\tau) \tag{5.45}$$

where:

$$R_{\tilde{u}_n\tilde{u}_m\tilde{u}_l\tilde{u}_k;\xi_1}(\tau) = E\left[\tilde{u}_1^{\frac{(n+1)}{2}-l}(t)\tilde{u}_1^{*\frac{(n-1)}{2}-l}(t)\tilde{u}_1^{\frac{(m-1)}{2}-k}(t+\tau)\right.$$
$$\left.\times \tilde{u}_1^{*\frac{(m+1)}{2}-k}(t+\tau)\tilde{u}_2^{l}(t)\tilde{u}_2^{*l}(t)\tilde{u}_2^{k}(t+\tau)\tilde{u}_2^{*k}(t+\tau)\right] \tag{5.46}$$

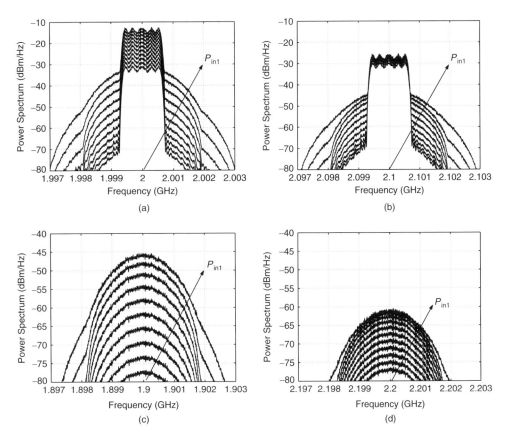

Figure 5.4 Simulated output spectrum: (a) first CDMA channel, (b) second CDMA channel, (c) lower intermodulation component; and (d) upper intermodulation component (Gharaibeh, 2004).

The PSD of the output signal $\tilde{z}_{\xi_1}(t)$ is obtained from the Fourier transform of the auto-correlation function as:

$$S_{\tilde{z}\tilde{z};\xi_1}(f) = \sum_{n=1}^{N}\sum_{m=1}^{N}\sum_{l=0}^{\frac{n-1}{2}}\sum_{k=0}^{\frac{m-1}{2}} b_{n,l}b_{m,k}^{*}S_{\tilde{u};nmlk}(f) \qquad (5.47)$$

where $\tilde{H}_{1;\xi_i}(f)$ is the baseband equivalent transfer functions of the input filter with respect to frequency ξ_i, and

$$S_{\tilde{u};nmlk;\xi_1}(f) = |\tilde{H}_{1;\xi_1}(0)|^{n+m-2l-2k}|\tilde{H}_{1;\xi_2}(0)|^{2l+2k}S_{\tilde{x};nmlk}(f) \qquad (5.48)$$

where $S_{\tilde{x};nmlk;\xi_1}(f)$ is the $(nmlk)$th-order PSD that can be found from the Fourier transform of the $nmlk$ autocorrelation function in Equation (5.39).

Note that Equation (5.47) is derived assuming that the input filter frequency response is constant within each signal bandwidth. The PSD at the output of the system is found by multiplying Equation (5.47) by the squared magnitude of the baseband equivalent of

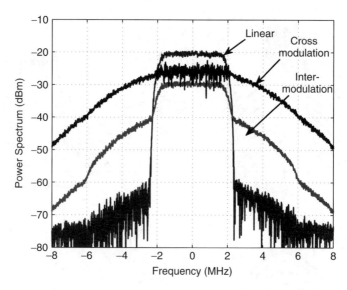

Figure 5.5 Output spectrum of a nonlinear model driven by two forward link WCDMA signals and partitioned into linear, cross-modulation, and intermodulation components (Gharaibeh and Steer, 2005).

the transfer function of the second filter:

$$S_{\tilde{y}\tilde{y};\xi_1}(f) = |\tilde{H}_{2;\xi_1}(f)|^2 S_{\tilde{z}\tilde{z};\xi_1}(f) \tag{5.49}$$

where $\tilde{H}_{2;\xi_1}(f)$ is the baseband equivalent transfer functions of the output filter with respect to frequency ξ_1.

A similar analysis can be developed for the intermodulation components with proper choice of the baseband equivalent of the input and output filters. Therefore, the output PSD of the intermodulation components is obtained in a similar way to Equation (5.47):

$$S_{\tilde{y}\tilde{y};\xi_{IM3}}(f) = |\tilde{H}_{2;\xi_{IM3}}(f)|^2 S_{\tilde{z}\tilde{z};\xi_{IM3}}(f) \tag{5.50}$$

The above results enable the autocorrelation function of the outputs centered at the carrier and intermodulation frequencies to be computed by combining the effects of the input and output filters as well as the nonlinear block. Note that with this formulation, the output spectrum is simply the spectrum of the memoryless nonlinearity multiplied by powers of the magnitude frequency response of the linear filters. This enables the easy implementation of the model in software where measured filter responses are multiplied by the spectrum generated from the Fourier transform of the autocorrelation function computed for the particular digitally modulated signal(s).

5.6 Response of Nonlinear Systems to a Random Signal and a Sinusoid

The case when the input to a nonlinear system consists of the sum of a single tone and a random signal appears when dealing with a receiver desensitization problem in

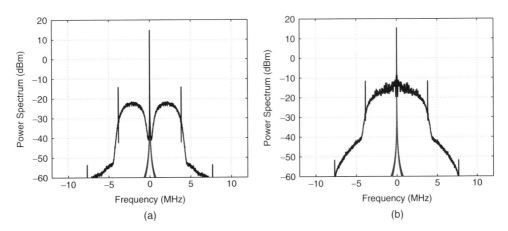

Figure 5.6 Spectrum of a single tone with cross-modulation from mixing with a WCDMA signal: (a) 1 DPCH signal and (b) 16 DPCH signal (Gharaibeh, 2004).

RF receivers as discussed in Chapter 1. The receiver desensitization problem can be treated as a special case of the above analysis. Let the input signal $x(t)$ be the sum of a single-tone jammer $x_1(t) = A_1 \cos(\omega_1 t + \theta)$ and a transmitter leakage CDMA signal $x_2(t) = A_2(t) \cos(\omega_2 t + \Phi(t))$, then $\tilde{x}_1(t) = A_1 e^{j\theta}$ and $\tilde{x}_2(t) = A_2(t) e^{j\Phi(t)}$. The output autocorrelation at the jammer center frequency can then be expressed as in Equation (5.38) with

$$R_{x_n x_m x_l x_k}(\tau) = A_1^{n+m-2l-2k} E[\tilde{x}_1^l(t)\tilde{x}_1^{*^l}(t)\tilde{x}_2^k(t+\tau)\tilde{x}_2^{*^k}(t+\tau)]. \tag{5.51}$$

Note that the above expression means that the jammer spectrum is widened by the cross-modulation distortion generated by the nonlinearity. The cross-modulation distortion acts as an interference to a weak desired signal adjacent to the jammer.

Figure 5.6 presents the output spectrum of a single tone when the power amplifier is driven by a single tone and a 1 DPCH WCDMA signal, (Figure 5.6(a)), and a single tone with a 16 DPCH WCDMA signal, (Figure 5.6(b)). The figure shows how the spectrum is widened because of the mixing with the modulated WCDMA carrier as a result the nonlinear interaction or cross-modulation. The figure also shows the cross-modulation spectrum depends on the nature of the WCDMA signal.

5.7 Summary

In this chapter, the autocorrelation function and the PSD at the output of a nonlinearity for the single- and multiple-channel cases was derived using the different nonlinear models. The analysis is useful when dealing with nonlinear distortion that results from real-world communication signals and enables nonlinear distortion to be quantified. In Chapter 7, the above analysis will be used to develop some distortion metrics that define the performance of communication systems under nonlinearity.

6

Nonlinear Distortion

Nonlinearity in wireless systems results in performance degradation because the response to a nonlinear system usually consists of undesirable signal components in addition to the desired signal. Unlike system noise, these undesirable signal components (known as distortion) cannot be removed by filtering since they usually exist within the frequency band of the desired signal. Therefore, nonlinearity, as will be seen in the following chapters, has daunting effects on the output waveform that inflect severe restrictions on system design, bandwidth, performance and the overall system power budget. In-band (or co-channel) distortion results in a degradation of SNR and ultimately as a degradation of BER while out-of-band distortion contributes to the degradation of the performance of the adjacent channel. From a constellation diagram point of view, the nonlinear behavior is manifested as compression and rotation of the signal constellation and this means that nonlinear distortion results in increasing system probability of error.

The quantification of nonlinear distortion when the input to the system consists of single or multiple digitally modulated signals is usually based on the nonlinear transformation of random processes that was discussed in Chapter 5. It was shown that with finite-bandwidth signals that represent digitally modulated signals, the nonlinear behavior of RF circuits results in two main impairments; the first is gain compression and the second is spectral regrowth and both limit system performance. Spectral regrowth consists of two components; the first lies inside the bandwidth of the signal and is called in-band (or co-channel) distortion while the second exists outside the bandwidth of the signal and is called Adjacent-Channel Interference (ACI). Nonlinear distortion is then predicted by estimating the output power spectrum of the nonlinearity that is defined as the Fourier transform of the output autocorrelation function. However, estimating the output spectrum does not mean that system performance can be predicted since in-band spectral regrowth may not be completely uncorrelated from the desired signal.

Characterizing the effective in-band distortion and its relation to the system figures of merit require the correlated and the uncorrelated components of the output spectrum to be identified. The correlated output component consists of an amplified version of the input waveform with gain compression/expansion and represents the useful part of the output that leads to correct detection of the received data. On the other hand, the uncorrelated

Nonlinear Distortion in Wireless Systems: Modeling and Simulation with MATLAB®, First Edition.
Khaled M. Gharaibeh.
© 2012 John Wiley & Sons, Ltd. Published 2012 by John Wiley & Sons, Ltd.

part adds to the system interference in a similar way to that of AWGN. In this context, an orthogonalization procedure is needed to partition the output spectrum into correlated and uncorrelated components so that the effective nonlinear distortion inside the signal bandwidth can be quantified.

In this chapter we present a generalized approach for the accurate estimation of nonlinear distortion introduced by nonlinear amplification in wireless systems (Gharaibeh *et al.*, 2007). We present a procedure based on Gram-Schmidt orthogonalization of the nonlinear model to determine the effective nonlinear distortion responsible for the degradation of system performance. The orthogonalization of the nonlinear model results in separating the correlated and uncorrelated distortion in the output spectrum of a nonlinearity and helps identify the effective distortion components responsible for the degradation of system performance.

6.1 Identification of Nonlinear Distortion in Digital Wireless Systems

The key to understanding the degradation of performance in wireless systems caused by nonlinear amplification is to recognize that the effective nonlinear distortion is uncorrelated with the desired signal. For example, SNR at the output of a nonlinear circuit is determined by the ratio of the desired signal to the uncorrelated in-band distortion and noise.

The identification of the uncorrelated distortion components is usually done by assuming that the input signal has Gaussian statistical properties. Then, using Bussgang theorem (Bussgang, 1952) or the Gaussian moment theorem (Miller, 1969) discussed in the previous chapter, the response of a general nonlinearity to a Gaussian process consists of an amplified replica of the input signal and an uncorrelated distortion component, see Bendat (1990), Gardner and Archer (1993), Mathews (1995) and Pedro and de-Carvalho (2001). The Gaussian assumption leads to a simplified analysis of distortion in communication systems, however, it is not always valid to model wireless communication signals that exhibit statistical properties that depend on the number of active users, user power profile, modulation and coding (Aparin, 2001). A theoretical analysis of the decomposition of the output spectrum into uncorrelated components without using the Gaussian assumption was studied in Aparin (2001), Blachman (1968), Gharaibeh *et al.* (2007), Kim and Powers (1993) and Raich *et al.* (2005), where based on the properties of the distribution function of the input signals, the output of a bandpass nonlinearity can be expressed as a sum of uncorrelated components.

To develop the concept of uncorrelated distortion, we consider a geometrical representation of a memoryless nonlinearity characterized by a third-order polynomial where the nonlinear output is the vector sum of the linear $a_1 x(t)$ and the third-order component $a_3 x(t)^3$ as shown in Figure 6.1 (Gharaibeh *et al.*, 2006). In this representation, the third-order output can be decomposed into two components: one in the direction of the linear output and the other orthogonal to it. The uncorrelated distortion output can now be identified in terms of a canceling signal, where a scaled replica of the input signal is subtracted from the total nonlinear output. Thus, for a third-order power series model, we define a canceling signal as a replica of the input signal used to cancel the correlated component of the output:

$$x_c(t) = a_1 x(t) + \alpha x(t) \qquad (6.1)$$

where the correlation coefficient α is defined (for a deterministic signal) as (Gharaibeh *et al.*, 2006)

$$\alpha = \frac{\int_{-\infty}^{\infty} y_3(t)x(t)dt}{\int_{-\infty}^{\infty} x(t)^2 dt} \tag{6.2}$$

The correlation coefficient α in this case represents the fraction of the cubic term that is correlated with the linear response. Thus, the correlated output can be defined as (Gharaibeh *et al.*, 2006)

$$y_c(t) = a_1 x(t) + \alpha x(t) \tag{6.3}$$

while the uncorrelated output is

$$y_d(t) = y(t) - y_c(t) = a_3 x(t)^3 - \alpha x(t) \tag{6.4}$$

Note that the correlated output $y_c(t)$ can be canceled by subtracting the signal $x_c(t)$ as shown in Figure 6.1(b), while the orthogonal component $y_d(t)$ cannot be canceled by a scaled replica of the input signal since it is orthogonal to the correlated output signal. Hence, the component $y_d(t)$ represents the effective uncorrelated distortion that contributes to the degradation of system performance, while $y_c(t)$ represents the correlated output that includes correlated distortion responsible for causing gain compression of the linear output.

This definition complies with the definition of nonlinear distortion in communication theory where the receiver is designed to distinguish between only two types of signals: the transmitted signal to which it is matched; and noise (Banelli, 2003; Dardari *et al.*, 2000; Eslami and Shafiee, 2002). The term correlation refers to the statistical resemblance between the output and the input signals and measures the ability of the receiver to recover useful information from the input signal. Therefore, if the signal is uncorrelated with the expected signal, then that part of the signal is considered as uncorrelated distortion noise that contributes to the degradation of the overall SNR. The effective SNR is then calculated as the ratio of the effective signal component (which includes the correlated distortion or

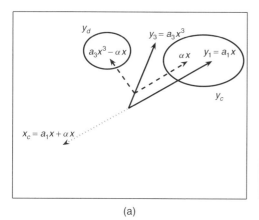
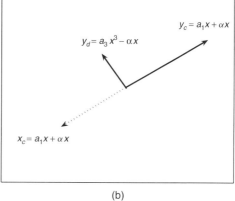

(a)   (b)

Figure 6.1 Geometrical interpretation of in-band distortion: (a) distortion vector and (b) orthogonal representations (Gharaibeh *et al.*, 2007).

gain compression) to the effective uncorrelated nonlinear distortion component. Therefore, the uncorrelated distortion is treated as an additive noise component in a similar way to the AWGN. This analysis of the nonlinear output greatly simplifies the analysis and simulation of nonlinear distortion and its impact on system performance and enables system performance metrics to be quantified easily.

6.2   Orthogonalization of the Behavioral Model

As discussed above, the orthogonalization of the behavioral model is needed for the prediction of nonlinear distortion in wireless communication systems where the objective is to extract the uncorrelated component of the nonlinear output that is responsible for the degradation of system performance. To this end, the behavioral model is developed by orthogonalizing the nonlinear response such that the output components of the model are orthogonal to each other. In the development presented here, the orthogonalization of the behavioral model is applied to models of a memoryless nonlinearity as well as and models of a nonlinearity with memory.

In order to clarify the concept of the identification of the effective nonlinear distortion, consider a general nonlinearity that can be characterized by:

$$y(t) = \sum_{n=1}^{N} y_n(t). \tag{6.5}$$

This form does not guarantee the representation of the output as pure linear and pure distortion terms. This is because different orders of nonlinearity ($y_n(t)$) may be correlated. For example, in a memoryless power-series model, the terms $y_n(t)$ that are represented by monomials of the form $x^n(t)$ are correlated.

The objective is to convert the nonlinear model in Equation (6.5) into a model with orthogonal output components of the form (Gharaibeh *et al.*, 2007)

$$y(t) = \sum_{n=1}^{N} s_n(t) \tag{6.6}$$

where $s_n(t)$ represent the n th-order orthogonal output. Orthogonality here is defined in the statistical sense as:

$$R_{s_n s_m}(\tau) = E[s_n(t)s_m(t+\tau)] = 0. \tag{6.7}$$

where $R_{s_n s_m}(\tau)$ is the cross-correlation function assuming that $s_n(t)$ and $s_m(t)$ are jointly WSS processes and E is the statistical expectation operator. As a result, the nonlinear output can be expressed as (Gharaibeh *et al.*, 2007)

$$y(t) = y_c(t) + y_d(t). \tag{6.8}$$

where

$$y_c(t) = s_1(t) \tag{6.9}$$

is the useful component of the output that is correlated with the input signal and

$$y_d(t) = \sum_{n=2}^{N} s_n(t) \tag{6.10}$$

is the uncorrelated component that represents nonlinear distortion.

To formulate the output of a nonlinearity as a sum of uncorrelated terms as in Equation (6.6), the Gram-Schmidt orthogonalization procedure can be used (Strang, 1988). Gram-Schmidt orthogonalization is a mathematical procedure by which a set of non orthogonal basis vectors is converted into an orthogonal set. To illustrate this procedure, let x_i be a set of non orthogonal basis vectors for a finite-dimensional vector space V then any vector $y \in V$ can be written as a linear combination of these basis vectors as

$$y = \sum_{n=1}^{N} b_n x_n \tag{6.11}$$

where $< x_n, x_m > \neq 0$ the $< >$ indicates inner product. The objective is to write the vector y in terms of a new set of orthogonal basis u_n as:

$$y = \sum_{n=1}^{N} c_n u_n \tag{6.12}$$

which means that the vector y consists of the sum orthogonal vectors $c_n u_n$. The orthogonal basis vectors u_n are produced using the Gram–Schmidt procedure as:

$$u_n = x_n - \sum_{m=1}^{n-1} \alpha_{mn} u_m \tag{6.13}$$

where

$$\alpha_{nm} = \frac{< x_n, u_m >}{||u_m||^2} \tag{6.14}$$

and the new coefficients c_n can then be found from the new coefficients b_n as:

$$c_n = b_n - \sum_{m=n}^{N} c_m \alpha_{mn} \tag{6.15}$$

Therefore, if x and y represent the input and output of a nonlinear model respectively, the Gram-Schmidt orthogonalization procedure leads to completely uncorrelated output terms. This procedure can be applied to any nonlinear model with correlated outputs and results in representing the output of a nonlinearity as a sum of uncorrelated terms. Note that Gram-Schmidt orthogonalization is more general than using orthogonal polynomials to model nonlinearity as in Chang (1990) and Zhou (2000) because it does not impose a certain probability distribution on the input waveform.

In the following subsections, this procedure is applied to a number of nonlinear models with the objective of quantifying the effective uncorrelated nonlinear distortion produced by nonlinear systems.

6.2.1 Orthogonalization of the Volterra Series Model

The orthogonalization of Volterra series model using Gaussian inputs was discussed in Schetzen (1981) where the Volterra kernels are converted to G-functionals that result in a nonlinear model with uncorrelated outputs. Here, the Gaussian assumption is not used and the Gram-Schmidt orthogonalization procedure is applied to the general Volterra series directly. however, some restrictions on the statistical properties of the input signal apply, as will be seen in the next sections.

Applying Gram-Schmidt orthogonalization to the Volterra model in Equation (3.1), the new set of orthogonal output components $s_n(t)$ in Equation (6.6) can be obtained by replacing the input basis functions $(x_1(t), x_2(t), \ldots, x_N(t))$ by a new set of orthogonal basis $(u_1(t), u_2(t), \ldots, u_N(t))$. Therefore, the orthogonal output components can be written as a multi dimensional convolution of the orthogonalized Volterra kernels with the powers of the orthogonalized inputs as

$$s_n(t) = \int_{-\infty}^{\infty} \cdots \int_{-\infty}^{\infty} g_n(\lambda_1, \ldots, \lambda_n) \prod_{i=1}^{n} u(t - \lambda_i) d\lambda_i. \tag{6.16}$$

where $g_n(\lambda_1, \ldots, \lambda_n)$ represents the orthogonalized Volterra kernels and $u_n(t) = \prod_{i=1}^{n} u(t - \lambda_i)$ represents a new set of orthogonal inputs. Therefore, using the Gram-Schmidt procedure the new set of orthogonal inputs can be obtained as

$$\prod_{i=1}^{n} u(t - \lambda_i) = \prod_{i=1}^{n} x(t - \lambda_i) - \sum_{m=1}^{n-1} \alpha_{mn}(\lambda_1, \ldots, \lambda_{m+n}) \prod_{i=1}^{m} u(t - \lambda_i) \tag{6.17}$$

where $\alpha_{nm}(\lambda_1, \ldots, \lambda_{m+n})$ is a time function that represents the correlation coefficient between the x_n and u_m terms:

$$\alpha_{nm}(\lambda_1, \ldots, \lambda_{m+n}) = \frac{E\left[\prod_{i=1}^{n} x(t - \lambda_i) \prod_{i=1}^{m} u(t - \lambda_i)\right]}{E\left[\prod_{i=1}^{m} u(t - \lambda_i) \prod_{i=1}^{m} u(t - \lambda_i)\right]} \tag{6.18}$$

and hence, the original set $x_n(t)$ can then be written as a linear combination of the orthogonal set $u_n(t)$ as

$$\prod_{i=1}^{n} x(t - \lambda_i) = \sum_{m=1}^{n} \alpha_{nm}(\lambda_1, \ldots, \lambda_{m+n}) \prod_{i=1}^{m} u(t - \lambda_i) \tag{6.19}$$

The new kernels $g_n(\lambda_1, \ldots, \lambda_n)$ are derived from the Volterra kernels as

$$g_n(\lambda_1, \ldots, \lambda_n) = \sum_{m=n}^{N} \int_{-\infty}^{\infty} \cdots \int_{-\infty}^{\infty} \alpha_{mn}(\lambda_1, \ldots, \lambda_{m+n}) h_m(\lambda_1, \ldots, \lambda_m) d\lambda_{n+1} \ldots d\lambda_{n+m}. \tag{6.20}$$

Note that there is always a difficulty in the evaluation of the orthogonal kernels and the associated correlation coefficients in the above analysis. The analysis can be greatly simplified if the input is assumed to have a Gaussian distribution or if a simplified version of Volterra series is used. These special cases result in similar analysis to the ones presented in Bendat (1990) and Schetzen (1981) that lead to closed-form expressions of the correlation coefficients and the uncorrelated outputs of a nonlinearity.

6.2.2   Orthogonalization of Wiener Model

A Wiener model is a simplified version of the Volterra series model where the n-dimensional Volterra kernel takes the form (Chen, 1989):

$$h_n(\lambda_1, \ldots, \lambda_n) = a_n h(\lambda_1) \ldots h(\lambda_n), \tag{6.21}$$

where a_n are a real coefficient and $h(t)$ is the impulse response of a linear filter as shown in Figure 6.2(a). The nth-order output of this model can be written as:

$$y_n(t) = \left(\int_{-\infty}^{\infty} h(\lambda)x(t-\lambda)d\lambda \right)^n . \tag{6.22}$$

Using the same development with the Volterra series model and utilizing the kernel relationship in Equation (6.21), the correlation coefficient between the mth- and the nth-order terms of the nonlinear output is found using Equation (6.18) as

$$\alpha_{nm}(\lambda) = \frac{E\left[x^n(t-\lambda)u^m(t-\lambda)\right]}{E\left[u^m(t-\lambda)u^m(t-\lambda)\right]} \tag{6.23}$$

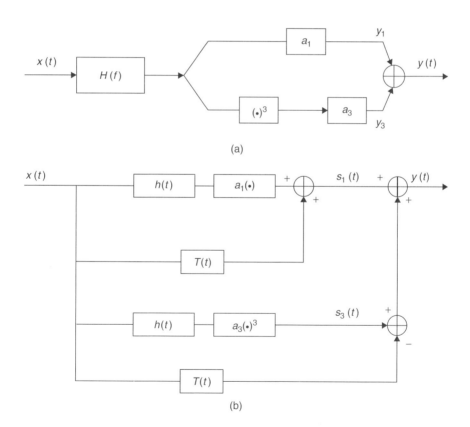

(a)

(b)

Figure 6.2   A third-order Wiener model and (b) orthogonal third-order Wiener model.

and hence, the orthogonalized kernels can be found by substituting Equation (6.23) in Equation (6.20):

$$g_n(\lambda_1, \ldots, \lambda_n) = \sum_{m=n}^{N} \int_{-\infty}^{\infty} \cdots \int_{-\infty}^{\infty} \alpha_{mn}(\lambda_1 + \cdots + \lambda_{m+n}) a_m \prod_{i=1}^{m} h(\lambda_i) \prod_{i=n+1}^{n+m} d\lambda_i$$

$$= \sum_{m=n}^{N} a_m T_{mn}(\lambda_1, \ldots, \lambda_n) \tag{6.24}$$

where

$$T_{mn}(\lambda_1, \ldots, \lambda_n) = \int_{-\infty}^{\infty} \cdots \int_{-\infty}^{\infty} \alpha_{mn}(\lambda_1 + \cdots + \lambda_{m+n}) \prod_{i=1}^{m} h(\lambda_i) \prod_{i=n+1}^{n+m} d\lambda_i. \tag{6.25}$$

represents a linear filter. The orthogonal basis functions can be written in terms of the original basis as

$$u^n(t - \lambda) = x^n(t - \lambda) - \sum_{m=1}^{n-1} \alpha_{nm}(\lambda) u^m(t - \lambda). \tag{6.26}$$

For a third-order nonlinearity and considering odd-order terms only, we have:

$$\alpha_{31}(\lambda) = \frac{E\left[x^3(t - \lambda)x(t - \lambda)\right]}{E\left[x(t - \lambda)x(t - \lambda)\right]} \tag{6.27}$$

and the orthogonalized kernels can be written using Equation (6.24) as

$$g_1(\lambda) = a_1 h(\lambda) + a_3 T_{31}(\lambda)$$

$$g_3(\lambda_1, \lambda_2, \lambda_3) = a_3 h(\lambda_1) h(\lambda_2) h(\lambda_3), \tag{6.28}$$

where $T_{31}(\lambda)$ represents a linear filter and can be found using Equation (6.25):

$$T_{31}(\lambda) = \int_{-\infty}^{\infty} \int_{-\infty}^{\infty} \int_{-\infty}^{\infty} \alpha_{31}(\lambda_1 + \lambda_2 + \lambda_3 + \lambda_4) h(\lambda_1) h(\lambda_2) h(\lambda_3) d\lambda_2 d\lambda_3 d\lambda_4 \tag{6.29}$$

The orthogonal set of basis functions can be found using Equation (6.26):

$$u_1(t - \lambda) = x(t - \lambda)$$

$$u_3(t - \lambda) = x^3(t - \lambda) - \alpha_{31}(\lambda) u_1(t - \lambda) \tag{6.30}$$

Therefore, the orthogonal outputs $s_n(t)$ are

$$s_1(t) = \int_{-\infty}^{\infty} [h(\lambda) + a_3 T_{31}(\lambda)] x(t - \lambda) d\lambda$$

$$s_3(t) = \left(\int_{-\infty}^{\infty} h(\lambda) x(t - \lambda) d\lambda\right)^3 - \int_{-\infty}^{\infty} a_3 T_{31}(\lambda) x(t - \lambda) d\lambda. \tag{6.31}$$

An orthogonalized third-order Wiener model is shown in Figure 6.2(b). The figure shows that the correlated output $s_1(t)$ consists of the linear input $y_1(t) = x(t) * h(t)$ in addition to a new component that represents the correlated part of the third-order output $y_3(t)$ (which is in this case represented by a new linear filter with impulse response $a_3 T_{31}(t)$). Thus, the correlated output is shaped in the frequency domain by a linear filter response represented by the Fourier transform of $(h(t) + a_3 T_{31}(t))$ which interprets the asymmetry in the shape of the output spectrum when the nonlinear system exhibits memory effects.

6.2.3 Orthogonalization of the Power-Series Model*

In this subsection, the orthogonalization of the behavioral model is applied to an envelope power-series model that represents a quasi-memoryless nonlinearity. The envelope version of the power-series model is expressed as in Equation (4.42):

$$\tilde{y}(t) = \sum_{\substack{n=1 \\ n \, \text{odd}}}^{N} b_n \tilde{x}_n(t) = \sum_{\substack{n=1 \\ n \, \text{odd}}}^{N} b_n |\tilde{x}(t)|^{n-1} \tilde{x}(t), \tag{6.32}$$

where the coefficients b_n are complex and represent the envelope coefficients that can be obtained by polynomial fitting of the measured AM–AM and AM–PM characteristics.

Using the orthogonalization procedure, the new set of orthogonal output components $\tilde{s}_n(t)$ in Equation (6.6) can be written as (Gharaibeh *et al.*, 2007)

$$\tilde{s}_n(t) = c_n \tilde{u}_n(t) \tag{6.33}$$

Therefore, the new set of orthogonal basis are

$$\tilde{u}_n(t) = \tilde{x}_n(t) - \sum_{m=1}^{n-2} \alpha_{nm} \tilde{u}_m(t) \tag{6.34}$$

where, in this case, the correlation coefficients α_{nm} are complex coefficients and can be expressed as

$$\alpha_{nm} = \frac{E[\tilde{x}_n(t)\tilde{u}_m^*(t)]}{E[\tilde{u}_m(t)\tilde{u}_m^*(t)]} \tag{6.35}$$

Hence, the original basis $\tilde{x}_n(t)$ can be written as a linear combination of the orthogonal basis $\tilde{u}_n(t)$ as

$$\tilde{x}_n(t) = \sum_{m=1}^{n} \alpha_{nm} \tilde{u}_m(t). \tag{6.36}$$

The new set of coefficients c_n that represents the orthogonalized model are derived from the original model coefficients b_n as

$$c_n = \sum_{m=n}^{N} b_m \alpha_{mn} \tag{6.37}$$

* This entire section is reproduced from Microwave Antenna and Propogation, 1, 1078–1085, Gharaibeh *et al.*, 2007 published by the IET, © IET 2007.

Therefore, the new set of coefficients depends on the original envelope coefficients and the input signal power level represented by the correlation coefficient α_{mn}. Special cases of this procedure can be developed when the distribution of the input process is known, such as using the Gaussian assumption as discussed in Bendat (1990). A 5th-order orthogonalized power-series model is shown in Figure 6.3.

Note that, with the power-series model, Gram-Schmidt orthogonalization leads to new polynomial representation with orthogonal terms. As discussed in Schetzen (1981) and Shirayaev (1996), the orthogonalization of the polynomial model leads to a new type of polynomial based on the probability distribution of the input process. For example, if the input process has a Gaussian distribution, the orthogonalization procedure leads to Hermite polynomial representation of the nonlinear model. Another example is when the input process has a Poisson distribution where the orthogonalization procedure leads to the Poisson–Charlier polynomial representation of the nonlinear model. Therefore, the orthogonalized model provides a generalized approach that is not tied to the probability distribution of the input signal.

For illustration, consider a 5th-order orthogonalized envelope power-series model, the orthogonal inputs are found using Equation (6.34):

$$\tilde{u}_1(t) = \tilde{x}_1(t)$$

$$\tilde{u}_3(t) = \tilde{x}_3(t) - \alpha_{31}\tilde{u}_1(t)$$

$$\tilde{u}_5(t) = \tilde{x}_5(t) - \alpha_{51}\tilde{u}_1(t) - \alpha_{53}\tilde{u}_3(t) \qquad (6.38)$$

and the new set of coefficients of the orthogonal model are obtained using Equation (6.37) as

$$c_1 = b_1 + \alpha_{31}b_3 + \alpha_{51}b_5$$

$$c_3 = b_3 + \alpha_{53}b_5$$

$$c_5 = b_5 \qquad (6.39)$$

where the correlation coefficients α_{mn} are found as in Equation (6.35) as

$$\alpha_{31} = \frac{E[\tilde{x}_3(t)\tilde{u}_1^*(t)]}{E[\tilde{u}_1(t)\tilde{u}_1^*(t)]}$$

$$\alpha_{51} = \frac{E[\tilde{x}_5(t)\tilde{u}_1^*(t)]}{E[\tilde{u}_1(t)\tilde{u}_1^*(t)]}$$

$$\alpha_{53} = \frac{E[\tilde{x}_5(t)\tilde{u}_3^*(t)]}{E[\tilde{u}_3(t)\tilde{u}_3^*(t)]} \qquad (6.40)$$

6.3 Autocorrelation Function and Spectral Analysis of the Orthogonalized Model

The objective now is to derive the autocorrelation function and the PSD of the output of the orthogonal behavioral model in order to identify and estimate the effective in-band distortion. The autocorrelation function is derived using Equations (6.8) that states

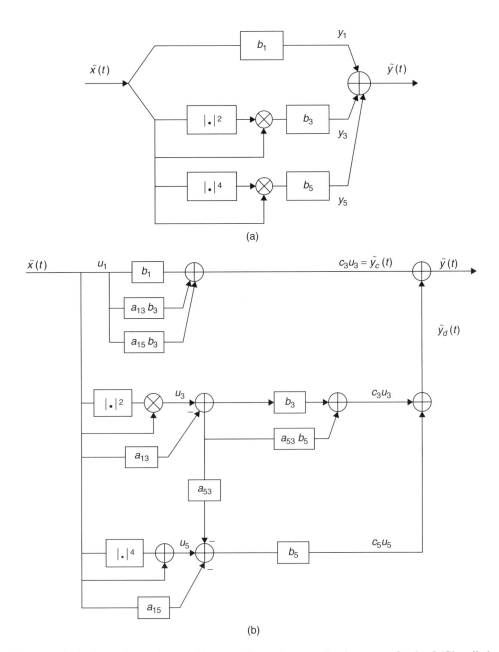

Figure 6.3   Orthogonal envelope nonlinear model with uncorrelated outputs of order 5 (Gharaibeh *et al.*, 2007).

that the cross-correlation function of the correlated and uncorrelated components is zero. Therefore, the output autocorrelation function (and hence, the output PSD) can be written as the sum of the autocorrelation functions (or PSDs) of correlated and uncorrelated components.

6.3.1 Output Autocorrelation Function

Using Equations (6.8) and the orthogonality of the output terms of the orthogonal model, the output autocorrelation function can now be written as:

$$R_{yy}(\tau) = E[y(t)y(t+\tau)] = R_{y_c y_c}(\tau) + R_{y_d y_d}(\tau) \tag{6.41}$$

where $R_{y_c y_c}(\tau)$ is the autocorrelation of the correlated component of the nonlinear output and $R_{y_d y_d}(\tau)$ is the autocorrelation of the uncorrelated distortion component. These autocorrelation functions can be expressed using Equations (6.9) and (6.10) as

$$R_{y_c y_c}(\tau) = R_{s_1 s_1}(\tau)$$

$$R_{y_d y_d}(\tau) = \sum_{n=2}^{N} R_{s_i s_i}(\tau) \tag{6.42}$$

Therefore, the orthogonalization of the nonlinear model results in the formulation of the output autocorrelation as a sum of the autocorrelation functions of the orthogonal components of the output. However; and as pointed out in Blachman (1968), this is not always true because even if $E[s_n(t)s_m^*(t)] = 0$ it is not necessary that $E[s_n(t)s_m^*(t+\tau)] = 0$. As discussed in Gharaibeh $et\ al.$ (2007) and based on the work done by Blachman in Blachman (1968), for a zero mean random process, $x(t)$, a sufficient and necessary condition for the output autocorrelation function to be written as a sum of uncorrelated components is that the process, $x(t)$, be a separable random process (has a separable distribution function) in the Nuttal sense (Nuttal, 1958). This implies that the cross-correlation function of the input signal $(x_1(t) = x(t))$ and its nth-order power $(x_n(t) = x^n(t))$ be written in the form:

$$E[x_n(t)x_1(t+\tau)] = \alpha_{n1} R_{xx}(\tau) \tag{6.43}$$

This identity holds for separable random processes such as Gaussian processes, however, a separable process does need to be Gaussian (Blachman, 1968). For wireless communication signals, this condition can be proved by using their statistical properties without assuming a Gaussian distribution.

6.3.2 Power Spectral Density

The PSD of the output of the nonlinearity is obtained from the Fourier transform of Equation (6.41):

$$S_{yy}(f) = S_{y_c y_c}(f) + S_{y_d y_d}(f), \tag{6.44}$$

where

$$S_{y_c y_c}(f) = S_{s_1 s_1}(f)$$
$$= |c_1|^2 S_{u_1 u_1}(f) \tag{6.45}$$

and

$$S_{y_d y_d}(f) = \sum_{n=2}^{N} S_{s_i s_i}(f)$$
$$= \sum_{n=2}^{N} |c_n|^2 S_{u_n u_n}(f) \tag{6.46}$$

which means that the PSD of the output of a nonlinearity is composed of two spectral components that represent the correlated output and the uncorrelated distortion.

Figure 6.4 shows the output spectrum of a forward-link WCDMA signals partitioned into correlated and uncorrelated spectral components. Note that the shape and level of each of the spectral components depends on the statistical properties of the input signal. Other signals may have different shapes of these spectral components. In general, for nonlinearities where memory effects are not present, the correlated spectrum usually follows the shape of the spectrum of the input signal, while the uncorrelated distortion spectrum has a shape that changes depending on the nonlinear order and on input signal statistics. Examples of the output spectrum of different communication signals will be given in Chapter 9.

It is also worth noting that the correlated spectrum is always confined to the bandwidth of the input signal, while the uncorrelated spectrum spreads outside the bandwidth of the input signal. Thus, the uncorrelated distortion spectrum, which represents the uncorrelated part of spectral regrowth, consists of two components; an in-band distortion component

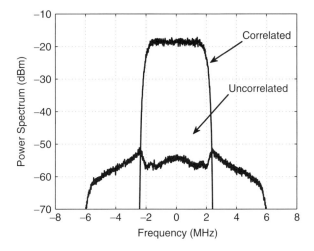

Figure 6.4 Output spectrum of a nonlinearity partitioned into correlated and uncorrelated components.

responsible for the degradation of system SNR in the main channel and an out-of-band component responsible for the degradation of SNR in the adjacent channel. The in-band distortion power can be found from the uncorrelated distortion spectrum as

$$P_{\text{In-band}} = \int_{f_0-B/2}^{f_0+B/2} S_{y_d y_d}(f) df \tag{6.47}$$

where f_0 is the carrier frequency and B is the bandwidth of the input signal. The Adjacent Channel Power (ACP) can be found as

$$P_{\text{ACP}} = \int_{f_1}^{f_2} S_{y_d y_d}(f) df \tag{6.48}$$

where f_1 and f_2 are frequency limits within the band of the adjacent channel.

6.4 Relationship Between System Performance and Uncorrelated Distortion

The formulation of the output of a nonlinearity as the sum correlated and uncorrelated components enables the effect of nonlinear distortion on the performance communication systems to be quantified. To clarify the impact of nonlinear distortion on system performance, consider a received signal at a communication receiver that consists of an amplified transmitted signal and an AWGN component. Therefore, using Equation (6.8) the received signal can be written in complex envelope form as

$$\tilde{r}(t) = \tilde{y}_c(t) + \tilde{y}_d(t) + \tilde{n}(t) \tag{6.49}$$

where $\tilde{n}(t)$ is an AWGN component. Note that the three signal components on the right-hand side of this equation are now uncorrelated and therefore, the uncorrelated output distortion is treated as an additive noise similar to AWGN. System performance is defined in terms of the effective SNR at the receiver that is defined as the ratio of the signal power to the total system noise power including nonlinear distortion.

In general, both the correlated and uncorrelated components of the output have different impacts on the system performance. To establish the relationship between system performance and nonlinear distortion, it is useful to analyze the impact of each of the output components in the orthogonalized model on the effective system SNR. From Equations (6.9) and (6.33), the correlated output, $\tilde{y}_c(t)$, of an envelope power-series model can be written as

$$\tilde{y}_c(t) = \tilde{s}_1(t) = c_1 \tilde{u}_1(t) = \sum_{m=1}^{N} b_m \alpha_{m1} \tilde{x}(t)$$

$$= b_1 \tilde{x}(t) + \sum_{m=3}^{N} b_m \alpha_{m1} \tilde{x}(t) \tag{6.50}$$

This equation means that the correlated output consists of two components; the first is a linearly amplified version of the input signal ($b_1 \tilde{x}(t)$) and the second is the correlated part of the nonlinear terms (the remaining terms in Equation (6.50)). The latter, known as gain compression, is represented as a scaled version of the input signal by a complex factor when nonlinearity includes AM–PM conversion and a real factor when AM–PM conversion is absent. Gain compression does not contribute to distortion noise but rather affects the signal level in a similar manner to gain saturation of discrete tones. With digitally modulated signals that involve linear modulation, gain compression causes the rotation and compression of the constellation points which results in increasing system BER. In systems that employ Automatic Gain Control (AGC), the complex scaling that results from gain compression can be removed and hence, system performance can be improved.

On the other hand, it is evident from Equation (6.49) that the uncorrelated distortion component of the output ($\tilde{y}_d(t)$) is an additive noise component where its effect on system performance is similar the effect of AWGN. For example, the uncorrelated distortion noise results in scattering of the constellation points of a digitally modulated signal in a similar manner to the effect of AWGN and this results in increasing system BER. The effect of this additive noise component can be removed by using linearization techniques such as predistortion that are widely used in wireless communication systems. Therefore, the

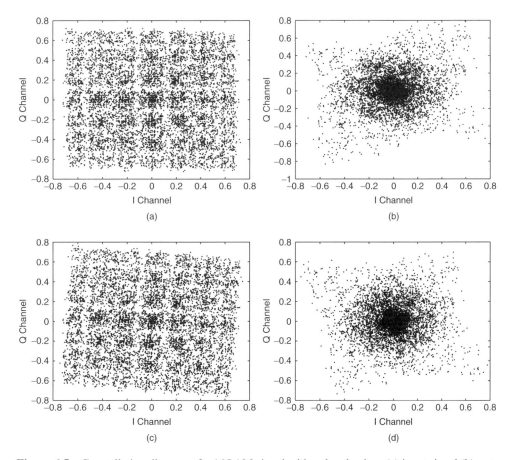

Figure 6.5 Constellation diagram of a 16QAM signal with pulse shaping; (a) input signal (b) output signal (c) correlated output and (d) uncorrelated output.

design of a predistorter based on removing the uncorrelated distortion noise result in improving system performance.

To clarify the impact of nonlinear distortion on received signal constellation, consider a linearly modulated input signal (16 QAM) applied to a power-series model with complex coefficients. Using the above analysis, the output of the nonlinearity is decomposed into correlated and uncorrelated components using the orthogonalization of the power-series model in the previous section and the constellation diagram of the input, total output, correlated output and uncorrelated output are plotted as shown in Figures 6.5(a)-(d). Figure 6.5(b) shows how the constellation diagram of the input signal is corrupted by nonlinearity. This corruption can be analyzed into two effects, the first is the effect of the correlated distortion that is included in the correlated output $\tilde{y}_c(t)$ and the second is the effect of the uncorrelated distortion. Figure 6.5(c) shows the constellation diagram of the correlated output where it is clear that it is a replica of the input with rotation and compression of the constellation points while Figure 6.5(d) shows the constellation of the uncorrelated output where it is clear that it is similar to the case when AWGN is present.

6.5 Examples

In the following subsections, examples are given on the orthogonalization of the behavioral model with different types of signals. This includes the case when the input signal is a NBGN signal and the cases of different multisine signals.

6.5.1 Narrowband Gaussian Noise[†]

As an example, we apply the above analysis to a NBGN process that is a very widely used approximation to CDMA signals (Gharaibeh et al., 2007). The complex envelope of a NBGN process is a complex Gaussian process with a flat spectrum over its bandwidth (see Appendix B). Hence, its autocorrelation function can be written as

$$R_{\tilde{x}\tilde{x}}(\tau) = E[\tilde{x}(t)\tilde{x}^*(t+\tau)] = P_x \text{sinc}(B\tau) \tag{6.51}$$

where $P_x = R_{\tilde{x}\tilde{x}}(0)$ represents the average power of the process and B is its bandwidth.

Now considering a 5th-order orthogonalized power-series model in Section 6.2.3, the correlation coefficients α_{mn} can be evaluated using Equation (6.40) and utilizing the properties of Gaussian moments presented in Chapter 5 (Equation (5.28)) (Gharaibeh et al., 2007):

$$\alpha_{31} = \frac{E[\tilde{x}_3(t)\tilde{u}_1^*(t)]}{E[\tilde{u}_1(t)\tilde{u}_1^*(t)]} = \frac{R_{\tilde{x}_3\tilde{u}_1}(0)}{R_{\tilde{u}_1\tilde{u}_1}(0)} = 2R_{\tilde{x}\tilde{x}}(0) = 2P_x$$

$$\alpha_{51} = \frac{E[\tilde{x}_5(t)\tilde{u}_1^*(t)]}{E[\tilde{u}_1(t)\tilde{u}_1^*(t)]} = \frac{R_{\tilde{x}_5\tilde{u}_1}(0)}{R_{\tilde{u}_1\tilde{u}_1}(0)} = 6R_{\tilde{x}\tilde{x}}^2(0) = 6P_x^2$$

$$\alpha_{53} = \frac{E[\tilde{x}_5(t)\tilde{u}_3^*(t)]}{E[\tilde{u}_3(t)\tilde{u}_3^*(t)]} = \frac{R_{\tilde{x}_5\tilde{u}_3}(0)}{R_{\tilde{u}_3\tilde{u}_3}(0)} = 6R_{\tilde{x}\tilde{x}}(0) = 6P_x. \tag{6.52}$$

[†] This entire section is reproduced from Microwave Antenna and Propogation, 1, 1078–1085, Gharaibeh et al., 2007 published by the IET, © IET 2007.

The coefficients of the orthogonal model are obtained by inserting Equation (6.52) into Equation (6.39) (Gharaibeh *et al.*, 2007):

$$c_1 = b_1 + \alpha_{31} b_3 + \alpha_{51} b_5 = b_1 + 2P_x b_3 + 6P_x^2 b_5$$

$$c_3 = b_3 + \alpha_{53} b_5 = b_3 + 6P_x b_5$$

$$c_5 = b_5 \tag{6.53}$$

The orthogonal basis are found by inserting Equation (6.53) in Equation (6.38):

$$\tilde{u}_1(t) = \tilde{x}_1(t)$$

$$\tilde{u}_3(t) = \tilde{x}_3(t) - \alpha_{31}\tilde{u}_1(t) = \tilde{x}_3(t) - 2P_x \tilde{u}_1(t)$$

$$\tilde{u}_5(t) = \tilde{x}_5(t) - \alpha_{51}\tilde{u}_1(t) - \alpha_{53}\tilde{u}_3(t) = \tilde{x}_5(t) - 6P_x^2 \tilde{u}_1(t) - 6P_x \tilde{u}_3(t) \tag{6.54}$$

It is easy to show that a Gaussian process is a separable random process because of the separability of the Gaussian distribution function, and therefore the condition in Equation (6.43) holds. For example, it is easy to show that on using the properties of Gaussian moments that (Gharaibeh *et al.*, 2007):

$$E[\tilde{s}_1(t + \tau)\tilde{s}_3(t)] = 0. \tag{6.55}$$

The cross-correlation between the other terms can be evaluated similarly. The autocorrelation of the correlated and uncorrelated components can now be evaluated by applying Equation (6.54) in Equation (6.42) as

$$R_{\tilde{y}_c \tilde{y}_c}(\tau) = R_{\tilde{s}_1 \tilde{s}_1}(\tau) = |c_1|^2 R_{\tilde{u}_1 \tilde{u}_1}(\tau)$$

$$= |b_1 + 2b_3 P_x + 6b_5 P_x^2|^2 R_{\tilde{x}\tilde{x}}(\tau) \tag{6.56}$$

and

$$R_{\tilde{y}_d \tilde{y}_d}(\tau) = R_{\tilde{s}_3 \tilde{s}_3}(\tau) + R_{\tilde{s}_5 \tilde{s}_5}(\tau)$$

$$= |c_3|^2 R_{\tilde{u}_3 \tilde{u}_3}(\tau) + |c_5|^2 R_{\tilde{u}_5 \tilde{u}_5}(\tau) \tag{6.57}$$

where, by using the properties Gaussian moments, we have:

$$R_{\tilde{u}_3 \tilde{u}_3}(\tau) = 2R_{\tilde{x}\tilde{x}}^3(\tau) \tag{6.58}$$

and

$$R_{\tilde{u}_5 \tilde{u}_5}(\tau) = 12R_{\tilde{x}\tilde{x}}^5(\tau). \tag{6.59}$$

The PSD of the third- and fifth-order orthogonal outputs can be found as

$$S_{\tilde{s}_3 \tilde{s}_3}(f) = |c_3|^2 S_{\tilde{u}_3 \tilde{u}_3}(f) = 2|c_3|^2 S_{\tilde{x}\tilde{x}}^{3*}(f) \tag{6.60}$$

and

$$S_{\tilde{s}_5 \tilde{s}_5}(f) = |c_5|^2 S_{\tilde{u}_5 \tilde{u}_5}(f) = 12|c_5|^2 S_{\tilde{x}\tilde{x}}^{5*}(f) \tag{6.61}$$

where $S^{k*}(f) = S(f) * S(f) * \ldots * S(f)$ and the $*$ denotes convolution. Hence, the uncorrelated distortion spectrum can be evaluated using as

$$S_{\tilde{y}_d \tilde{y}_d}(f) = S_{\tilde{s}_3 \tilde{s}_3}(f) + S_{\tilde{s}_5 \tilde{s}_5}(f)$$
$$= 2|c_3|^2 S_{\tilde{x}\tilde{x}}^{3*}(f) + 12|c_5|^2 S_{\tilde{x}\tilde{x}}^{5*}(f) \tag{6.62}$$

6.5.2 Multisines with Deterministic Phases

In the following subsections, multisine input signals with equal amplitudes A and deterministic phases of the form

$$x(t) = \sum_{i=1}^{K} A \cos(\omega_i t + \phi_i). \tag{6.63}$$

are considered. The output of a memoryless nonlinearity at the fundamental tones is derived from their complex envelopes using the analysis presented in Section 4.5. The uncorrelated distortion is then quantified using the basic concepts developed in Section 6.1 in terms of a canceling process to the correlated output.

6.5.2.1 Single Tone

Consider the case where the input to a memoryless power-series model is a single-tone with frequency ω_c as in Equation (4.118), the output at the fundamental frequency (the first zonal output) can be derived from the complex envelope in Equation (4.119) as

$$y(t) = g_1(A) \cos(\omega_c t + \phi + g_2(A)) \tag{6.64}$$

where g_1 and g_2 represent the AM–AM and AM–PM characteristics. In this case, the nonlinear response to a single tone is therefore a replica of the input signal with modified amplitude and phase. A canceling signal can therefore be designed to cancel the whole output signal regardless of the phase of the input signal. This means that the response to a single tone is a single component that is correlated with the input signal and hence, the uncorrelated distortion component is absent. This is intuitive because the response of a nonlinear system to a single tone results in gain compression and not distortion.

6.5.2.2 Two Tones

Now consider the response of a memoryless power-series model to a two-tone input with equal amplitudes as shown in Equation (4.121), the complex envelope of the output of the nonlinearity at each of the fundamental frequencies ω_1 and ω_2 can be found from Equations (4.122) and (4.123) as (Gharaibeh et al., 2006)

$$\tilde{y}_{\xi_1}(t) = \sum_{n=1}^{N} \sum_{l=0}^{\frac{n-1}{2}} b_{n,l} A^n e^{j\phi_1} \tag{6.65}$$

and for the second tone:

$$\tilde{y}_{\xi_2}(t) = \sum_{n=1}^{N} \sum_{l=0}^{\frac{n-1}{2}} b_{n,l} A^n e^{j\phi_2} \tag{6.66}$$

then the bandpass output at the fundamental frequencies (the first zonal output) can be derived from the complex envelops as

$$y(t) = g_1(A) \cos[\omega_1 t + \phi_1 + g_2(A)]$$

$$+ g_1(A) \cos[\omega_2 t + \phi_2 + g_2(A)] \tag{6.67}$$

where $g_1(A)$ and $g_2(A)$ are real functions of the amplitudes and that represent the AM–AM and AM–PM conversions, respectively, and can be easily evaluated from Equations (6.65) and (6.66). This form means that with a two-tone signal the output tones that lie within the input band have the same phase change as the linear output regardless of the initial phase of the input tones and the nonlinear order. Therefore, effective in-band distortion is absent, as considering Equation (6.67), the output signal has a phase that is totally correlated with the input signal. Thus, a canceling signal can be designed to cancel the output signal at the fundamental frequencies completely regardless of the initial phase. The uncorrelated output, therefore, is only present in the out-of-band component which is evident from the forms of the intermodulation components as in Equation (4.125) and Equation (4.126).

6.5.2.3 Four Tones

For a four-tone input with equal amplitudes A and deterministic phases, the input $x(t)$ to a nonlinearity is given by

$$x(t) = \sum_{i=1}^{4} A \cos(\omega_i t + \phi_i). \tag{6.68}$$

For illustration, consider a third-order power series, then the output complex envelope at the fundamental frequencies can be found using the general form of the complex envelope of the response of a power-series model to multiple inputs in Equation (4.108) with $K = 4$. It is easy to show that the complex envelope at each of the fundamental frequencies is (Gharaibeh *et al.*, 2006)

$$\tilde{y}_{\omega_1}(t) = b_1 A e^{j\theta_1} + b_3 A^3 \left(7 e^{j\theta_1} + e^{j(2\theta_2 - \theta_3)} + 2 e^{j(\theta_2 + \theta_3 - \theta_4)} \right)$$

$$\tilde{y}_{\omega_2}(t) = b_1 A e^{j\theta_2} + b_3 A^3 \left(7 e^{j\theta_2} + e^{j(2\theta_3 - \theta_4)} + 2 e^{j(\theta_4 + \theta_1 - \theta_3)} + 2 e^{j(\theta_3 + \theta_1 - \theta_2)} \right)$$

$$\tilde{y}_{\omega_3}(t) = b_1 A e^{j\theta_3} + b_3 A^3 \left(7 e^{j\theta_3} + e^{j(2\theta_2 - \theta_1)} + 2 e^{j(\theta_4 + \theta_1 - \theta_2)} + 2 e^{j(\theta_4 + \theta_2 - \theta_3)} \right)$$

$$\tilde{y}_{\omega_4}(t) = b_1 A e^{j\theta_4} + b_3 A^3 \left(7 e^{j\theta_4} + e^{j(2\theta_3 - \theta_2)} + 2 e^{j(\theta_3 + \theta_2 - \theta_1)} \right) \tag{6.69}$$

Therefore, the total output at the fundamental frequencies (the first zonal output) can then be derived from these complex envelops as

$$y(t) = \sum_{i=1}^{4} g_{1i}(A)\cos(\omega_i t + \Theta_i(\phi_1, \ldots \phi_4) + g_{2i}(A))$$

(6.70)

where $g_{1i}(A)$ and $g_{1i}(A)$ are real functions of the input amplitudes and the nonlinear coefficients and $\Theta_i(\phi_1, \ldots \phi_4)$ is a linear function of the initial phases ϕ_i. Note that a four-tone signal produces distortion components at the fundamental frequencies that can be either correlated or uncorrelated with the linear output depending on the initial phases of the input tones. This is evident from Equation (6.70) where the phase term of the bandpass signal is a function of all the phases of the input tones (not a constant phase) and hence, it is not possible to design a canceling signal that cancels all the signal components at all frequencies. This means that the uncorrelated distortion is present inside the band of the multisine signal.

For the special case where the initial phases of the input tones are zero, different terms in each of the output components in Equation (6.69) add up in the following way:

$$\tilde{y}_{\omega_1}(t) = \tilde{y}_{\omega_4}(t) = B_1$$

$$\tilde{y}_{\omega_2}(t) = \tilde{y}_{\omega_3}(t) = B_2$$

(6.71)

where $B_1 = b_1 A + 10 b_3 A^3$ and $B_2 = b_1 A + 12 b_3 A^3$. Hence, Equation (6.70) becomes

$$y(t) = |B_1|\cos(\omega_1 t + \angle B_1) + |B_2|\cos(\omega_2 t + \angle B_2)$$

$$+ |B_2|\cos(\omega_3 t + \angle B_2) + |B_1|\cos(\omega_4 t + \angle B_1)$$

(6.72)

Note that the amplitude of the output at each of the four tones is not equal and therefore, a canceling signal will not cancel the whole signal at the fundamental frequencies. This means that the remainder from the canceling process is what is referred to as uncorrelated in-band distortion. The coefficient α for a four-tone signal can be computed from Equation (6.2) as:

$$\alpha = \frac{\int_{-\infty}^{\infty} y_3(t)x(t)dt}{\int_{-\infty}^{\infty} x(t)^2 dt} = 11 b_3 A^2$$

(6.73)

and hence, the in-band distortion component is given by (Gharaibeh et al., 2006)

$$y_{IB}(t) = y(t) - x_c(t)$$

$$= 2 b_3 A^3 [\cos(\omega_1 t) - \cos(\omega_2 t) - \cos(\omega_3 t) + \cos(\omega_4 t)]$$

(6.74)

Note that the in-band distortion components have equal powers at the four fundamental tones in the case of phase-aligned tones (zero initial phases). The same analysis can be performed when the initial phases are not zero. Figure 6.6 shows a time-domain representation of the phase aligned four-tone signal, a canceling signal and the resulting in-band distortion. The frequency-domain representation is depicted in Figure 6.7(a) where the total output spectrum and the total uncorrelated distortion spectrum are simulated. Figure 6.7(b) shows the effective in-band distortion of the four-tone signal as a function

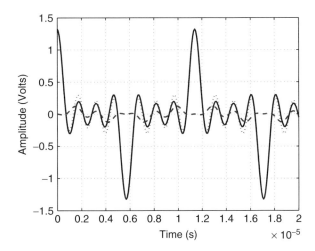

Figure 6.6 Time-domain representation of a phase-aligned four-tone signal: solid: output signal, dotted: canceling signal, and dashed: in-band distortion.

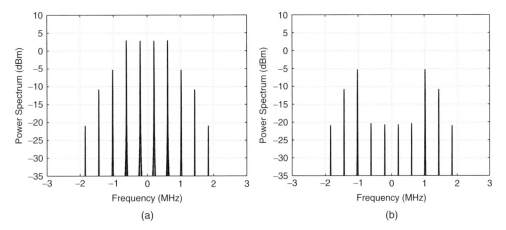

Figure 6.7 Phase aligned four-tone signal: (a) output spectrum and (b) uncorrelated distortion spectrum.

of input power and for different initial phase combinations. Note that the minimum in-band distortion occurs when the initial phases are zero (phase-aligned multisines).

The above analysis provides a basis for the identification of in-band distortion of multisine signals. However, it is not adequate for identifying distortion when the input tones have random phases and/or amplitudes. A multisine signal with random amplitudes and phases models a digitally modulated carrier that does not have a deterministic representation. Therefore multisines are a useful tool in understanding distortion in communication signals. In the following section, the orthogonalization of the nonlinear model is done with multisine signals with random phases where uncorrelated distortion is quantified.

6.5.3 Multisines with Random Phases

The analysis of multisines with random phases is done under the assumption that multisines have uniformly distributed random phases and hence the estimated distortion is found as a statistical average. The statistical average is estimated in simulation by computing the average distortion of a large number of realizations of the multisines having phases generated using a random number generator for each realization, as will be shown next. It is worth emphasizing that the analysis presented here does not require the input signal to have Gaussian distribution, which is a common assumption for a large number of tones with random phases. Therefore, it is valid with any number of tones provided that their phases are completely random.

In order to estimate the uncorrelated distortion of multisines with random phases we follow a similar procedure to the one used for communication signals in the previous sections with a third-order power-series model. The following subsections discuss special cases of multisine signals with random phases and derive the uncorrelated output distortion terms for single, two and four tones.

6.5.3.1 Single Tone

For a single-tone input with a uniformly distributed random phase, the complex envelope of the input $\tilde{x}(t)$ can be expressed as:

$$\tilde{x}(t) = A e^{j\phi} \tag{6.75}$$

where ϕ is a random phase uniformly distributed in $[0, \pi]$. The correlation coefficient α is found using Equation (6.35):

$$\alpha_{13} = b_3 \frac{E[\tilde{x}_3(t)\tilde{x}_1^*(t)]}{E[\tilde{x}_1(t)\tilde{x}_1^*(t)]} = b_3 A^2 \tag{6.76}$$

and the autocorrelation function of the correlated and uncorrelated outputs can be derived using Equation (6.42) as

$$R_{\tilde{y}_c \tilde{y}_c}(\tau) = |b_1 + b_3 A^2|^2 R_{\tilde{x}\tilde{x}}(\tau)$$

$$R_{\tilde{y}_d \tilde{y}_d}(\tau) = 0 \tag{6.77}$$

where $R_{\tilde{x}\tilde{x}}(\tau) = A^2$. Therefore, if the input to a nonlinear system is a single tone with a random phase, then the first zonal output is a single tone with a compressed amplitude regardless of the initial phase. Therefore, the output is completely correlated with the input and thus, a nonlinear system behaves as a distortion less system with single-tone inputs.

6.5.3.2 Two Tones

For a two-tone input with frequencies ω_1 and ω_2 each with amplitude $A/2$, the input $x(t)$ can be expressed in complex envelope form using as (Gharaibeh *et al.*, 2006)

$$\tilde{x}(t) = A \cos\left(\frac{\omega_{\mathrm{m}} t}{2} + \phi\right) \tag{6.78}$$

where $\omega_m = \omega_2 - \omega_1$ and ϕ is a random phase uniformly distributed in $[0, \pi]$. Now using the orthogonalization procedure, the correlation coefficients α can be found as:

$$\alpha_{13} = b_3 \frac{E[\tilde{x}_3(t)\tilde{x}_1^*(t)]}{E[\tilde{x}_1(t)\tilde{x}_1^*(t)]} = \frac{3A^2 b_3}{4} \tag{6.79}$$

and it follows that

$$R_{\tilde{y}_c\tilde{y}_c}(\tau) = \frac{|b_1 + (3A^2/4)b_3|^2 A^2}{2} \cos\left(\frac{\omega_m \tau}{2}\right) \tag{6.80}$$

and

$$R_{\tilde{y}_d\tilde{y}_d}(\tau) = \frac{|b_3|^2 A^6}{32} \cos\left(\frac{3\omega_m \tau}{2}\right) \tag{6.81}$$

Therefore, the output autocorrelation function of a two-tone input consists of components at integer multiples of intermodulation frequency ω_m. The uncorrelated part of the output consists of the out-of-band intermodulation products. Note that a two-tone test cannot predict the uncorrelated in-band distortion since the only distortion terms that result as uncorrelated components are the uncorrelated intermodulation components that lie outside the frequency band of the input tones.

6.5.3.3 Four Tones

A four-tone input with frequencies ω_1, ω_2, ω_3 and ω_4 and equal amplitudes $A/2$ can be expressed in a complex envelope form as (Gharaibeh *et al.*, 2006)

$$\tilde{x}(t) = A\cos\left(\frac{\omega_m t}{2} + \phi_1\right) + A\cos\left(\frac{3\omega_m t}{2} + \phi_2\right) \tag{6.82}$$

where it is assumed that the tones are equally spaced in frequency such that $2\omega_m = \omega_4 - \omega_3 = \omega_3 - \omega_2 = \omega_2 - \omega_1$. The phases ϕ_1 and ϕ_2 are independent random phase uniformly distributed in $[0, \pi]$. The correlation coefficients for a third-order power-series model can be found from Equation (6.35) as:

$$\alpha_{13} = b_3 \frac{E[\tilde{x}_3(t)\tilde{x}_1^*(t)]}{E[\tilde{x}_1(t)\tilde{x}_1^*(t)]} = \frac{9A^2 b_3}{4} \tag{6.83}$$

and hence

$$\tilde{y}_c(t) = b_1 + \frac{9A^2 b_3}{4}\tilde{x}_1(t)$$

$$\tilde{y}_d(t) = b_3\tilde{x}_3(t) - \frac{9A^2 b_3}{4}\tilde{x}_1(t) \tag{6.84}$$

which results in writing the autocorrelation functions of the correlated and uncorrelated outputs as

$$R_{\tilde{y}_c\tilde{y}_c}(\tau) = \frac{|b_1 + (9A^2/4)b_3|^2 A^2}{2}\left[\cos\left(\frac{\omega_m \tau}{2}\right) + \cos\left(\frac{3\omega_m \tau}{2}\right)\right] \tag{6.85}$$

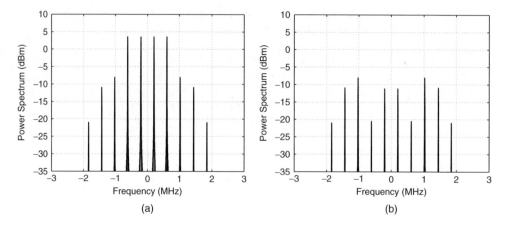

Figure 6.8 Four-tone signal with random phases; (a) output spectrum and (b) uncorrelated distortion spectrum.

and

$$R_{\tilde{y}_d \tilde{y}_d}(\tau) = \frac{|b_3|^2 A^6}{32} [9 \cos\left(\frac{\omega_m \tau}{2}\right) + \cos\left(\frac{3\omega_m \tau}{2}\right)$$

$$+ 18 \cos\left(\frac{5\omega_m \tau}{2}\right) + 9 \cos\left(\frac{7\omega_m \tau}{2}\right)$$

$$+ \cos\left(\frac{9\omega_m \tau}{2}\right)] \tag{6.86}$$

The uncorrelated distortion is manifested as frequency components at the fundamental in addition to the intermodulation frequencies. The in-band distortion is therefore represented by the components $9 \cos(\omega_m \tau/2) + \cos(3\omega_m \tau/2)$. Note that in comparison to the case of phase-aligned four-tone signal, the in-band distortion is not equal at all the fundamental frequencies, where there is a difference of about 9.5 dB between the components at ω_1 and ω_4 and those at ω_2 and ω_3.

Figure 6.8 shows a simulated output spectrum and the uncorrelated distortion spectrum of a four-tone signal with uniformly distributed random phases. The simulated spectrum matches the results obtained above.

6.6 Measurement of Uncorrelated Distortion

Measurement of the uncorrelated in-band distortion in WCDMA systems can be achieved using feed forward cancelation of the desired signal at the output of the nonlinear amplifier (Gharaibeh *et al.*, 2007). Figure 6.9 shows the measurement setup of a feed forward cancelation loop used for measuring the uncorrelated distortion spectrum. The input signal, generated using a vector signal generator, is split using a power splitter into two branches. The first is amplified by the nonlinear PA under test and the second is used after phase reversal to cancel the correlated component of the output. The variable phase shifter

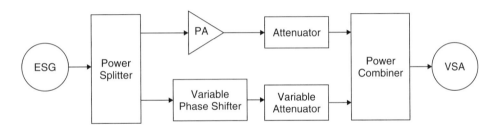

Figure 6.9 Feed forward cancelation measurement setup (Gharaibeh *et al.*, 2007).

and the variable attenuator are used to adjust the linear input level and phase to ensure a perfect cancelation of the linear output of the amplifier. The resulting output at the power combiner consists of only the uncorrelated distortion. The in-band distortion is then measured within the signal bandwidth using a vector signal analyzer.

6.7 Summary

In this chapter the orthogonalization of the behavioral model, which leads to the accurate estimation of nonlinear distortion using different signal models has been presented. The orthogonalization of the behavioral model enables the effective SNR to be directly related to nonlinear distortion. The uncorrelated co-channel distortion was obtained from the output spectrum estimated using signal realizations and measured nonlinear characteristics, which means that system metrics and in-band distortion can be estimated directly and accurately from simple measurements.

By considering a number of discrete tones with random phases, it was shown that-single and two-tone tests are inadequate to model in-band uncorrelated distortion. Single-tone excitation of a nonlinear system results only in gain compression and not distortion. With two-tone excitation, the output consists of compressed outputs at the fundamental frequencies and uncorrelated out-of-band intermodulation components. With K tones where $K > 2$, the output consists of uncorrelated components at the fundamental frequencies. In particular, we have shown that with a four-tone signal with random phases, the amount of in-band distortion is higher than if the phases are aligned, for example, zero initial phases. This is because the output components add to the linear output, whereas with random phases the probability of having components that are orthogonal to the linear output is higher, and hence these components are considered as in-band distortion.

7

Nonlinear System Figures of Merit

Wireless communication system performance is usually characterized by the effective SNR that can be related to system BER. SNR is measured as the ratio of the useful signal power to the AWGN. However, the effective SNR depends on many contributors including thermal noise, phase noise, nonlinear distortion, spurious signals, etc. Intermodulation distortion is a significant contributor to the degradation of system SNR when operating at high output power levels in systems utilizing power efficient nonlinear amplifiers. With nonlinear distortion, the effective SNR is defined as the ratio of the useful signal power to the sum of AWGN and nonlinear distortion powers. Other system metrics are also used to quantify system performance and can directly be related to the effective SNR. Transmitter signal quality in digital wireless communication systems is specified by the EVM of the transmitted signal and the waveform quality factor (ρ). Other system metrics such as NPR and NF are also measures of the system performance and they quantify nonlinear distortion in a different way.

Predicting system performance metrics such as SNR, EVM and the waveform quality factor (ρ) is usually a difficult task when the signal is processed by a nonlinear amplifier. A high dynamic range receiver and demodulator are required to detect the baseband symbols necessary for calculating waveform quality metrics. It is therefore desirable to accurately estimate these metrics from measurements that do not require a sophisticated digital receiver.

As discussed in the previous chapter, the output spectrum can be partitioned into correlated and uncorrelated components. This partition enables the effective in-band (or co-channel) distortion, which contributes to the degradation of system metrics, to be quantified. The correlated output component consists of an amplified version of the input waveform with gain compression/expansion and represents the useful part of the output that leads to correct detection of the received data. On the other hand, the uncorrelated part adds to the system interference in a similar way to that of AWGN. The effective in-band distortion is defined as the component of the nonlinear output that shares the same frequency band as the input signal, but is "uncorrelated" with the ideal transmitter signal.

Nonlinear Distortion in Wireless Systems: Modeling and Simulation with MATLAB®, First Edition. Khaled M. Gharaibeh.
© 2012 John Wiley & Sons, Ltd. Published 2012 by John Wiley & Sons, Ltd.

From a communications point of view, the receiver is designed to distinguish between only two types of signals: the transmitted signal to which it is matched; and noise. Here, the term correlation refers not only to the statistical resemblance between the output and the input signals, but also to the ability of the receiver to recover useful information from the transmitter signal. Therefore, if part of the transmitted signal is uncorrelated with the expected waveform, that part of the signal is considered as uncorrelated distortion noise, which contributes to the degradation of system SNR.

In this chapter, we utilize the techniques developed in the previous chapter to develop the relationship between nonlinear distortion and the most common metrics of system performance; SNR, EVM, ρ, NPR and NF.

7.1 Analogue System Nonlinear Figures of Merit

7.1.1 Intermodulation Ratio

The nonlinear performance of an amplifier with analogue signals is generally quantified using two-tone testing. The troublesome tones at the output of the amplifier are commonly called the third-order intermodulation components IM3. This terminology is loose as other intermodulation orders are involved. Third-Order Intermodulation Ratio (IMR3) is determined by considering the response of two equal-amplitude input tones, of frequencies ξ_1 and ξ_2. IMR3 is the ratio of the power of the lower (IM3$_L$) intermodulation tone, here at $2\xi_1 - \xi_2$ (or $2\xi_2 - \xi_1$) to the power in one of the fundamental tones (ξ_1 or ξ_2):

$$\text{IMR3}_L = \frac{P_{\text{IM3L}}}{P_{\text{fund}}} = \frac{P_{2\xi_1 - \xi_2}}{P_{\xi_1}} \tag{7.1}$$

and IM3$_U$ is defined in the same way. Generalized power-series analysis (Rhyne and Steer, 1987) can be used to model intermodulation distortion of a two-tone signal. The output at a particular frequency ($\xi_i = n_1\xi_1 + n_2\xi_2$) is then the vector addition of a number of Intermodulation Products (IP's). Therefore, using the memoryless version of the GPS in Equation (4.143), then for IM3$_L$ ($n_1 = 2$ and $n_2 = -1$) we have (Rhyne and Steer, 1987):

$$\text{IMR}_{3L} = K[1 + T] \tag{7.2}$$

where:

$$K = 2a_3 \frac{3!}{2^3} \frac{X_1^2}{2} X_1^* \tag{7.3}$$

is the intermodulation term, $*$ indicates complex conjugation, X_1 and X_2 are the phasors of the tones at ξ_1 and ξ_2 and a_i is the ith-order term of the behavioral model that represents a memoryless nonlinear system as a power series as in Equation (7.2). The saturation term T is defined as (Rhyne and Steer, 1987):

$$T = \sum_{\alpha=1}^{\infty} \left\{ \left(\frac{(3 + 2\alpha)!}{3!2^{2\alpha}} \right) \frac{a_{3+2\alpha}}{a_3} \right\} \sum_{s_1+s_2=\alpha} \Phi \tag{7.4}$$

where

$$\Phi = \frac{|X_1|^{2s_1}}{s_1!(2+s_1)!} \frac{|X_2|^{2s_2}}{s_2!(1+s_2)!}. \tag{7.5}$$

The intermodulation term K, has a simple relationship to the input tones and varies as the cube of the level of one of the input tones yielding the classic 3:1 slope. However, this is valid only when the saturation term T is zero (for small signals) but as the signal levels become larger, this term grows because of the contribution of the fifth- and higher-order components.

7.1.2 Intercept Points

Traditionally the third-order Intercept Point (IP3) describes nonlinearity of a power amplifier as it is a quantity specified by the manufacturer. By definition, the ith-order Input Intercept Point (IIPi) is the intercept point of extrapolated output power-input power $(P_{in} - P_{out})$ curve and extrapolated intermodulation power-input power $(P_{in} - P_{IMi})$ curve where (P_{IMi}) denotes the ith-order intermodulation distortion power calculated from a two-tone (separated by small frequency difference) test of the nonlinear device.

Using the model in Rhyne and Steer (1987), the output of the nonlinear device for a single-tone input is expressed as

$$Y = a_1 X + \sum_{\alpha=1}^{\alpha_m} \left\{ \left(\frac{(1+2\alpha)!}{2^{2\alpha}\alpha!(1+\alpha)!} \right) \right\} a_{1+2\alpha} |X|^{2\alpha} X, \tag{7.6}$$

that is as

$$Y = a_1 X + \frac{3}{4}|X|^2 X + \frac{5}{8}|X|^4 X + \dots . \tag{7.7}$$

By considering a pure third-order nonlinearity, the value of X in Equation (7.7) (neglecting terms of order >3) at the intercept point can be expressed as

$$X_{IP3} = \sqrt{\frac{4}{3} \left| \frac{a_1}{a_3} \right|} \tag{7.8}$$

Intercept points represent a good figure of merit for amplifier linearity because they are independent of the input and output powers and are characteristics of the amplifier alone. However, IIPi are extrapolated values and indicate the nonlinear response only in the region where the P_{IMi} has a slope equal to i. IIP3, for example, is not a reliable indicator of performance outside the region where the P_{IM3} has a slope equal to 3, see Figure 7.1, and it cannot predict over the entire power range. Therefore, IIPi may not be the proper parameter to characterize nonlinearity with a nonconstant envelope modulated signal. In addition, an intercept point of order i is constant only when that particular ith intermodulation order is present alone. Since single-order nonlinearity in real nonlinear systems is rarely present, the intercept point concept does not accurately represent nonlinearity.

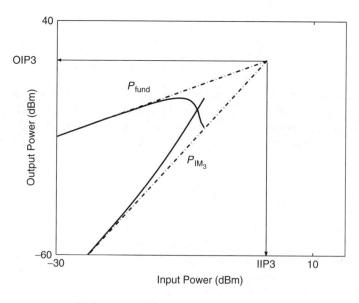

Figure 7.1   Third-order intercept point.

7.1.3   1-dB Compression Point

A very common parameter of nonlinearity is the 1-dB compression point. The 1-dB compression point is the input power at which the extrapolated linear response is greater than that power by 1 dB. Using Equation (7.7), the 1-dB compression point correspond to a factor of $10^{-1/20} = 0.89$ change in voltage gain. Therefore, the corresponding value of X where the power drops to 1 dB below the linear gain is found by solving

$$Y = a_1 X + \frac{3}{4}|X|^2 X + \frac{5}{8}|X|^4 X + \ldots = 0.89X. \tag{7.9}$$

Note that the value of the 1-dB compression point is also dependent on other intermodulation products above-third order. The relation between the 1 dB compression point and the third-order intercept point can be understood by considering a pure third-order nonlinearity where the value of X in Equation (7.9) (neglecting terms of order > 3) can be expressed as

$$X_{1\,\mathrm{dB}} = \sqrt{0.11\frac{4}{3}\left|\frac{a_1}{a_3}\right|} \tag{7.10}$$

and the value of X at the intercept point is

$$X_{\mathrm{IP3}} = \sqrt{\frac{4}{3}\left|\frac{a_1}{a_3}\right|} \tag{7.11}$$

Therefore, $\mathrm{IP3} - \mathrm{P_{1\,dB}} = 9.6\,\mathrm{dB}$ is a fundamental lower limit for the difference. This difference becomes higher when higher-order terms are considered because of the decrease in $P_{1\,dB}$ as the influence of fifth- and higher-order intermodulation products is becoming more significant. This imposes a restriction on the applicability of IIP3 as a measure of nonlinearity, where it becomes more applicable when $P_{1\,dB}$ is higher. Figure 7.1 illustrates the relationship between IIP3 and $P_{1\,dB}$.

7.2 Adjacent-Channel Power Ratio (ACPR)

Adjacent-Channel Interference (ACI) is spectral broadening (spectral regrowth) that results from nonlinear amplification in wireless communication systems. ACI is mitigated in many wireless systems by not using adjacent channels in the same geographical area. However, it is still necessary to establish a lower limit on the ratio of the power in the main channel to the amount of power induced in the adjacent channel. This ratio is known as the Adjacent-Channel Power Ratio (ACPR) (sometimes also termed adjacent channel leakage ratio–ACLR) and is defined in terms of the PSD of the output of a nonlinearity as

$$\text{ACPR} = \frac{\int_{f_1}^{f_2} S_{yy}(f)df}{\int_{f_3}^{f_4} S_{yy}(f)df}. \tag{7.12}$$

The frequencies f_1 and f_2 are the frequency limits of the main channel, while f_3 and f_4 are the limits of the upper adjacent channel. Figure 7.2 clarifies the definition of ACPR and the frequency limits included in the definition.

The definition of the frequency limits is system dependent. For example, in the IS-95 CDMA mobile system standard (IS95 Standard, 1993) ACPR is defined as the ratio of the adjacent channel power in a 30-kHz resolution bandwidth ($f_4 - f_3 = 30\,kHz$), swept over the adjacent channel and measured at 885 k Hz offset from the carrier, to the total power in the main channel ($f_2 - f_1 = 1.23\,MHz$). In WCDMA systems, ACPR is defined with the main channel and adjacent channels measured in a 3.84-MHz bandwidth ($f_2 - f_1 = f_4 - f_3 = 3.84\,MHz$) and the adjacent channel centered at 5 MHz offset from the carrier (3GPP2 Standard, 1999).

7.3 Signal-to-Noise Ratio (SNR)

System SNR is defined as the ratio of signal power to total noise power and is usually estimated assuming a linear AWGN channel. However, nonlinear distortion increases the system BER for a fixed AWGN power and hence an alternative definition of SNR is needed to develop the relationship between nonlinear distortion and system performance. Therefore, system performance is expressed in terms of the effective SNR (also known as Signal-to-Noise and Distortion Ratio (SNDR)) that is defined as the ratio of signal power to the AWGN and nonlinear distortion power in the band of interest.

In order to develop the relationship between the co-channel (in-band) distortion and the effective SNR (and hence system BER), recall the formulation of the PSD of the output of nonlinearity as the sum of the PSDs of the correlated and uncorrelated components in Equation (6.44):

$$S_{\tilde{y}\tilde{y}}(f) = S_{\tilde{y}_c\tilde{y}_c}(f) + S_{\tilde{y}_d\tilde{y}_d}(f) \tag{7.13}$$

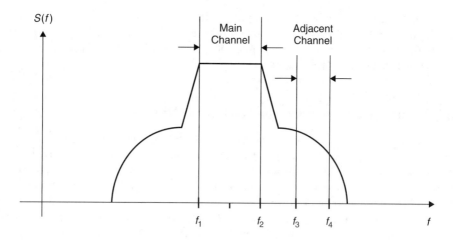

Figure 7.2   Definition of ACPR.

The effective in-band distortion can now be expressed in terms of the PSD of the uncorrelated distortion spectrum as

$$P_{\text{In-band}} = \int_{-B/2}^{B/2} S_{\tilde{y}_d \tilde{y}_d}(f) df \qquad (7.14)$$

where B is the bandwidth of the input signal.

The effective system SNR can be expressed in terms of the PSDs of the correlated and uncorrelated output components of Equation (7.13) as

$$\text{SNR} = \frac{\int_{-B/2}^{B/2} S_{\tilde{y}_c \tilde{y}_c}(f) df}{\int_{-B/2}^{B/2} S_{\tilde{y}_d \tilde{y}_d}(f) df + N_0 B} \qquad (7.15)$$

Note that the effective system SNR is a function of both the nonlinear distortion and the PSD of AWGN represented by its PSD, N_0. Evaluating the effective SNR is important to determine the system BER and the system noise figure. These parameters are usually estimated assuming a linear AWGN channel, however, nonlinear distortion increases the system BER for a fixed AWGN power.

As discussed in the previous chapter, the in-band portion of the uncorrelated output distortion is responsible for the degradation of the system SNR, and hence system BER. The probability of bit error can then be evaluated, for example, from the error probability of QPSK or BPSK system as (Conti *et al.*, 2000)

$$P_e = Q\left(\sqrt{2\text{SNR}}\right) \qquad (7.16)$$

Note that SNR is a function of both the nonlinear distortion and the energy per bit-to-AWGN ratio E_b/N_0.

7.4   CDMA Waveform Quality Factor (ρ)

The waveform quality factor is a measure of the correlation between a scaled version of the input and the total in-channel output waveforms. The waveform quality factor is defined as (Gharaibeh *et al.*, 2004):

$$
\begin{aligned}
\rho &= \frac{E[\tilde{y}(t)\tilde{y}_c^*(t)]^2}{E[|\tilde{y}(t)|^2]E[|\tilde{y}_c(t)|^2]} \\
&= \frac{\int_{-B/2}^{B/2} S_{\tilde{y}_c\tilde{y}_c}(f)df}{\int_{-B/2}^{B/2} S_{\tilde{y}_c\tilde{y}_c}(f)df + \int_{-B/2}^{B/2} S_{\tilde{y}_d\tilde{y}_d}(f)df + N_0 B}
\end{aligned}
\tag{7.17}
$$

where the in-band portion of the signals $y(t)$, $y_c(t)$ and $y_d(t)$ is used in this expression. Note that ρ can be directly related to the effective SNR and it measures the fraction of the useful part of the signal at the receiver.

Comparing Equation (7.15) and Equation (7.17), the relationship between ρ and SNR can be written as

$$
\rho = \frac{\text{SNR}}{\text{SNR}+1}.
\tag{7.18}
$$

The evaluation of ρ in the first line of Equation (7.17) can be done using the properties of Gaussian processes. In IS-95 systems, ρ is usually measured when only the pilot channel is transmitted (Aparin, 2001). In this case, the NBGN assumption of the signal model is not valid and hence, the estimated ρ using the properties of the Gaussian higher-order moments does not lead to an accurate estimation of ρ. Hence, the orthogonalization of the model which leads to the second line in Equation (7.17) is needed to accurately determine ρ.

7.5   Error Vector Magnitude (EVM)

The Error Vector Magnitude (EVM) is a common figure of merit for system linearity in digital wireless communication standards (including GSM, NADC, IS-95 and WCDMA systems) where a maximum level of EVM is specified. EVM is defined in the context of digitally modulated signals, where it is a measure of the departure of signals constellation from its ideal reference because of nonlinearity.

EVM can be defined in terms of the signal and noise power as (Gharaibeh *et al.*, 2004)

$$
\begin{aligned}
\text{EVM} &= \sqrt{\frac{E[\tilde{y}_d^2(t)] + E[\tilde{n}(t)^2]}{E[y_c^2(t)]}} \\
&= \sqrt{\frac{\int_{-B/2}^{B/2} S_{\tilde{y}_d\tilde{y}_d}(f)df + N_0 B}{\int_{-B/2}^{B/2} S_{\tilde{y}_c\tilde{y}_c}(f)df}}.
\end{aligned}
\tag{7.19}
$$

Using Equations (7.15), (7.18) and (7.19), EVM can be related to SNR and ρ as follows

$$
\text{EVM} = \sqrt{\frac{1}{\text{SNR}}}
\tag{7.20}
$$

and

$$\text{EVM} = \sqrt{\frac{1}{\rho} - 1}. \tag{7.21}$$

Note that EVM and ρ are directly related to SNR and hence, to the ability of the receiver to perform reliable detection of the transmitted data.

7.6 Co-Channel Power Ratio (CCPR)

CCPR is defined as the ratio between total linear output power to the total distortion power within the input signal bandwidth (Pedro and de-Carvalho, 2001), therefore:

$$\text{CCPR} = \frac{\int_{-B/2}^{B/2} |b_1|^2 S_{\tilde{x}\tilde{x}}(f)df}{\int_{-B/2}^{B/2} S_{\tilde{y}_g \tilde{y}_g}(f)df + N_0 B} \tag{7.22}$$

where by using Equation (5.13)

$$S_{\tilde{y}_g \tilde{y}_g}(f) = \sum_{\substack{n=1 \\ n+m \neq 2}}^{N} \sum_{m=1}^{N} b_m b_n S_{\tilde{x}^n \tilde{x}^m}(f) \tag{7.23}$$

where $S_{\tilde{x}^n \tilde{x}^m}(f)$ is the Fourier transform of $R_{\tilde{x}^n \tilde{x}^m}(\tau)$ defined in Equation (5.14).

CCPR can be measured using a feed forward cancelation loop tuned to small-signal cancelation. While this approach is valid to measure the CCPR it faces the question of whether to consider signal correlated distortion as an additive noise component or as part of the useful signal, and hence it questions the applicability of CCPR as a useful measure of distortion. This rationale was studied in Pedro and de-Carvalho (2001) and Vanhoenacker et al. (2001) where it was shown that CCPR applies in certain situations while it does not represent a useful measure of in-band distortion in other situations.

7.7 Noise-to-Power Ratio (NPR)

Noise-loading techniques have been used to measure in-band distortion of nonlinear PAs (Fenton, 1977; Kuo, 1973; Koch, 1971; Reis, 1976). In noise loading a nonlinear system is excited by a NBGN signal (that has a flat spectrum over a band of frequencies) after passing it into a notch filter. NPR is then defined as the ratio of the output power of a nonlinear system with the notch present and measured within the notch bandwidth to the output power without the notch in the same notch bandwidth (Fenton, 1977; Gharaibeh, 2010; Kuo, 1973; Koch, 1971; Pedro and de-Carvalho, 2001; Reis, 1976; Vanhoenacker et al., 2001), see Figure 7.3. Therefore, NPR represents the ratio of the output power to the effective in-band distortion power within the notch bandwidth.

To establish the concept of NPR, let the complex envelope of the input signal to the notch filter be $\tilde{x}(t)$ then, from Figure 7.3, the signal at the output of the notch filter can be expressed as

$$\hat{x}(t) = \tilde{h}(t) * \tilde{x}(t) \tag{7.24}$$

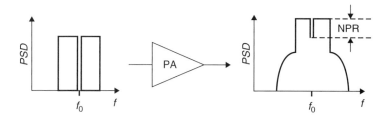

Figure 7.3 Definition of NPR.

where the $*$ indicates convolution, $\hat{x}(t)$ is the output of the notch filter and $\tilde{h}(t)$ is the low pass equivalent of the notch filter having a notch at frequency f_0. The baseband equivalent of the transfer function of the notch filter can be represented as:

$$H(f) = 1 - P(f) \tag{7.25}$$

where $P(f)$ is the transfer function of a rectangular-shaped filter with bandwidth B_n:

$$P(f) = \begin{cases} 1, & |f| \leq B_n; \\ 0, & |f| > B_n. \end{cases} \tag{7.26}$$

Let $\hat{y}(t)$ be the output of the nonlinear PA when $\hat{x}$ is applied and $\tilde{y}(t)$ be the output of the nonlinear PA when $\tilde{x}(t)$ is applied, then from Figure 7.3, NPR can be defined as (Gharaibeh, 2010; Kuo, 1973)

$$\mathrm{NPR} = \frac{S_{\tilde{y}\tilde{y}}(0)}{S_{\hat{y}\hat{y}}(0)} \tag{7.27}$$

where $S_{\hat{y}\hat{y}}(f)$ is the PSD of the output of the nonlinear PA when the notch is present and $S_{\tilde{y}\tilde{y}}(f)$ is the PSD of the output when the notch is not applied. If the system is linear and noise free then NPR approaches infinity.

7.7.1 NPR of Communication Signals

The main problem with NPR is that it is highly dependent on the characteristics of the stimulus. NPR is usually defined in terms of white Gaussian noise that has a flat spectrum and hence it does not represent distortion in a general communication signal that may not be Gaussian (Aparin, 2001). Therefore, if the PDF of the input process is not Gaussian, NPR does not represent the effective in-band distortion.

The problem now is to derive NPR for a general signal and to find its relation to the effective uncorrelated in-band distortion. Using the formulation in Equation (6.8), the output of the nonlinear PA when the notch is applied to the input signal can be written as

$$\hat{y}(t) = \hat{y}_c(t) + \hat{y}_d(t). \tag{7.28}$$

Therefore, using Equation (7.13), NPR in Equation (7.27) can be defined in terms of the correlated and uncorrelated spectra as

$$\text{NPR} = \frac{S_{\tilde{y}\tilde{y}}(0)}{S_{\hat{y}\hat{y}}(0)} = \frac{S_{\tilde{y}\tilde{y}}(0)}{S_{\hat{y}_c\hat{y}_c}(0) + S_{\hat{y}_d\hat{y}_d}(0)}. \tag{7.29}$$

Assuming that the notch is very narrow, the notch filter will eliminate all signal correlated component within the notch bandwidth, hence

$$S_{\hat{y}_c\hat{y}_c}(0) = |H(0)|^2 S_{\tilde{y}_c\tilde{y}_c}(0) = 0. \tag{7.30}$$

Hence, NPR can now be expressed as (Gharaibeh, 2010)

$$\text{NPR} = \frac{S_{\tilde{y}\tilde{y}}(0)}{S_{\hat{y}_d\hat{y}_d}(0)}. \tag{7.31}$$

Therefore, NPR is now defined in terms of the uncorrelated spectrum of the notched output $S_{\hat{y}_d\hat{y}_d}(0)$ and not the uncorrelated spectrum of the original output $S_{\tilde{y}_d\tilde{y}_d}(0)$ that represents the effective in-band distortion. In general $S_{\hat{y}_d\hat{y}_d}(0)$ is not equal $S_{\tilde{y}_d\tilde{y}_d}(0)$ and hence, NPR does not always represent the effective uncorrelated in-band distortion characterized by $S_{\tilde{y}_d\tilde{y}_d}(0)$ unless the input is a Gaussian process, as will be shown in the next sections.

7.7.2 NBGN Model for Input Signal

The NPR method emulates the effective in-band distortion of many signals by loading the amplifier with a NBGN signal. Considering a NBGN process at the input of a third-order power-series model, the autocorrelation function of the uncorrelated component at the output of a nonlinearity for a NBGN input can be found using Equation (6.57) by replacing $\tilde{x}(t)$ by $\hat{x}(t)$ (Gharaibeh, 2010):

$$R_{\hat{y}_d\hat{y}_d}(\tau) = |c_3|^2 R_{\hat{u}_3\hat{u}_3}(\tau)$$
$$= 2|b_3|^2 R_{\hat{x}\hat{x}}^3(\tau) \tag{7.32}$$

The PSD of the uncorrelated output of the nonlinearity when the notch is applied can be found using Equation (6.62) (Gharaibeh, 2010):

$$S_{\hat{y}_d\hat{y}_d}(f) = 2|c_3|^2 S_{\hat{x}\hat{x}}^{*3}(f). \tag{7.33}$$

where $S_{\hat{x}\hat{x}}^{*3}(f)$ is the triple-time convolution of $S_{\hat{x}\hat{x}}(f)$. Note that for a narrow notch, the triple time convolution of a notched NBGN spectrum is approximately equal to that of the spectrum without the notch ($S_{\hat{x}\hat{x}}^{*3}(f) \simeq S_{\tilde{x}\tilde{x}}^{*3}(f)$), as shown in Figure 7.4. This means that the uncorrelated distortion spectrum is the same regardless of whether the notch is applied to the input or not ($S_{\hat{y}_d\hat{y}_d}(f) \simeq S_{y_dy_d}(f)$).

The output PSD without the notch can be calculated in a similar way as (Gharaibeh, 2010)

$$S_{\tilde{y}\tilde{y}}(f) = S_{\tilde{y}_c\tilde{y}_c}(f) + S_{\tilde{y}_d\tilde{y}_d}(f)$$
$$= |c_1|^2 S_{\tilde{x}\tilde{x}}(f) + 2|c_3|^2 S_{\tilde{x}\tilde{x}}^{*3}(f). \tag{7.34}$$

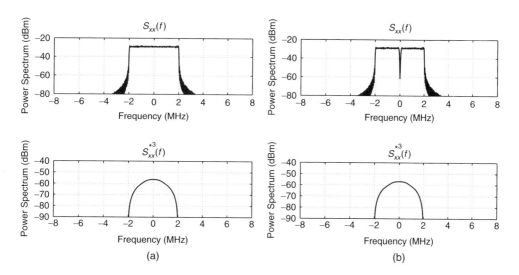

Figure 7.4 Triple-time convolution of $S_{\tilde{x}\tilde{x}}(f)$; (a) and $S_{\hat{x}\hat{x}}(f)$; (b) (Gharaibeh, 2010).

Therefore, NPR for NBGN input can now be found by inserting Equation (7.34) into Equation (7.31) (Gharaibeh, 2010):

$$\text{NPR} = \frac{S_{\tilde{y}\tilde{y}}(0)}{S_{\hat{y}_d\hat{y}_d}(0)} = \frac{|c_1|^2 S_{\tilde{x}\tilde{x}}(0) + 2|c_3|^2 S^{*3}_{\tilde{x}\tilde{x}}(0)}{2|c_3|^2 S^{*3}_{\tilde{x}\tilde{x}}(0)} \tag{7.35}$$

Now since $S_{\hat{y}_d\hat{y}_d}(0) \simeq S_{\tilde{y}_d\tilde{y}_d}(0)$, NPR in this case represents the uncorrelated in-band distortion within the notch bandwidth. Note that in the case of a general communication signal, the uncorrelated distortion spectrum cannot be written as the triple-time convolution of the input spectrum as in Equation (7.33) and hence $S_{\hat{y}_d\hat{y}_d}(0) \neq S_{\tilde{y}_d\tilde{y}_d}(0)$.

It is worth noting that the shape of $S_{\hat{y}_d\hat{y}_d}(f)$ depends on the statistical properties of the input signal. In WCDMA systems, the number and selection of the channelization codes (the Orthogonal Variable Spreading Factor (OVSF) codes) is an important factor in determining the validity of NPR as a measure of in-band distortion. Figures 7.5(a) and 7.5(b) show $(S_{\hat{y}\hat{y}}(f))$ and $(S_{\tilde{y}_d\tilde{y}_d}(f))$ for a lightly loaded WCDMA signal (1DPCH) and a heavily loaded WCDMA signal (8 DPCH). It is clear that in the case of a heavily loaded WCDMA signal, the NPR test matches the effective in-band distortion, while in the case of a lightly loaded signal, the NPR test underestimates the effective in-band distortion. This is due to the fact that the signal statistics of a heavily loaded WCDMA signal approach the Gaussian distribution (by the central limit theorem) and hence, the equality $S_{\hat{y}_d\hat{y}_d}(0) = S_{\tilde{y}_d\tilde{y}_d}(0)$ holds, as shown in the previous subsection.

7.8 Noise Figure in Nonlinear Systems

The Noise Figure (NF) is defined as the ratio of the signal-to-noise ratio at the input to the signal-to-noise ratio at the output of a system, and hence it quantifies the amount of noise added by a system. For linear systems, the evaluation of NF is straightforward and

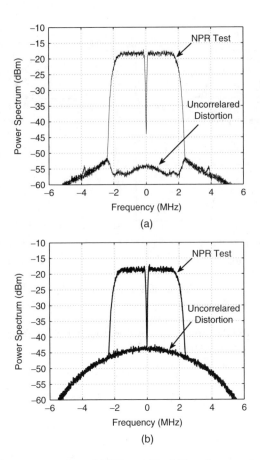

Figure 7.5 Simulated output spectrum of NPR test ($S_{\hat{y}\hat{y}}(f)$) and uncorrelated distortion spectrum ($S_{y_d y_d}(f)$) of WCDMA signals: (a) 1 DPCH and (b) 8 DPCH.

depends only on the noise level introduced by the system. In nonlinear RF circuits such as LNAs, which are usually nonlinear, the evaluation of the output SNR is more complicated than linear systems. This is because the interaction of signal and noise by nonlinearity results in many output signal components that may be correlated to each other and this complicates the evaluation of system NF (Gharaibeh, 2009).

To establish the analysis of NF in nonlinear systems, an LNA model is used. The model for the bandpass LNA is shown in Figure 7.6 where the input to the LNA $\tilde{x}(t) = \tilde{s}(t) + \tilde{n}(t)$ consists of a signal $\tilde{s}(t)$ and a Gaussian noise $\tilde{n}(t)$. The bandpass filter that precedes the amplifier passes the signal and band limits the input noise to a narrow bandwidth B_n. We will assume that the input noise bandwidth is larger than the signal bandwidth B_s. The bandpass filter that follows the LNA confines the output spectrum to the bandwidth of the signal B_s. In the following development, the LNA is assumed to exhibit a memoryless nonlinearity that can be modeled using an envelope power-series model as in Equation (4.42).

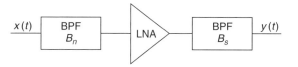

Figure 7.6   Bandpass LNA model.

7.8.1   Nonlinear Noise Figure

The system noise figure is defined as the ratio of the SNR at the input of the LNA to the SNR at its output:

$$NF = \frac{SNR_i}{SNR_o} \tag{7.36}$$

The problem now is to determine the effective output SNR given that nonlinearity produces distortion terms at the output of the LNA that contribute to the output system noise. To evaluate the effective output SNR, the response of the nonlinearity to an input $\tilde{x}(t) = \tilde{s}(t) + \tilde{n}(t)$ needs to be expressed in the form (Gharaibeh, 2009):

$$\tilde{y}_o(t) = \tilde{s}_o(t) + \tilde{n}_o(t) + \tilde{y}_d(t) \tag{7.37}$$

where $\tilde{s}_o(t)$ represents the useful output signal component, $\tilde{n}_o(t)$ represents the correlated output noise and $\tilde{y}_d(t)$ represents the uncorrelated nonlinear distortion component.

The input SNR is defined as the ratio of the input signal power to input noise power as:

$$SNR_i = \frac{P_{s_i}}{P_{n_i}} \tag{7.38}$$

where P_{n_i} is the average power of the input noise process within the noise bandwidth B_n and P_{s_i} is the average power of the input signal within the signal bandwidth B_s. The effective system output SNR is defined as the ratio of signal power to the total output noise power including the effective nonlinear distortion power:

$$SNR_o = \frac{P_{s_0}}{P_{n_o} + P_d + N_0} \tag{7.39}$$

where N_0 is the added system noise, P_{n_o} is the output noise power, P_{s_0} is the output signal power and P_d is the output uncorrelated in-band nonlinear distortion power, all computed within the signal bandwidth B_s. The identification of the uncorrelated distortion component ($\tilde{y}_d(t)$) that adds to system noise is the key here to the accurate evaluation of the output SNR and NF.

Using the analysis of the orthogonal power-series model in the Chapter 6, (Equation (6.42) with $\tilde{x}(t) = \tilde{s}(t) + \tilde{n}(t)$) and using the independence of signal and noise processes, the autocorrelation of the correlated component of the output can be written as (Gharaibeh, 2009)

$$R_{\tilde{y}_c \tilde{y}_c}(\tau) = |c_1|^2 R_{\tilde{u}_1 \tilde{u}_1}(\tau) = |c_1|^2 R_{\tilde{s}\tilde{s}}(\tau) + |c_1|^2 R_{\tilde{n}\tilde{n}}(\tau) \tag{7.40}$$

and for the uncorrelated component:

$$R_{\tilde{y}_d \tilde{y}_d}(\tau) = \sum_{i=3}^{N} |c_i|^2 R_{\tilde{u}_i \tilde{u}_i}(\tau) \tag{7.41}$$

The PSD of the correlated and uncorrelated output components can be found using Equations (7.45) and (7.46) (Gharaibeh, 2009):

$$S_{\tilde{y}_c \tilde{y}_c}(f) = |c_1|^2 S_{\tilde{s}\tilde{s}}(f) + |c_1|^2 S_{\tilde{n}\tilde{n}}(f) \tag{7.42}$$

and

$$S_{\tilde{y}_d \tilde{y}_d}(f) = \sum_{i=3}^{N} |c_i|^2 S_{\tilde{u}_i \tilde{u}_i}(f) \tag{7.43}$$

Note that the PSD of the uncorrelated output consists of signal–signal, signal–noise and noise–noise interactions and represents the effective output component that determines the noise performance of the LNA, and hence its NF.

Figure 7.7 shows the PSD of the response of a nonlinear amplifier to the sum of a WCDMA signal and a NBGN process partitioned into correlated and uncorrelated components. It is clear that the correlated output component consists of the correlated components of signal and noise, as is evident from the shape of the correlated spectrum. On the other hand, the uncorrelated spectrum consists of nonlinear distortion that results from all interactions of signal–signal and noise–noise and signal–noise by nonlinearity.

Now using the above analysis, the input SNR can be evaluated from the PSDs of the input signal and input noise as follows: $P_{n_i} = \int_{-B_n/2}^{B_n/2} S_{\tilde{n}\tilde{n}}(f)df$ and $P_{s_i} = \int_{-B_s/2}^{B_s/2} S_{\tilde{s}\tilde{s}}(f)df$ and hence, the input SNR can be evaluated as (Gharaibeh, 2009)

$$\text{SNR}_i = \frac{\int_{-B_s/2}^{B_s/2} S_{\tilde{s}\tilde{s}}(f)df}{\int_{-B_n/2}^{B_n/2} S_{\tilde{n}\tilde{n}}(f)df} \tag{7.44}$$

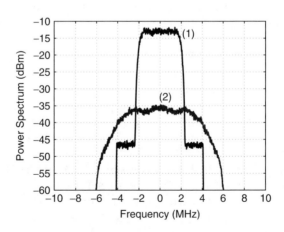

Figure 7.7 Output spectrum of a WCDMA signal plus NBGN, (1): correlated output spectrum and (2): uncorrelated distortion spectrum; $P_{in} = -6$ dBm and $\text{SNR}_i = 33$ dB (Gharaibeh, 2009).

The output power of the signal, noise and distortion are computed from their PSDs as follows: $P_{n_o} = |c_1|^2 \int_{-B_s/2}^{B_s/2} S_{\tilde{n}\tilde{n}}(f)df$, $P_{s_o} = |c_1|^2 \int_{-B_s/2}^{B_s/2} S_{\tilde{s}\tilde{s}}(f)df$ and (Gharaibeh, 2009)

$$P_d = \int_{-B_s/2}^{B_s/2} S_{\tilde{y}_d\tilde{y}_d}(f)df = \sum_{i=3}^{N} |c_i|^2 \int_{-B_s/2}^{B_s/2} S_{\tilde{u}_i\tilde{u}_i}(f)df \tag{7.45}$$

Thus, the output SNR can be evaluated from Equation (7.39) as (Gharaibeh, 2009)

$$\text{SNR}_o = \frac{|c_1|^2 \int_{-B_s/2}^{B_s/2} S_{\tilde{s}\tilde{s}}(f)df}{|c_1|^2 \int_{-B_s/2}^{B_s/2} S_{\tilde{n}\tilde{n}}(f)df + \int_{-B_s/2}^{B_s/2} S_{\tilde{y}_d\tilde{y}_d}(f)df + N_0}$$

$$\tag{7.46}$$

The system NF can now be evaluated in terms of the output PSDs as (Gharaibeh, 2009)

$$\begin{aligned}
\text{NF} &= \frac{\text{SNR}_i}{\text{SNR}_o} \\
&= \frac{|c_1|^2 \int_{-B_s/2}^{B_s/2} S_{\tilde{n}\tilde{n}}(f)df + \int_{-B_s/2}^{B_s/2} S_{\tilde{y}_d\tilde{y}_d}(f)df + N_0}{|c_1|^2 \int_{-B_n/2}^{B_n/2} S_{\tilde{n}\tilde{n}}(f)df}.
\end{aligned} \tag{7.47}$$

Note that the power levels of the output signal, noise and distortion components are all computed within the signal bandwidth B_s. This is because the output signals are bandlimited by the output filter to the signal bandwidth B_s.

The above definition of the NF in Equation (7.47) indicates that it is dependent on the input power embedded in the coefficients of the orthogonal power-series model (c_i). This is because the gain of the amplifier and the nonlinear distortion (and hence the output SNR) are dependent on the input power of the signal and noise components which is different from the original IEEE definition of NF for linear systems as pointed out in Pedro *et al.* (2001). Note also that the above analysis clearly separates different signal, noise and distortion components and their interactions at the output of nonlinearity and hence presents a more realistic approach to the evaluation of the effective output SNR and NF. The interaction between the signal and noise and their nonlinear distortion components are all included in the uncorrelated distortion spectrum $S_{\tilde{y}_d\tilde{y}_d}(f)$, which is the key here to the evaluation of a NF that realistically represents system noise performance (Gharaibeh, 2009).

It is also worth noting that the above analysis is done with the assumption that input noise has a larger bandwidth than the input signal. The output distortion that results from the interaction of the signal and noise components with different bandwidths is different from that when both signal and noise have the same bandwidth because of the mixing properties of nonlinearity. This is manifested by the integration of the PSD of noise in the denominator of Equation (7.47), which is done over the input noise bandwidth B_n, which is greater than the signal bandwidth B_s (Gharaibeh, 2009).

7.8.2 NBGN Model for Input Signal and Noise

In this subsection, the case when both the input signal and noise are modeled as NBGN processes with a flat spectrum over its bandwidth is considered. Therefore, for an input signal with power P_s and a bandwidth B_s, its autocorrelation function is given by (Gharaibeh, 2009)

$$R_{\tilde{s}\tilde{s}}(\tau) = P_s \text{sinc}(B_s \tau) \tag{7.48}$$

where $P_s = R_{ss}(0)$ and hence, its PSD is

$$S_{\tilde{s}\tilde{s}}(f) = \frac{P_s}{B_s} \text{rect}\left(\frac{f}{B_s}\right) \tag{7.49}$$

which is a flat spectrum over the band B_s.

The input noise $n(t)$ with power P_s and bandwidth B_n is also characterized by a NBGN process with an autocorrelation function

$$R_{\tilde{n}\tilde{n}}(\tau) = P_n \text{sinc}(B_n \tau) \tag{7.50}$$

where $P_n = R_{\tilde{n}\tilde{n}}(0)$, and its PSD is given by

$$S_{\tilde{n}\tilde{n}}(f) = \frac{P_n}{B_n} \text{rect}\left(\frac{f}{B_n}\right) \tag{7.51}$$

Considering a 5th-order orthogonalized power-series model and using the development in Section 6.5.2 with $\tilde{x}(t) = \tilde{s}(t) + \tilde{n}(t)$, we have (Gharaibeh, 2009)

$$S_{\tilde{y}_d \tilde{y}_d}(f) = |c_3|^2 S_{\tilde{u}_3 \tilde{u}_3}(f) + |c_5|^2 S_{\tilde{u}_5 \tilde{u}_5}(f) \tag{7.52}$$

where the coefficients c_i are obtained using Equation (6.53):

$$c_1 = b_1 + 2Pb_3 + 6P^2 b_5$$

$$c_3 = b_3 + 6Pb_5$$

$$c_5 = b_5 \tag{7.53}$$

with $P = P_s + P_n$ and

$$S_{\tilde{u}_3 \tilde{u}_3}(f) = 2S_{\tilde{s}\tilde{s}}^{3*}(f) + 2S_{\tilde{n}\tilde{n}}^{3*}(f) + 6S_{\tilde{s}\tilde{s}}^{2*}(f) * S_{\tilde{n}\tilde{n}}(f)$$
$$+ 6S_{\tilde{n}\tilde{n}}^{2*}(f) * S_{\tilde{n}\tilde{n}}(f) \tag{7.54}$$

and

$$S_{\tilde{u}_5 \tilde{u}_5}(f) = 12S_{\tilde{s}\tilde{s}}^{5*}(f) + 12S_{\tilde{n}\tilde{n}}^{5*}(f) + 60S_{\tilde{s}\tilde{s}}^{4*}(f) * S_{\tilde{n}\tilde{n}}(f)$$
$$+ 120S_{\tilde{s}\tilde{s}}^{3*}(f) * S_{\tilde{n}\tilde{n}}^{2*}(f) + 120S_{\tilde{s}\tilde{s}}^{2*}(f) * S_{\tilde{n}\tilde{n}}^{3*}(f)$$
$$+ 60S_{\tilde{s}\tilde{s}}(f) * S_{\tilde{n}\tilde{n}}^{4*}(f) \tag{7.55}$$

where $S^{k*}(f) = S(f) * S(f) * \ldots * S(f)$ and the $*$ denotes convolution. NF can now be evaluated in a closed form by inserting the last three equations into Equation (7.47).

Note that with the Gaussian assumption all the higher-order spectra can be computed from the first-order spectra $S_{\tilde{s}\tilde{s}}(f)$ and $S_{\tilde{n}\tilde{n}}(f)$. Hence, the NBGN assumption enables a closed-form evaluation of system NF in terms of the input power of the signal and noise components.

7.9 Summary

In this chapter, the definition of communications system figures of merit in terms of nonlinear distortion has been presented. The techniques developed in the previous chapter were used to develop the relationship between in-band nonlinear distortion and the most common metrics of system performance; SNR, NPR, NF, EVM and ρ. Closed-form expressions for these metrics have been developed when the input signal to a nonlinearity is modeled as a NBGN process with flat spectrum over its bandwidth. The definitions developed in this chapter will be used in the following chapters to develop simulation models in MATLAB® and Simulink® where these metrics are estimated from simulated signal spectra at the output of nonlinearity.

8

Communication System Models and Simulation in MATLAB®

The growing complexity of wireless communication systems imposes the use of computer simulations to evaluate system performance that leads to efficient system design and reduces the development time. MATLAB® has been one of the most effective and powerful software tools for simulation of wireless system performance. The efficient way that MATLAB® handles array and matrix manipulations provides system designers with a fast tool for the simulation of a complex communication systems. In addition, MATLAB® has a huge number of ready-to-use functions in its toolboxes that eases the implementation of complex system models.

In simulation of communication systems, MATLAB® has the capability of random signal generation that simulates wireless communication signals through its embedded random number generators and the functions that generate modulation formats. Through its RF and Communications toolboxes, MATLAB® represents an efficient tool for the design, modeling and simulation of nonlinear systems. On the other hand, MATLAB® optimization toolbox offers ready to use tools for nonlinear model optimization and model embedding in system simulations. Together, these tools make MATLAB® an optimum tool for the efficient simulation of nonlinearity in wireless communication systems. In addition, MATLAB® offers the capability of spectrum estimation and system metric simulation including BER simulations.

In this chapter, the basic MATLAB® tools used for the simulation of wireless communication systems are presented. These include deterministic and random signal generation, simulation of a wide variety of modulation formats, simulation of wireless communication standards, and the simulation of performance of a wireless system. In addition, Simulation of the performance of an overall communication system in Simulink® is presented. Simulink® tools are used as a verification platform for the concepts presented in this book.

Nonlinear Distortion in Wireless Systems: Modeling and Simulation with MATLAB®, First Edition.
Khaled M. Gharaibeh.
© 2012 John Wiley & Sons, Ltd. Published 2012 by John Wiley & Sons, Ltd.

8.1 Simulation of Communication Systems

Simulation of communication systems is usually done with the objective of estimating system performance by computing the response of a system to random process samples using a computer program. In general, a computer simulation is used to simulate signals that flow through a real system and to estimate its response. Therefore, a simulation process must include the generation of random process samples, and then using computer calculations to estimate the output of the system based on a given system model (Jeruchem *et al.*, 2000). Thus, a computer simulation involves three main processes that are random signal generation, a mathematical or empirical system model and the calculations of the system response using a computer algorithm. These processes are usually referred to as Monte Carlo (MC) simulations that are basically based on the simulation of a random phenomena without fully repeating the underlying physical experiment.

8.1.1 Random Signal Generation

The generation of samples of a random process is done using random number generation algorithms based on various models of random processes, see Appendix B. The efficiency of these algorithms in producing the statistical properties of a certain random process controls the accuracy of the overall simulation process. Since a random process can be viewed as a collection of random variables with certain probability distribution and correlation properties, the generation of a random process with a given probability distribution is based on random number generators. In many computer software platforms, the generation of random numbers with a given probability distribution is based on transformation of uniformly distributed random variables by a memoryless transformation. Independent uniformly distributed random numbers can be generated using simple recursive algorithms. To establish the correlation properties of the generated random process, a linear transformation with memory is used to produce the required correlation and spectral properties of a random process.

In simulation of random processes, the important property of *Ergodicity* is used to estimate the statistical properties of the generated random process. Ergodicity implies that the statistical properties of a random process such as expected value, variance, moments and correlation functions can be estimated from the time averages of the simulated process, which means that a single realization of a process is enough to fully characterize the process. Ergodicity does not have a rigorous mathematical basis but is rather based on practical observation of real communication signals. In computer simulations, the assumption that a process is ergodic greatly simplifies the models for random processes and hence, it represents the basis of computer simulation of communication systems.

8.1.2 System Models

System models in a simulation environment can either be generated mathematically (exact mathematical operation) or empirically (using block models derived from measured data). As discussed in Chapter 3, the choice of the model for a given phenomena depends on two main factors: firstly complexity that in turn affects the quality of the simulations and secondly accuracy that affects the capability of the simulations to faithfully reproduce system response. Unfortunately, these are conflicting requirements where increasing the level of details in a model results in increased computer calculations. However, increasing the level of detail of a given model enables finer design space parameters to be explored than is usually achieved by measurements or formula-based approaches to the evaluation of system performance. The expense of increased computer calculations is overcome by the advances in computer processing software and hardware technologies that improve day after day.

8.1.3 Baseband versus Passband Simulations

One important issue in the simulation of communication systems is the problem of simulating bandpass systems and signals, that is, systems that involve modulation of a baseband signal. Bandpass signals have a frequency spectrum that is concentrated around a carrier frequency. Similarly, a bandpass system is a system whose frequency response is limited to a frequency band that is centered around a carrier frequency. The very high carrier frequency of wireless communication systems implies that a very high sampling rate is needed when simulating such signals or systems since the sampling rate is commensurate with the carrier frequency by Nyquist theorem. Nyquist's theorem states that the sampling rate of a bandlimited signal or system must be equal to at least twice the bandwidth in order not to introduce aliasing distortion. Therefore, if a linear system to be simulated has a spectrum that spreads between $\pm B/2$, that is, has a bandwidth of B, then the sampling rate must be greater than or equal to B. This means that a large number of samples are needed to fully represent the signal in a simulation environment, which means a large number of computations.

In nonlinear systems, the situation is different because nonlinearity usually results in widening the spectrum of the input signal. For example, if a nonlinear system is represented by a polynomial model, the system results in an increase in the bandwidth of the signal by a factor N. Hence, in order to avoid aliasing, the sampling rate must be at least NB. A simple illustration is a square-law device $y(t) = x^2(t)$ that multiplies the signal by itself in the time domain. By Fourier transform properties, multiplication in the time domain is equivalent to convolution in the frequency domain, hence the output spectrum of the square law device is $X(f) * X(f)$ where $*$ indicates convolution. In this case, it can be shown that if $x(t)$ has a bandwidth B, then $Y(f) = X(f) * X(f)$ has a bandwidth of $2B$. In general, the contribution of higher-order nonlinearities to the output of the system become minor for large values of N. Therefore, a lower sampling rate than NB can be used if a certain aliasing error can be tolerated. The optimum sampling rate in this case can be obtained experimentally by running simulation of the spectrum at the output of nonlinearity at different sampling rates and then estimating the sampling rate that keeps the aliasing error at an acceptably low level (Jeruchem et al., 2000).

The solution to the problem of high sampling rate requirements of the simulation is to use complex envelope simulations. As shown in Chapter 4, the complex envelope of a signal or the baseband equivalent of a system contains all the useful information, and hence a system can be simulated using only its baseband equivalent and the complex envelope of the input and output signals. In this case, the sampling rate is proportional to the bandwidth of the complex envelope and not to the carrier frequency, which means much lower computations than the case of simulating the signal at the passant.

8.2   Choosing the Sampling Rate in MATLAB® Simulations

The sampling rate is the rate at which the message signal is sampled during the simulation. Therefore, each signal simulated in MATLAB® must be associated with a sampling rate Fd that associates entries of the signal value with time increments of 1/Fd. For example, for a signal taken from an alphabet of *M* values, the signal consists of integers in the range [0, M-1]. If the sampling rate is specified to be Fd, then the signal values represent samples taken at time increments 1/Fd. In general, the sampling rate of the signal can be found from the knowledge of the time vector associated with the signal:

```
Fd = size(x,1) / (max(t)-min(t));
```

for a signal x sampled at times t.

Baseband digital modulation functions in MATLAB® do not have a built-in notion of time, hence, there is no need to include a sampling rate in the argument of functions that perform modulation. However, when the signal is to be plotted versus time, a time vector based on the sampling rate must be specified to associate signal values with time increments. For example, in *M*-ary modulation, the mapping process increases the sampling rate of the signal from Fd to Fs, whereas the demapping process decreases the sampling rate from Fs to Fd. If the baseband signal is a binary signal with sampling rate, say Fd=8kHz, then the output of an 8-PSK modulator would be a signal with sampling rate Fs=1kHz. That is, the modulated symbols are integers drawn from an alphabet [0,1,...,7] with time increments of 1 ms. The input and output of the modulator can be plotted vs. time using a time vector generated as

```
tx=[0:length(x)-1]*1/Fd
ty=[0:length(x)-1]*1/Fs
```

On the other hand, when designing a pulse-shaping filter, the sampling rates at the input and output of the filter must be specified since these filters perform up-sampling of data and hence the ratio of the sampling rate at the input and output of the filter Fs/Fd (which must be an integer) must be specified. One exception where the sampling rate needs to be specified in the argument of the modulation function, is nonlinear modulation functions (FSK and its variants) where individual values of Fs need to be specified.

8.3   Random Signal Generation in MATLAB®

Random signal generation in MATLAB® Communications toolbox is performed using a number of functions that are used to generate random sequences with given statistical properties. Random signal generation in MATLAB® is used to simulate either noise signal or signal sources, compute the error rates or generate scatter plots or eye diagrams of digital modulation formats.

8.3.1   White Gaussian Noise Generator

Perhaps the most common random signal is the White Gaussian Noise (WGN) that is used to model noise in communication systems. The *wgn* function generates either real or complex white Gaussian noise process with a specified power in dB, dBm or linear units. For example, the command below generates a complex white Gaussian noise column vector of length N whose power is K dBm in a load impedance of R Ohms.

```
x = wgn(N,1,K,R,'complex','powertype');
```

Another function is the *awgn* which can be used to add additive WGN (AWGN) to an input signal.

8.3.2 Random Matrices

The *randsrc* function generates random matrices whose entries are chosen independently from a specified alphabet A where each element of the alphabet is equally probable. For example, the command

```
x = randsrc(N,M,A)
```

generates an $N \times M$ matrix whose entries are chosen from an alphabet A. For example, to generate an Independent Identically Distributed (i.i.d.) sequence of bipolar data, the alphabet A is chosen as $A = [1 - 1]$.

If the elements of the alphabet are to be chosen with nonequal probabilities, then the vector that contains the alphabet is concatenated with the probabilities of the elements:

```
x = randsrc(N,M,[A; P])
```

For example, the following command produces an 3×4 random matrix with elements chosen from an alphabet [123] with probabilities [0.1 0.3 0.6]

```
x = randsrc(3,4,[1,3,5; .1,.3,.6])
```

8.3.3 Random Integer Matrices

The *randint* function generates a matrix of random integers chosen from a range of values. For example, the command

```
x = randint(N,M,[a,b])
```

generates an $N \times M$ matrix whose entries are chosen from the values between a and b. A special case is the generation of random binary matrices when $a = 0$ and $b = 1$.

The `randerr` function generates binary matrices with a specified number of zeros and ones and is meant for testing error-control coding. The command

```
x = randerr(N,M,[0,1,2,...; p0,p1,p2,...])
```

generates an $N \times M$ binary matrix having the property that each row contains zero '1's with probability p_0 or one '1' with probability p_1, etc. For example, the command

```
x = randerr(5,4,[1,2; 0.4,0.6])
```

produces a 5-by-4 matrix where each row has one '1' with probability 0.4 or two '1's with probability 0.6.

8.4   Pulse-Shaping Filters

Pulse-shaping filters can be constructed using `fdesign.pulseshaping` specification objects from the Signal Processing Toolbox (The Signal Processing Toolbox, 2009) where the properties of the filter such as the shape, order, stop-band attenuation, roll-off factor and frequency response can be specified. The command

$$d = fdesign.pulseshaping(sps,shape,...spec,value1,value2,...,fs,$$
$$magunits) \qquad (8.1)$$

constructs the filter d which has specifications as shown in Table 8.1.
  The command

```
H = design(d);
```

produces the frequency response/impulse response of the filter. The frequency response can be plotted using the command

```
fvtool(H)
```

and the impulse response can be plotted using the command:

```
fvtool(H, 'impulse').
```

8.4.1   Raised Cosine Filters

Pulse-shaping filters can be constructed using `fdesign.pulseshaping` specification objects by setting the "shape" property to 'Raised Cosine'. The values (`value1`, `value2,...`) in Equation (8.1) define the values of the string `spec` and are defined as follows:

- `Ast`: stopband attenuation (in dB).
- `Beta`: roll-off factor expressed as a real-valued scalar ranging from 0 to 1.
- `Nsym`: filter order in symbols.
- `N`: filter order (must be even).
- The length of the impulse response is $N + 1$.

The string `magunits` specifies the units for magnitude specification of the input arguments that take one of the following entries:

- linear: specify the magnitude in linear units;
- dB: specify the magnitude in dB (decibels);
- squared: specify the magnitude in power units.

Table 8.1   Pulse shaping

String	Description
sps	oversampling
shape	'Raised Cosine', 'Square Root Raised Cosine', 'Gaussian'
spec	'Ast,Beta' 'Nsym,Beta', 'Nsym,BT', 'N,Beta'
fs	sampling frequency of the signal to be filtered (Hz)
magunits	units for magnitude specification of the input arguments

When the `magunits` argument is omitted, all magnitudes are assumed to be in decibels. For example, the command:

```
d = fdesign.pulseshaping(8,'Raised Cosine','Ast,Beta',30,0.3)
H = design(d);
fvtool(H)
```

produces a pulse shaping filter with over sampling of 8, stop-band attenuation of 30 dB and a roll-off factor of 0.3. Figure 8.1 shows the impulse response and the frequency response of the resulting pulse-shaping filter.

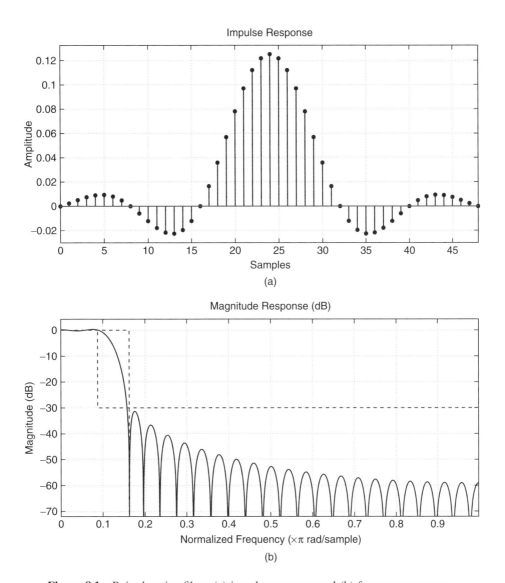

Figure 8.1 Raised cosine filter; (a) impulse response and (b) frequency response.

Digital data can also be filtered by raised cosine pulse-shaping filters using the function `rcosflt` that applies a raised cosine filter with specified characteristic to input data. The command

```
y=rcosflt(x,Fd,Fs,type,rolloff,delay);
```

filters the digital data x whose sampling rate is `Fd` by a raised cosine filter and produces filtered data y whose sampling rate is `Fs`. The command up samples the input data by a factor `Fs/Fd` before filtering where the ratio must be an integer. The function up samples the input data by inserting `Fs/Fd-1` zeros between input data samples. The up sampled data consists of `Fs/Fd` samples per symbol and has sampling rate `Fs`.

The string `type` specifies the type of the RC filter. The available types of the raised cosine filter are 'fir', 'fir/sqrt' that produces FIR square-root RC filter, 'iir' that produces an IIR raised cosine filter and 'iir/sqrt' that produces an IIR square-root RC filter.

The value `rolloff` specifies the roll-off factor of the filter that determines the bandwidth of the filter. For example, a roll-off factor of r means that the bandwidth of the filter is $1 + r$ times the input sampling frequency, `Fd`. This also means that the transition band of the filter extends from r`Fd` to $(1 + r)$`Fd`.

The value `delay` specifies the group delay taps of the filter that defines the time between the filter's initial response and its peak response. The group delay influences the size of the output, as well as the order of the filter.

Another way of filtering the data by an RC filter is to first specify the filter transfer function using the `rcosine` function and then applying `rcosflt` to the input data. For example

```
[num,den] = rcosine(Fd,Fs,'iir',rolloff,delay);
y = rcosflt(x,Fd,Fs,'iir/filter',num,den,delay);
```

where `[num,den]` contain the filter transfer function coefficients in ascending order of powers of z^{-1}.

8.4.2   Gaussian Filters

A Gaussian filter can be constructed if the shape property in Equation (8.1) is specified as 'Gaussian'. In this case, the entries for *spec* are chosen as `'Nsym,BT'` where the string entries are defined as follows:

- `Nsym`: filter order in symbols.
- `BT`: the 3 dB bandwidth–symbol time product.

For example the following command:

```
d = fdesign.pulseshaping(8,'Gaussian','Nsym,BT',4,0.2)
H = design(d);
fvtool(H)
```

produces a Gaussian filter with oversampling $= 8$, filter order of 4 symbols and $BT = 0.2$. Figure 8.2 shows the frequency response and the impulse response of the designed Gaussian pulse-shaping filter.

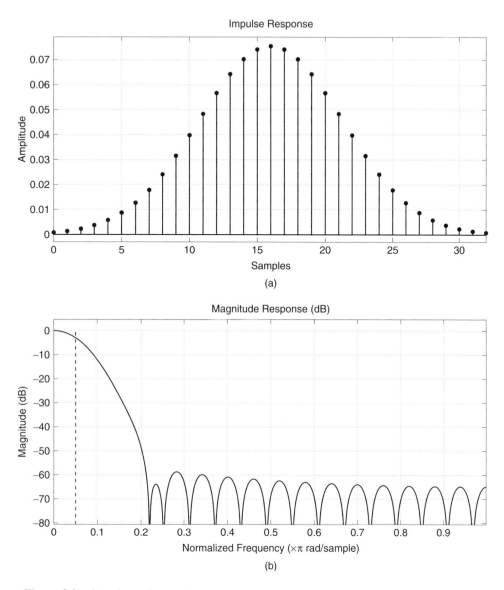

Figure 8.2 Gaussian pulse-shaping filter; (a) impulse response and (b) frequency response.

8.5 Error Detection and Correction

Error detection and correction coding is used to combat errors that result from channel impairments. MATLAB® has a number of functions and objects for generation of coding techniques to perform encoding and decoding of data. Table 8.2 lists the main functions used in implementing Block Codes, Convolutional Codes and Cyclic Redundancy Check codes. Since coding techniques are out of the scope of this book, the reader is referred to MATLAB® documentation (The Communication Toolbox, 2009) for more information.

Table 8.2 Block coding techniques: toolbox functions and objects

Code Type	Functions
Linear	block, encode, decode, gen2par, syndtable
Cyclic	encode, decode, cyclpoly, cyclgen, gen2par, syndtable
BCH	bchenc, bchdec, bchgenpoly
LDPC	fec.ldpcenc, fec.ldpcdec
Hamming	encode, decode, hammgen, gen2par,syndtable
Reed–Solomon	rsenc, rsdec, rsgenpoly, rsencof,rsdecof
Convolutional Codes	convenc vitdec, poly2trellis

8.6 Digital Modulation in MATLAB®

Baseband simulation of digital modulation formats in MATLAB® is based on producing the complex envelope of the output of a modulator given the complex envelope of a message signal. The complex envelope of the modulated signal can be related to the passband output of the modulator using Equation (4.8). M-ary modulation of a digital signal is implemented in two steps: firstly; the input binary bit stream is mapped into a symbol stream where each symbol represents an n-tuple of bits and takes the values $[0, 1, .., M - 1]$, secondly; the resulting symbol stream is then modulated using one of the modulation schemes such as Amplitude Shift Keying (ASK), Phase Shift Keying (PSK) or Frequency Shift Keying (FSK) and their variants.

8.6.1 Linear Modulation

Baseband simulation of linear digital modulation formats such as PSK, DPSK, QPSK, OQPSK, PAM and QAM is done using the *modem* objects. A `modem` object is a MATLAB® variable that contains information about the modulation class, M-ary number, and the constellation mapping (The Communication Toolbox, 2009). The `modem` object takes a symbol stream as input and produces the modulation object. The modulated signal can be generated using the `modulate` function that operates on the `modem` variable. Table 8.3 shows the modem object of different modulation formats.

Table 8.3 The modem objects used to generate different modulation formats (The Communication Toolbox, 2009)

Modulation Type	Modem Object
DPSK	modem.dpskmod and modem.dpskdemod
General QAM	modem.genqammod and modem.genqamdemod
MSK	modem.mskmod and modem.mskdemod
OQPSK	modem.oqpskmod and modem.oqpskdemod
PAM	modem.pammod and modem.pamdemod
PSK	modem.pskmod and modem.pskdemod
QAM	modem.qammod and modem.qamdemod

Table 8.4 Argument description of the modem object (The Communication Toolbox, 2009)

Argument	Description
M	*M*-ary value.
PhaseOffset	Phase offset of ideal signal constellation in radians.
Constellation	Ideal signal constellation.
SymbolOrder	Type of mapping employed for mapping symbols to ideal constellation points. The choices are: -'binary': for binary mapping -'gray': for gray mapping -'user-defined': for custom mapping
SymbolMapping	Symbol mapping is a list of integer values from 0 to $M-1$ that correspond to ideal constellation points.
InputType	Type of input to be processed by QAM modulator object. The choices are: -'bit': for bit/binary input -'integer': for integer/symbol input

The modem object of any of the modulation formats in Table 8.3 can be constructed as follows:

```
H = modem.xxxmod('M', value, 'PHASEOFFSET', value,...
 'SYMBOLORDER', value,'INPUTTYPE', value)
```

where xxx takes one of the values shown in Table 8.3 and defines the modulation type. The variables in the argument of the modem object are defined as shown in Table 8.4.

The command

```
y = modulate(H, x)
```

where H is the handle to a modulator object and x is a message signal, produces the complex envelope of the modulated signal.

As an example, the following script produces the complex envelope of a 16QAM modulated signal. The script starts by generating a random bit stream, maps the bit stream to 4-bit symbols using the bi2de function and then modulates the resulting symbol stream using 16 QAM with rectangular pulse shaping.

```
%File Name: QAM.m
%% Setup
% Define parameters.
M = 16; % Size of signal constellation
k = log2(M); % Number of bits per symbol
n = 10000; % Number of bits to process
nsamp = 4; % Oversampling rate
hMod = modem.qammod(M); % Create a 32-QAM modulator
%% Signal Source
% Create a binary data stream as a column vector.
```

```
x = randint(n,1); % Random binary data stream
% Plot first 40 bits in a stem plot.
stem(x(1:40),'filled');
title('Random Bits');
xlabel('Bit Index'); ylabel('Binary Value');
%% Bit-to-Symbol Mapping
% Convert the bits in x into k-bit symbols.
xsym = bi2de(reshape(x,k,length(x)/k).','left-msb');
%% Stem Plot of Symbols
% Plot first 10 symbols in a stem plot.
figure; % Create new figure window.
stem(xsym(1:10));
title('Random Symbols');
xlabel('Symbol Index'); ylabel('Integer Value');
%% Modulation
hMod=modem.qammod(M);
y = modulate(hMod,xsym); % Modulate using 32-QAM.
%% Follow with rectangular pulse shaping.
ypulse = rectpulse(y,nsamp);
```

Figure 8.3 shows the bit/symbol mapping and the resulting 16QAM modulated waveform.

8.6.2 Nonlinear Modulation

For FSK and MSK modulation formats, the function FSKMOD can be used. The command

```
y = FSKMOD(x,M,FREQ_SEP,NSAMP,Fs,PHASE_CONT,SYMBOL_ORDER);
```

produces the complex envelope of an M-ary FSK modulated signal given a message signal x where

- M: is the alphabet size and must be an integer power of two.
- x: is the message signal that must consist of integers between 0 and $M-1$.
- FREQ_SEP: is the desired separation between successive frequencies, in Hz.
- NSAMP: denotes the number of samples per symbol and must be an integer greater than 1.
- Fs: specifies the sampling frequency (Hz). The default sampling frequency is 1.
- PHASE_CONT: specifies the phase continuity across FSK symbols. PHASE_CONT can be either 'cont' for continuous phase, or 'discont' for discontinuous phase. The default is 'cont'.
- SYMBOL_ORDER: specifies how the function assigns binary words to corresponding integers. If SYMBOL_ORDER is set to 'bin' (default), then the function uses a natural binary-coded ordering. If SYMBOL_ORDER is set to 'gray', then the function uses a gray-coded ordering.

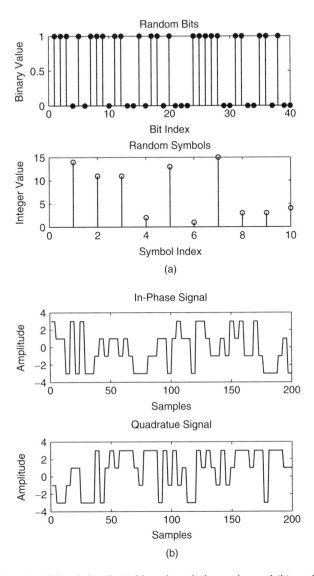

Figure 8.3 16-QAM modulated signal; (a) bit and symbol mapping and (b) modulated waveform.

For example, the following code generates a 16-ary FSK signal for an input binary signal x:

```
File Name: FSK.m
%% Setup
% Define parameters.
M = 16; % Size of signal constellation
```

```
k = log2(M); % Number of bits per symbol
n = 1000; % Number of bits to process
nsamp = 4; % Oversampling rate
freq_sep=.2;
Fs=1; %Sampling rate
%% Signal Source
% Create a binary data stream as a column vector.
x = randint(n,1); % Random binary data stream
%% Bit-to-Symbol Mapping
% Convert the bits in x into k-bit symbols.
xsym = bi2de(reshape(x,k,length(x)/k).','left-msb');
% Generation of a continuous phase 16-ary FSK with gray coding
y = FSKMOD(xsym,M,freq_sep,nsamp,Fs,'cont','gray')
```

Figure 8.4 shows the resulting 16-FSK modulated waveform.

For MSK modulation, the modem object `modem.mskmod` can be used as shown in the previous section (Table 8.3).

8.7   Channel Models in MATLAB®

Communication channels introduce impairments to the transmitted signal that cause data errors at the receiver. Different channel models can be assumed when simulating system performance in MATLAB®. These include additive white Gaussian noise channels, fading channels including many models such as Raleigh and Rician fading channel models, MIMO channels and binary symmetrical channels. Each of these models is based on mathematical description of the channel, which is based on the properties of the channel and is dealt with differently in MATLAB®."

Table 8.5 lists the main functions used in implementing different channel types. Since, the main objective of this book is to study the system performance under nonlinear channels and not the above channel models, these models will not be discussed in details. MATLAB® models used to model the degradation of system performance due to nonlinear amplification are discussed in details in the next chapter.

8.8   Simulation of System Performance in MATLAB®

Simulation of communication system performance is usually based on calculating the error rates which result from corruption of a random sequence of data processed by a channel model, by additive noise. Other performance metrics that can be simulated include the eye diagrams, scatter plots, EVM and ACPR, which provide different views for system performance and signal quality at the receiver.

This section provides the necessary tools to simulate these metrics in MATLAB® using different functions such as `biterr`, `symerr` `eyediagram` and `scatterplot`. These functions are designed to estimate system performance of different wireless standards under various channel models.

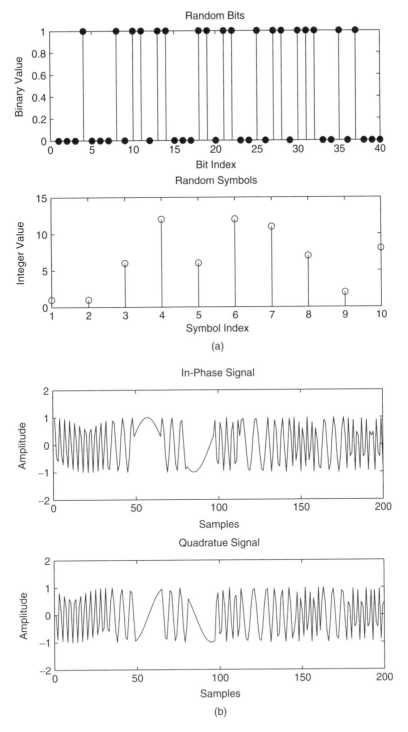

Figure 8.4 16-FSK modulated signal; (a) bit and symbol mapping and (b) modulated waveform.

Table 8.5   Main functions used in implementing different channel types

Channel Type	Function
Additive White Gaussian Noise (AWGN) channel	awgn
Multiple-Input Multiple-Output (MIMO) channel	mimochan
Fading channel	rayleighchan, ricianchan, doppler
Binary symmetric channel	bsc

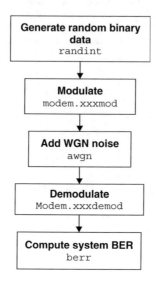

Figure 8.5   Simulation of BER in MATLAB® (The Communication Toolbox, 2009; The Communication Blockset, 2009).

8.8.1   BER

Simulation of BER of a communication system using MC simulations in MATLAB® is based on simulating the transmission of random data through a channel model and then comparing the resulting data with the data before transmission. The channel model is usually assumed to be an AWGN channel where white Gaussian noise is artificially added to the transmitted signal while the other parts of the communication system (transmitter and receiver) are simulated using different system specifications as shown in the previous sections. Figure 8.5 shows a flowchart for the simulation of BER in MATLAB® (The Communication Blockset, 2009).

The functions `biterr` and `symerr` compute the BER and SER by comparing the data entering the transmitter (the original data (bit or symbol levels)) to the data at the output of the receiver. An error is counted if there is a discrepancy between the corresponding bits in the two sequences of bits. The bit or symbol error rate is the number of errors divided by the total number (of bits or symbols) transmitted.

The following script simulates BER of the 16PSK modulation with rectangular pulse shaping. Figure 8.6 shows the result of the simulations.

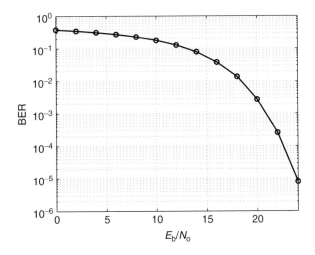

Figure 8.6 Simulated BER of a 16 PSK signal with rectangular pulse shaping.

```
File Name: BER_Example
clear all
%% Setup
% Define parameters.
M = 16; % Size of signal constellation
k = log2(M); % Number of bits per symbol
n = 1e6; % Number of bits to process
nsamp = 4; % Oversampling rate
hMod = modem.pskmod(M); % Create a 16PSK modulator
EbNo=0:2:24; % Range of EbNo
for p=1:length(EbNo)
%% Signal Source
% Create a binary data stream as a column vector.
txbits = randint(n,1); % Random binary data stream
% Plot first 40 bits in a stem plot.
% stem(x(1:40),'filled');
% title('Random Bits');
% xlabel('Bit Index'); ylabel('Binary Value');
%% Bit-to-Symbol Mapping
% Convert the txbits into k-bit symbols.
txsym = bi2de(reshape(txbits,k,length(txbits)/k).','left-msb');
%% Modulation
y = modulate(hMod,txsym); % Modulate using 16PSK.
% Apply rectangular pulse shaping.
ypulse = rectpulse(y,nsamp);
% Transmitted Signal
ytx = ypulse;
%% AWGN Channel
% Apply AWGN channel.
SNR = EbNo(p) + 10*log10(k) - 10*log10(nsamp);
```

```
ynoisy = awgn(ytx,SNR,'measured');
% Received Signal
yrx=downsample(ynoisy,nsamp);
%% Demodulation
% Demodulate signal using 16PSK.
rxsym = demodulate(modem.pskdemod(M),yrx);
% Symbol-to-Bit Mapping
% Undo bit-to-symbol mapping
rxbits = de2bi(rxsym,'left-msb'); % Convert integers to bits.
% Convert rxbits to a vector
rxbits = reshape(rxbits.',numel(rxbits),1);
% BER Computation
% Compare txbits and rxbits to obtain the number of errors and
% the bit error rate.
[NumErrors,BER(p)] = biterr(txbits,rxbits)
% the symbol error rate.
[NumErrors,SER(p)] = symerr(txsym,rxsym);
end
semilogy(EbNo,BER,'o-')
```

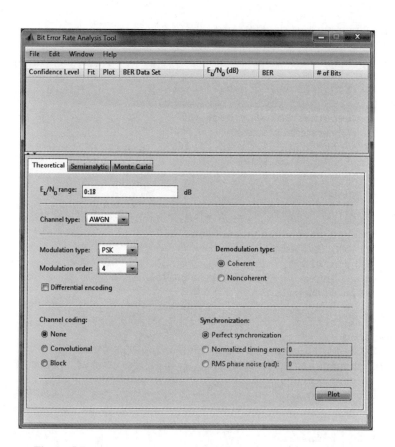

Figure 8.7 BERtool GUI (The Communication Toolbox, 2009).

BER can also be simulated using the Bit Error Rate Analysis Tool (BERTool) in the Communications Toolbox. BERTool is a Graphical User Interface (GUI) that enables BER performance of a communication system to be analyzed using theoretical, semi analytic, or Monte Carlo simulation-based approach. The command `bertool` launches the GUI of the BERTool in MATLAB®. Figure 8.7 shows the BERTool GUI where the type of BER simulations can be chosen from the tabs "Theoretical", "Semi-analytic", or "Monte Carlo".

The BERtool provides simulation of BER vs. E_b/N_o of various modulation schemes under various channel models, types of demodulation, coding schemes and types of synchronization. The theoretical BER is calculated using a set of theoretical BER data for systems that use an AWGN channel. Monte Carlo simulations are done based on Simulink® models or MATLAB® functions that can be called through the GUI. These models can be modified to include specifications of a given communication system model. The semi analytic technique uses a combination of simulation and analytical models to determine the error rate of a communication system. The semi analytic technique is based on quasi-analytic simulation methods where some a priori knowledge (or assumptions) that allows the construction of an abstraction (that can be dealt with analytically) are implied (Jeruchem *et al.*, 2000).

The following code is used by the *Monte Carlo Simulation* as a simulation function in the BERTool (The Communication Toolbox, 2009). Figure 8.8 shows the results of MC simulations compared to theoretical BER.

```
% File Name: bertool_simfcn.m
function [ber, numBits] = bertool_simfcn(EbNo,...
maxNumErrs, maxNumBits)
% Import Java class for BERTool.
import com.mathworks.toolbox.comm.BERTool;
% Initialize variables related to exit criteria.
totErr = 0;  % Number of errors observed
numBits = 0; % Number of bits processed
% - Set up parameters. - % Set up initial parameters.
% Define parameters.
M = 16; % Size of signal constellation
k = log2(M); % Number of bits per symbol
siglen = 10000; % Number of bits to process
hMod = modem.qammod(M); % Create a 32-QAM modulator
hDemod = modem.qamdemod(hMod); % Create a DPSK
ntrials = 0; % Number of passes through the loop
% Simulate until number of errors exceeds maxNumErrs
% or number of bits processed exceeds maxNumBits.
while((totErr < maxNumErrs)  && (numBits < maxNumBits))
    % Check if the user clicked the Stop button of BERTool.
    if (BERTool.getSimulationStop)
        break;
    end
% - Proceed with simulation.
% Modulate the message signal
% Create a binary data stream as a column vector.
msg = randint(siglen, 1); % Generate message sequence.
% msg =randint(siglen, 1, 2, 9973);
```

```
% Bit-to-Symbol Mapping
% Convert the bits in msg into k-bit symbols.
xsym = bi2de(reshape(msg,k,length(msg)/k).','left-msb');
% Modulation-Transmitted Signal
txsig = modulate(hMod,xsym); % Modulate using 32-QAM.
% Channel
% Send signal over an AWGN channel.
snr = EbNo+10*log10(k);
txsignoisy = awgn(txsig,snr,'measured');
% Received Signal
rxsig=txsignoisy;
% Demodulate signal using 16-QAM.
decodmsg = demodulate(hDemod, rxsig); % Demodulate.
% Undo the bit-to-symbol mapping performed earlier.
zsym = de2bi(decodmsg,'left-msb'); % Convert integers to bits.
% Convert z from a matrix to a vector.
zbit = reshape(zsym.',numel(zsym),1);
newerrs = biterr(msg,zbit); % Errors in this trial
ntrials = ntrials + 1; % Update trial index.
% Update the total number of errors.
totErr = totErr + newerrs;
% Update the total number of bits processed.
numBits = ntrials * siglen;
    % - Be sure to update totErr and numBits.
    % - INSERT YOUR CODE HERE.
end % End of loop
% Compute the BER.
ber = totErr/numBits
```

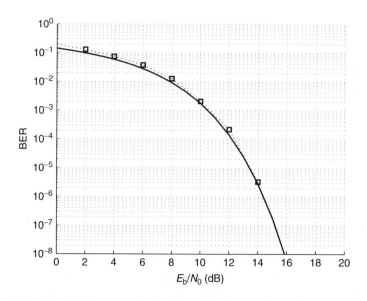

Figure 8.8 BER of 16 QAM system obtained from BERtool; solid: theoretical and □: simulated.

8.8.2 Scatter Plots

The function `scatterplot` plots the signal constellation of the transmitted or the received signals where the in-phase vs. the quadrature components of the modulated signals are plotted. To plot the signal constellation of a modulated signal generated using the `modem.xxxmod` object, the `Constellation` property of the object can be used. For example, the following script:

```
% File Name: ScatPlot.m
% Number of points in constellation
M = 16;
% Modulator object
hMod=modem.pskmod(M);
pt = hMod.Constellation; % Vector of points in constellation
% Plot the constellation.
scatterplot(pt);
```

retrieves the points in a 16 PSK modulated signal constellation and plots the signal constellation using `scatterplot`. Figure 8.9 shows the resulting constellation diagram.

A scatter plot can also be generated using the `commscope` object. For example, the command:

```
hScope = commscope.ScatterPlot
update(h,x)
```

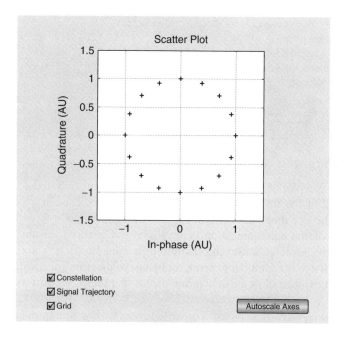

Figure 8.9 Scatter plot of a 16 PSK modulated signal with RRC pulse shaping.

produces a constellation diagram of the transmitted signal x. The following script produces the constellation diagram of a 16PSK signal with RC pulse shaping using the `commscope` object.

```
% File Name: ScatPlot_CommScope.m
% Number of points in constellation
M = 16;
% Modulator object
hMod=modem.pskmod(M); % Modulator object
% Design a pulse shaping filter
nsamp = 16; % up sampling rate
hFilDesign = fdesign.pulseshaping(nsamp,'Raised Cosine',...
...'Nsym,Beta',nsamp,0.50);
hFil = design(hFilDesign);
% Create a scatter plot
hScope = commscope.ScatterPlot
% Set the samples per symbol to the upsampling rate
% of the signal
hScope.SamplesPerSymbol = nsamp;
% Generate data symbols
d = randi([0 hMod.M-1], 100, 1);
% Generate modulated symbols
sym = modulate(hMod, d);
% Apply pulse shaping
xmt = filter(hFil, upsample(sym, nsamp));
% Set the constellation value of the scatter plot to
% the expected constellation
hScope.Constellation = hMod.Constellation;
% Set MeasurementDelay to the group delay of the filter
groupDelay = (hFilDesign.NumberOfSymbols/2);
hScope.MeasurementDelay = groupDelay /hScope.SymbolRate;
% Update the scatter plot with transmitted signal
update(hScope, xmt)
% Display the ideal constellation
hScope.PlotSettings.Constellation = 'on';
```

8.8.3 Eye Diagrams

An eye diagram is a plot of the transmitted signal against time on a fixed-interval axis (T) where at the end of the fixed interval, the signal is wrapped around to the beginning of the time axis. The eye diagram is useful for studying signal quality and in particular the effects of ISI in digital communication systems. Several quality measures of the received signal such as the level of noise and distortion and quality of synchronization can be obtained from an eye diagram. In general, an open eye diagram means that signal distortion is minimum, while a closed eye diagram means that the signal suffers from distortion due to ISI.

The command `eyediagram(x,n,period)` creates an eye diagram for the signal x, plotting n samples in each trace with the horizontal axis range between – `period`/2 and `period`/2. For example, the following script produces an eye diagram for a 16QAM signal with a square-root raised cosine pulse shaping.

```
% File Name:
%% Setup
% Define parameters.
M = 16; % Size of signal constellation
k = log2(M); % Number of bits per symbol
n = 10000; % Number of bits to process
% Oversampling rate
nsamp = 4;
% Create a 16-QAM modulator
hMod = modem.qammod(M);
%% Signal Source
% Create a binary data stream as a column vector.
x = randint(n,1); % Random binary data stream
%% Bit-to-Symbol Mapping
% Convert the bits in x into k-bit symbols.
xsym = bi2de(reshape(x,k,length(x)/k).','left-msb');
%% Modulation
hMod=modem.qammod(M);
y = modulate(hMod,xsym); % Modulate using 32-QAM.
% Upsample and apply square root raised cosine filter.
ytx = rcosflt(y,1,nsamp,'fir/sqrt');
% Create eye diagram for part of filtered signal.
eyediagram(ytx(1:1000),nsamp*2);
```

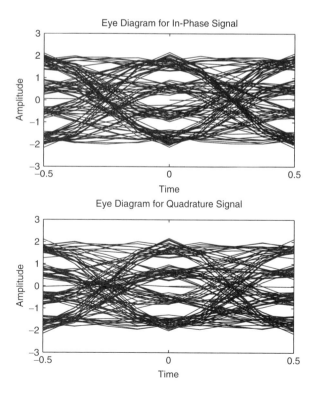

Figure 8.10 Eye diagram of a 16 PSK modulated signal with RRC pulse shaping.

Figure 8.10 shows the resulting eye diagram. Note that the `eyediagram` command creates an eye diagram for part of the filtered signal before the addition of noise where only the effect of the pulse shaping is studied. The figure shows that the signal suffers from significant ISI because the filter is a square root raised cosine filter, not a full raised cosine filter.

The eye diagram can also be generated using the `commscope` object. For example, the command:

```
hScope = commscope.eyediagram
update(h,x)
```

produces an eye diagram of the transmitted signal x in a similar manner as the constellation diagram described above.

8.9 Generation of Communications Signals in MATLAB®

8.9.1 Narrowband Gaussian Noise

A narrowband Gaussian noise signal can be generated by passing a white Gaussian noise signal through either a lowpass or a bandpass filter. The following code generates a narrowband Gaussian noise signal at baseband.

```
% File Name: NBGNSpecPlot.m
clear all
% Set noise bandwidth and sampling rate
NoiseBW=4;
Ts=4.8828e-2;
% Calculate filter impulse response
t0=-1e1:Ts:1e1;
FiltCoeff=sinc(NoiseBW*t0);
%% Generate a complex white Gaussian noise signal
% with power=0 dBm
Noise0= wgn(1,132000,0,1,'complex','dBm');
% Apply an ideal  bandpass filter to the white noise signal
Noise=filter(FiltCoeff,1,Noise0);
% Normalize the power to 0 dBm
P=10*log10(mean(abs(Noise).^2)/(50*.001));
Pscale=10^(P/10);
Noise=sqrt(1/Pscale)*(Noise);
%% Calculate the ACF and PSD of the noise signal
W=2^11;
Rn=xcorr(Noise',Noise',W,'biased');
len=length(Rn);
Sn=fftshift(fft(hanning(len).*Rn/len));
% Generate a frequency vector
m=len/2;
delf=2/(Ts*len);
x=-(m):(m-1);
f=delf*x;
% Scale the power of the signal
```

```
Pin=5;
ps=sqrt(10^(Pin/10));
SndB=10*log10(ps^2*Sn/.05);
P=10*log10(sum(abs(ps^2*Sn))/0.05);
%% Plot PSD of NBGN process
plot(f,SndB,'b')
```

Figure 8.11 shows the resulting spectrum of a NBGN signal at baseband.

8.9.2 OFDM Signals

Figure 8.12 shows a flow chart for the procedure used to generate a OFDM signal according to the transmitter model in Figure 2.12 (a). A random data source generates baseband user data that is then entered to a channel encoder and interleaver. Coded data is then mapped using a constellation mapper that generates one of the following modulation formats BPSK, QPSK, or QAM. The serial modulated data symbols are then entered to a Serial-to-Parallel converter (S/P) that produces N data symbol streams with rate $R = Rs/N$. The parallel data streams are then mapped to time domain using the IFFT block. This block generates a vector of N elements, where each complex number element represents one sample of the OFDM symbol.

The complex baseband OFDM signal can be expressed as:

$$s(n) = \sum_{i=0}^{N-1} d_i e^{\frac{j2\pi i}{N}}, 0 \leq n \leq N - 1 \tag{8.2}$$

This sequence corresponds to samples of the multicarrier signal: that is, the multicarrier signal consists of linearly modulated sub channels, and the right-hand side of

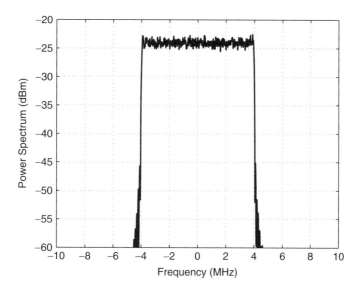

Figure 8.11 PSD of a NBGN signal.

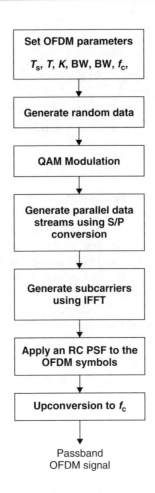

Figure 8.12 A flow chart for OFDM signal simulations (Hasan, 2007).

Equation (8.2) corresponds to samples of a sum of QAM symbols d_i each modulated by carrier frequency $e^{\frac{j2\pi i}{N}}$, $i = 0, ..., N - 1$. The cyclic prefix is then added to the OFDM symbol, and the resulting time samples $x[n] = x[-\mu], ..., .x[N - 1] = x[N - \mu], ..., x[0], ..., x[N - 1]$ are ordered by the parallel-to-serial converter and passed through a D/A converter, resulting in the baseband OFDM signal $x(t)$, which is then up converted to frequency f_0. The following script generates an OFDM signals in MATLAB®. Figures 8.13 to 8.15 show the resulting OFDM signal and its PSD.

```
% File Name: OFDMGen.m
clear all
close all
%% OFDM Signal Parameters
K=2000; % number of subcarriers
BWsub=5e3;
BW=K*BWsub; % BW of the modulated signal
```

```
R=2*BW; % Symbol rate of BB QAM data
T=1/R; % baseband elementary period
FS=4096; % IFFT length
Tofdm=FS*T/2; % useful OFDM symbol period
delta=256*T; % guard band duration
Ts=delta+Tofdm; % Total OFDM symbol period
q=10; % carrier period to elementary period ratio
fc=2*q*BW; % carrier frequency
Rs=4*fc; % simulation period
t=0:1/Rs:Tofdm; % Time Variable
% Set up paramters of RC pulse shaping filter
GroupDelay=10;
Alpha=0.4
%% QAM Modulated Data generation
n=K+1;   % Number of QAM symbols
M=16;   % Number of bits per symbol of QAM
% Generate 16 QAM symbols
k = log2(M); % Number of bits per symbol
% Create a 16-QAM modulation object
hMod = modem.qammod(M);
% Create a binary data stream as a column vector.
rand('state',0);
bits = randint(k*n,1); % Random binary data stream
% Convert bits into k-bit symbols.
syms = bi2de(reshape(bits,k,length(bits)/k).','left-msb');
% Modulate using 16-QAM.
bits = modulate(hMod,syms);
figure
plot(real(syms))
%% Generation of parallel data streams of one symbol
A=length(syms);
FFTData=zeros(FS,1);
%Zero padding to match the IFFT length
FFTData(1:(A/2)) = [ bits(1:(A/2)).'];
FFTData((FS-((A/2)-1)):FS) = [ syms(((A/2)+1):A).'];
figure;
plot(real(FFTData));
%% Generation of subcarriers using IFFT
Subcarr=FS.*ifft(FFTData,FS);
t=0:T/2:Tofdm; % Time vector of subcarriers
figure;
plot(t(401:900),real(Subcarr(401:900)));
%% Apply an RC pulse shaping filter to the OFDM symbols
OFDM_BB=rcosflt(Subcarr,1,2*q,'fir',Alpha,GroupDelay);
OFDM_BB=OFDM_BB(1:length(t));
%% Plot the real and imaginary parts of the Baseband signal
figure;
subplot(211);
plot(t(401:900),real(OFDM_BB(401:900)));
subplot(212);
plot(t(401:900),imag(OFDM_BB(401:900)));
```

```
%% Plot the PSD of the baseband signal
Rx=xcorr(OFDM_BB,OFDM_BB,2^11,'biased'); % Autocorrelation function
len=length(Rx);
Sx=fftshift(fft(Rx/len)); % Power spectral density
SxdB_BB=10*log10(Sx/0.05);  % Power spectral density in dBm
len=length(Rx);
m=len/2;
delf=Rs*1e-6/len;
xx=-(m):(m-1);
f=delf*xx;
figure
plot(f,(SxdB_BB))
%% Upconversion and generation passband signal
OFDM=(real(OFDM_BB)').*cos(2*pi*fc*t)-(imag(OFDM_BB)').
    *sin(2*pi*fc*t);
% Plot the modulated signal in time
figure;
plot(t(80:480),OFDM(80:480));
%% Generation and plotting of PSD of passband signal
Rx=xcorr(OFDM,OFDM,2^11,'biased'); % Autocorrelation function
len=length(Rx);
Sx=fftshift(fft(hanning(len)'.*Rx/len)); % Power spectral density
SxdB=10*log10(Sx/0.05);  % Power spectral density in dBm
len=length(Rx);
m=len/2;
delf=Rs*1e-6/len;
xx=-(m):(m-1);
f=delf*xx; % frequency vector
figure
plot(f,(SxdB))
```

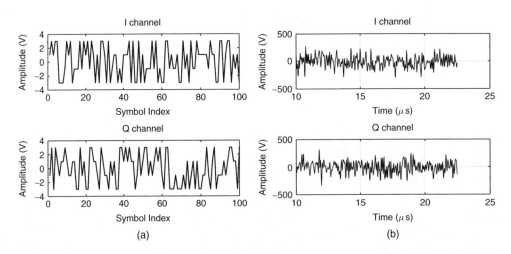

Figure 8.13   OFDM modulation; (a) QAM signal and (b) subcarriers.

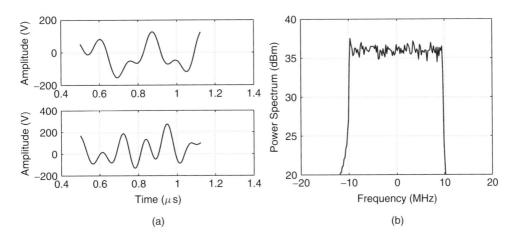

Figure 8.14 OFDM baseband signal; (a) time domain and (b) frequency domain.

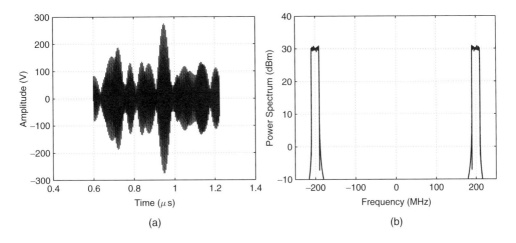

Figure 8.15 OFDM passband signal; (a) time domain and (b) frequency domain.

8.9.3 DS-SS Signals

Various models for DS-SS signal generation in simulation exist. The basic idea in the generation of DS-SS signals is to perform spreading of a randomly generated baseband sequence using a spreading sequence. Spreading sequences can be generated using random sequence generators. In multiuser CDMA systems, orthogonal spreading sequences (such as orthogonal Walsh codes) are used to produce a multiuser coded signal that consists of the sum of spread user data. Figure 8.16 shows a block diagram for a basic CDMA signal generator. $2n$ random streams that represent user data (each user has I and Q data streams) are first generated and then oversampled in order to perform spreading by a code. The oversampling rate is equal to spreading factor since each bit is

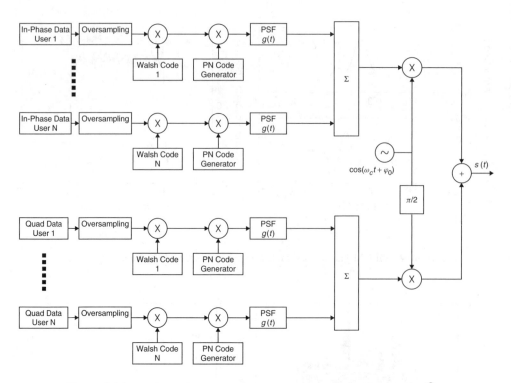

Figure 8.16   A block diagram of CDMA signal generation in MATLAB®.

represented by N chips that represent a Walsh code. Then each user data is multiplied by a unique repeated Walsh sequence to provide orthogonal covering. The resulting user-coded sequence is further multiplied by a PN sequence of the same length. Each coded user data is then pulse shaped using a raised cosine filter and then added together to produce a composite CDMA signal. The script below produces a CDMA signal for a 64-user system where Walsh codes are generated using a Hadamard matrix generator. The script also generates and plots the PSD of the resulting signal.

```
% File Name: DS_CDMA_Gen.m
clear all
% Set the number of user data symbols
n=(2^6);
% Oversampling
nsamp=4;
% Length of Walsh code
N=64;
% Generate Walsh codes of length N
H=hadamard(N);
% Generate repeated Walsh codes of length N*n
WalshCode=repmat(WalshCode,1,n);
% Initialize variables
Ichanc=zeros(1,N*n*nsamp+80);
```

```
Qchanc=Ichanc;
% Generate a PN sequence for the I and Q channels
PNI=randsrc(n*N,1,[1 -1])';
PNQ=randsrc(n*N,1,[1 -1])';
for d=1:N
    bitsI=randsrc(n,1,[1 -1]);
    bitsQ=randsrc(n,1,[1 -1]);
    % Match the sequence length to be multiple of N (1 bit length=N
    % samples)
    I =rectpulse(bitsI,N)';
    Q =rectpulse(bitsQ,N)';
    % Walsh code spreading of the I and Q data
    walshbitsi=WalshCode(d,:).*I;
    walshbitsq=WalshCode(d,:).*Q;
    % Randomize the spreaded I and Q sequences using PN codes
    walshbitsI=PNI.*walshbitsi;
    walshbitsQ=PNQ.*walshbitsq;
    % Filter the I and Q data with an
    % RC pulse shaping FIR filter
    Ichan=rcosflt(walshbitsI,1,nsamp,'fir',0.4,10);
    Qchan=rcosflt(walshbitsQ,1,nsamp,'fir',0.4,10);
    % Generate the composite signal (multiuser signal)
    Ichanc=Ichanc+Ichan';
    Qchanc=Qchanc+Qchan';
end
%%%%%%%%%%%%%%%%%%%%%%%%%%%%%%%%%%%%%%%%%%%%%%%%%%%%%%%%%%%%%%%
% Normalize the power to zero dBm
Pin=10*log10(mean(abs(Ichanc+1i*Qchanc).^2)/(50*.001));
Pscale=10^(Pin/10);
% Complex DS-CDMA signal generation
CDMA=sqrt(1/Pscale)*(Ichanc+1i*Qchanc);
% Compute the autocorrelation estimate
y=transpose(CDMA);
ry=xcorr(y,y,2^11,'biased');
% Calculate the power spectral density of the CDMA signal
len=length(ry);
Sy=fftshift(fft(hanning(len).*ry/len));
% Create a frequency vector
len=length(Sy);
m=len/2;
delf=10/(len);
xx=-(m):(m-1);
f=delf*xx;
% Plot the PSD vs. frequency
SydB=10*log10(Sy/.05);
figure
plot(f,SydB);
```

Figure 8.17(a) shows the I and Q data of user k, Figure 8.17(b) shows the user data after spreading by Walsh code k, Figure 8.17(c) shows the resulting sequence after further

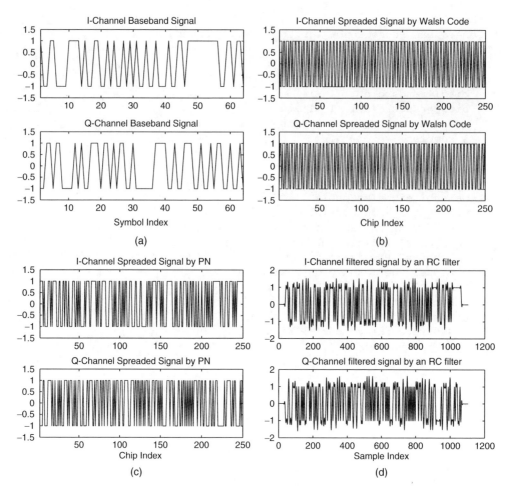

Figure 8.17 DS-CDMA signal generation; (a) I and Q data of user k, (b) after spreading by Walsh code, (c) after PN spreading and (d) after baseband filtering.

spreading by a PN sequence and Figure 8.17(d) shows the resulting signal after baseband filtering (RC pulse shaping). Figure 8.18 shows the PSD of the resulting CDMA signal.

8.9.4 Multisine Signals

A multisine signal $x(t)$ consisting of the sum of an even number of tones with a uniform frequency separation can be represented as (Gharaibeh *et al.*, 2006)

$$x(t) = \sum_{k=-K}^{K} \frac{A}{2} \cos \left(\omega_c t + (k - 1/2) \, \omega_m t + \phi_k \right) \tag{8.3}$$

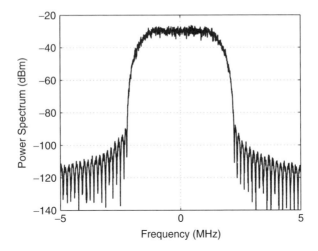

Figure 8.18 PSD of a CDMA signal.

where K is an even number and $K \neq 0$, $\omega_m = \omega_k - \omega_{-k}$ is the frequency separation between the input tones and ϕ_k are the phases of the input tones which are assumed to be independent random phases uniformly distributed in $[0, 2\pi]$.

The multisine signal in Equation (8.3) can be written as a sum of complex conjugate pairs as (Gharaibeh *et al.*, 2006)

$$x(t) = \sum_{k=-K}^{K} \frac{1}{2} \tilde{x}_k(t) e^{j\omega_c t} \tag{8.4}$$

it follows that $\tilde{x}_k = A \cos\left(\left(k - \frac{1}{2}\right) \omega_m t + \theta_{1k}\right) e^{j\theta_{2k}}$ where $\theta_{2k} = (\phi_k + \phi_{-k})/2$ and $\theta_{1k} = (\phi_k - \phi_{-k})/2$ and hence the complex envelope of Equation (8.4) is

$$\tilde{x}(t) = \sum_{k=1}^{K} A \cos\left(\left(k - \frac{1}{2}\right) \omega_m t + \theta_{1k}\right) e^{j\theta_{2k}}. \tag{8.5}$$

For the special case where $\phi_k = -\phi_{-k}$ the above equation can be written as

$$\tilde{x}(t) = \sum_{k=1}^{K} A \cos\left(\left(k - \frac{1}{2}\right) \omega_m t + \phi_k\right). \tag{8.6}$$

This form can be easily simulated under the above assumptions where the only random variable is the phase of each of the tones. Uniformly distributed random phases can be generated using a uniform random number generator in MATLAB®.

The following script generates the complex envelope of a multisine signal that consists of 8 tones with random phases and plots its frequency spectrum. Figure 8.19 shows the time-domain representation of an 8-tone signal and Figure 8.20 shows its PSD.

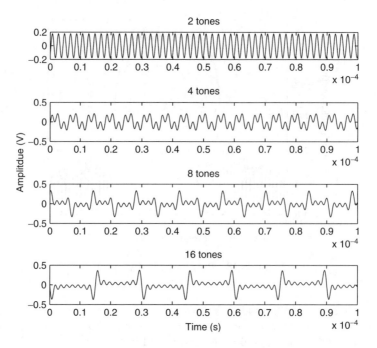

Figure 8.19 A phase-aligned multisine signal for different number of tones.

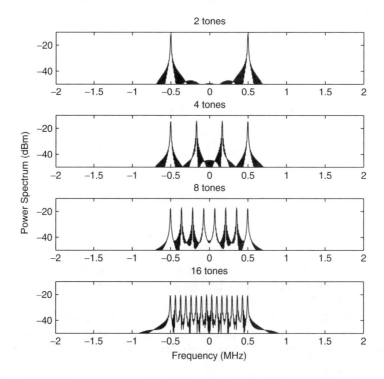

Figure 8.20 Power spectrum of multisine signals for different number of tones.

```
% File Name: MulltiSine_Gen.m
clear all
n=2;    % Number of tones
bw=1e6; % Bandwidth of multisine signal
freq_sep=bw/(n-1); % Frequency separation between tones (Constant)
% Generate the frequency vector
for i=1:n/2
    f(i)=(freq_sep/2)+freq_sep*(i-1);
end
% Set the sampling frequency and generate a time vector
fs=20*f(n/2);   % Sampling frequency
Ts=1/fs; % Sampling time
t=0:Ts:1e3*Ts; % Time vector
% Generate tones with random phases.
Msine=zeros(size(t)); % Initialize multisine signal with zeros
for k=1:(n)/2
    rand('seed',sum(100*clock));
    rand('state',sum(100*clock));
    phi=-pi+(2*pi)*(rand(1)); % Random phases generation for tone k
    Msine=Msine+cos(2*pi*f(k)*t+phi); % Multisine signal
end
% Normalize the power of the multisine signal to zero dB
Pin=10*log10(mean(abs(Msine).^2)/(50*.001));
Pscale=10^(Pin/10);
Msine=sqrt(1/Pscale)*(Msine);
%%%%%%%%%%%%%%%%%%%%%%%%%%%%%%%%%%%%%%%%%%%%%%%%%%%%%%%%%%%%%%%%%%%%%%%
% Compute the autocorrelation function and the power spectrum
% of the multisine signal at input power = pin (dBm).
pin=-5; % Input power =-5 dBm
A=sqrt(10^((pin)/10)); % Amplitude scaling
Msine_scaled=A*Msine;
Rx=xcorr(Msine_scaled,Msine_scaled,2^10,'biased'); % Autocorrelation
len=length(Rx);
Sx=fftshift(fft(Rx/len)); % Power spectral density
SxdB=10*log10(Sx/0.05);  % Power spectral density in dBm
% Generate a frequency vector to plot the spectrum
len=length(Rx);
m=len/2;
delf=fs/1e6/len;
xx=-(m):(m-1);
f1=delf*xx;
%%%%%%%%%%%%%%%%%%%%%%%%%%%%%%%%%%%%%%%%%%%%%%%%%%%%%%%%%%%%%%
% Plot the time domain multisine signal
figure
plot(t,Msine_scaled)
axis([0 1e-4 -1.5 1.5])
hold on
```

```
% Plot the power spectrum of multisine signal
figure
plot(f1,SxdB)
axis([-2 2 -50 -10])
```

Note that the above scripts generates a single realization of a multisine signal with a random phase. If a full representation of the random signal is required, then, the script needs to be repeated to generate other realizations. When dealing with simulation of nonlinear distortion of multisines, distortion is estimated by averaging a large number of phase realizations. The number of realizations required depends on the number of tones since as the number of tones increases the probability of having a uniformly distributed phases increases.

8.10   Example

The following script simulates overall communication system operations including data generation, pulse shaping, modulation and demodulation. It also plots an eye diagram and a scatter plot and conducts BER calculations.

```
% File Name: Example_ModDemodBER.m
clear all, close all
%% Setup
% Define parameters.
M = 16; % Size of signal constellation
k = log2(M); % Number of bits per symbol
n = k*1e6; % Number of bits to process
nsamp = 4; % Oversampling rate
hMod = modem.qammod(M); % Create a 32-QAM modulator
%% Data Generation
% Create a binary data stream as a column vector.
txbits = randint(n,1); % Random binary data stream
% Bit-to-Symbol Mapping
% Convert the bits in x into k-bit symbols.
txsym = bi2de(reshape(txbits,k,length(txbits)/k).','left-msb');
%% Modulation
hMod=modem.qammod(M);
QAM = modulate(hMod,txsym); % Modulate using 32-QAM.
%% RC pulse-shaping.
GroupDelay=10;
alpha=.8;
QAM_PS=rcosflt(QAM,1,nsamp,'fir',alpha,GroupDelay);
%% Create a scatter plot
hScope = commscope.ScatterPlot
% Set the samples per symbol to the upsampling rate of the signal
hScope.SamplesPerSymbol = nsamp;
% Set the constellation value of the scatter plot to the
% expected constellation
hScope.Constellation = hMod.Constellation;
% Update the scatter plot with transmitted signal
```

```
update(hScope, QAM)
%% Create an Eye Diagram
eyediagram(QAM_PS(1:1000),nsamp*2);
%% Send signal over an AWGN channel
% Transmitted Signal
txSig = QAM_PS;
% Signal to Noise Ratio
EbNo =15; % In dB
snr = EbNo + 10*log10(k) - 10*log10(nsamp);
txnoisy = awgn(txSig,snr,'measured');
%% Received Signal
rxSig=downsample(txnoisy,nsamp);
%% Demodulation
% Demodulate signal using 16-QAM.
rxsym = demodulate(modem.qamdemod(M),rxSig);
% Remove group delay of pulse shaping filter
rxsym=rxsym(GroupDelay+1:size(rxsym)-GroupDelay);
% Symbol-to-Bit Mapping
% Undo the bit-to-symbol mapping
rxbits = de2bi(rxsym,'left-msb'); % Convert integers to bits.
% Convert rxbits from a matrix to a vector.
rxbits = reshape(rxbits.',numel(rxbits),1);
%% BER Computation
% Compare x and z to obtain the number of errors and
% the bit error rate.
[number_of_errors,bit_error_rate] = biterr(txbits,rxbits)
% the symbol error rate.
[number_of_errors,sym_error_rate] = symerr(txsym,rxsym)
```

8.11   Random Signal Generation in Simulink®

In addition to the random data and noise generators in the main Simulink® sources library (including reading a measured random signal from a.mat file), random data sources and noise generators are available in Comm. Source library in the Communications Blockset library. The Comm. Sources library consists of three main sublibraries: Random Data Sources, Noise Generators and Random Sequence sublibraries (The Communication Blockset, 2009). Figure 8.21 shows the main sublibraries of the communication sources library within the Communications Blockset.

8.11.1   Random Data Sources

A number of random signals can be generated using this sublibrary such as Random Integer Generator, Poisson Integer Generator and Bernoulli Binary Generator. Figure 8.22 shows the Simulink® library of random data sources. The Bernoulli Binary Generator block generates independent random bits (binary integers) where each bit is generated from a Bernoulli random variable. Thus, the generated bit stream can represent binary data sources in real-world communication systems. Random Integer Generator generates vectors of non-negative random integers having uniform distribution, while

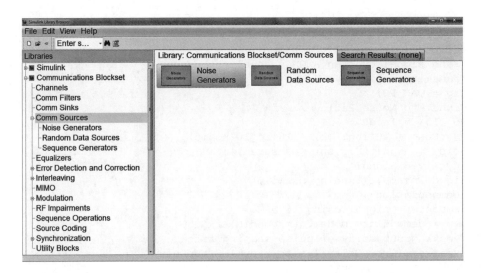

Figure 8.21   Simulink® communication sources (The Communication Blockset, 2009).

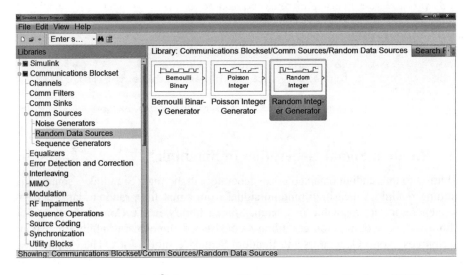

Figure 8.22   Simulink® data sources (The Communication Blockset, 2009).

the Poisson Integer Generator block generates random non-negative integers having a Poisson distribution.

8.11.2   Random Noise Generators

Noise Generators sublibrary contains a number of random real number generators with a given probability distribution that simulate channel noise. The sublibrary includes Gaussian Noise Generator, Rayleigh Noise Generator, Rician Noise Generator and

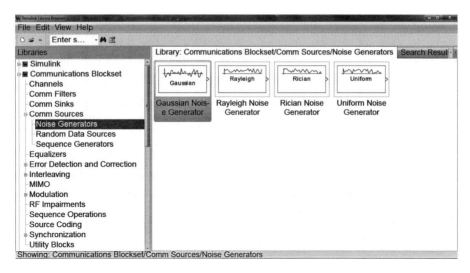

Figure 8.23   Simulink® noise sources (The Communication Blockset, 2009).

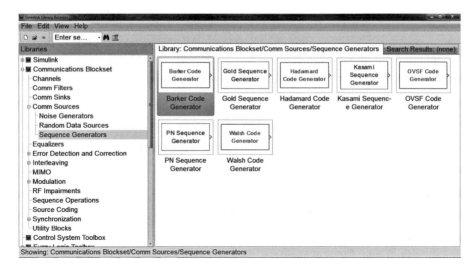

Figure 8.24   Simulink® random sequence sources (The Communication Blockset, 2009).

Uniform Noise Generator. Figure 8.23 shows the Simulink® sub-library of random noise sources (The Communication Blockset, 2009).

8.11.3   Sequence Generators

The Sequence Generators sublibrary contains blocks that generate random sequences using shift registers for spreading or synchronization in a communication system. These random sequences include Pseudo-random sequences, Synchronization codes

Table 8.6   Simulink® random sequence generators (The Communication Blockset, 2009)

Type	Random Sequence
Pseudorandom Sequences	Gold sequences, Kasami Sequence, Pseudonoise sequences
Synchronization codes	Barker Code
Orthogonal codes	Hadamard codes, OVSF codes, Walsh codes

and Orthogonal codes. These codes can be used for multiple-access spread spectrum communication systems, synchronization, and data scrambling (The Communication Blockset, 2009). Figure 8.24 shows the Simulink® sublibrary of random sequence sources. Table 8.6 shows the different sequence generator blocks in the sequence generator sublibrary.

8.12   Digital Modulation in Simulink®

The Digital Modulation library in the Communications Blockset consists of blocks that perform baseband simulation of digital modulation/demodulation schemes. The baseband modulation blocks generate the lowpass equivalent of common digital modulation schemes. Figure 8.25 shows the digital modulation sublibraries under the Digital Modulation library. These sublibraries include (The Communication Blockset, 2009):

- amplitude modulation: PAM, ASK, QAM;
- phase modulation PM: MPSK, DQPSK, OQPSK, CPSK;
- frequency modulation: FSK;
- Continuous Phase Modulation (CPM): CPFSK, MSK, GMSK;
- trellis-coded modulation: TCM, MPSK TCM, MQAM TCM.

8.13   Simulation of System Performance in Simulink®

System performance can be simulated in Simulink® using a number of approaches. The "Sinks" library in the "Communications Blockset" contains three types of signal scopes:

- Eye Diagrams: for plotting the eye diagram of a discrete signal.
- Scatter Plots: for plotting scatter plot of a discrete signal.
- Signal Trajectories: for plotting signal trajectory of a discrete signal.

These scopes can be used to facilitate viewing the performance of a digital communication system under channel or RF impairments. Figure 8.26 shows the communications sinks sublibrary.

On the other hand, error rates can be calculated in communication system model using the Error Rate Calculation block that compares input data from a transmitter with input data from a receiver and calculates the error rates or the number of errors at either the bit or symbol levels.

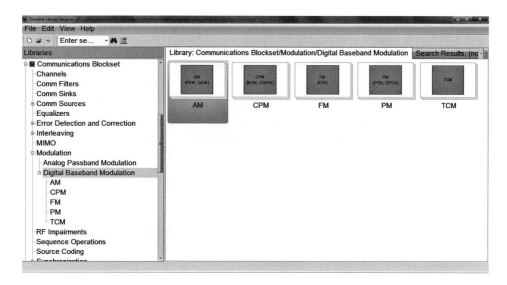

Figure 8.25 Simulink® digital modulation library (The Communication Blockset, 2009).

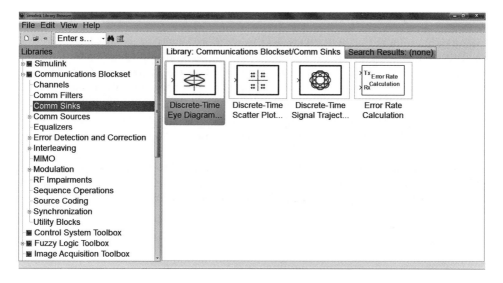

Figure 8.26 Simulink® communication sinks (The Communication Blockset, 2009).

Signal spectrum calculation blocks can be used to assess system performance by viewing the changes that occur to the input signal spectrum as a result of channel or RF impairments. There are sinks under the Signal Processing Blockset such as the Vector Scope, Spectrum Scope, Matrix Viewer, and Waterfall Scope blocks that can be used to view the signal in the frequency domain. The Spectrum Scope block displays the frequency spectrum of an input signal by computing its FFT.

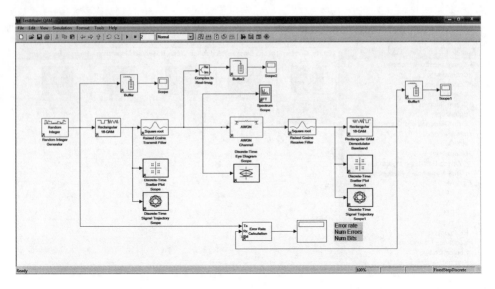

Figure 8.27   Example 1: QAM modulation/demodulation model in Simulink®.

8.13.1   Example 1: Random Sources and Modulation

In the simple example shown in Figure 8.27 (File Name: Example_Mod.mdl), the various blocks that can be used to assess the performance of communication systems are used. The example consists of a model for simulation of QAM transmitter and receiver system with an AWGN channel. The transmitter model consists of a random integer generator, a QAM modulator and a transmit pulse-shaping filer. The channel is modeled as an AWGN channel model using the AWGN channel block. The receiver model consist of a receive pulse-shaping filter and a QAM demodulator. The parameters of the various blocks are shown in Table 8.7.

The performance of the system in Example 1 is assessed using time scopes, spectrum scopes, signal constellation, signal trajectory and a BER calculator. Figure 8.28 shows the plots generated by various performance blocks where it is shown how noise affects the

Table 8.7   Parameters of various blocks of Example 1

Block	Parameters
Random Integer Generator	$M = 16$, Sample time = 1e-5
QAM modulator	$M = 16$, Contellation = Binary
AWGN channel	$E_s/N_o = 30$ dB
Transmit pulse shaping filer	Group delay = 8, $\alpha = 0.3$, Upsampling = 8
Receive pulse shaping filter	Group delay = 8, $\alpha = 0.3$, Downsampling = 8
QAM demodulator	$M = 16$, Contellation = Binary
Error Rate Calculation	Receive Delay = 6

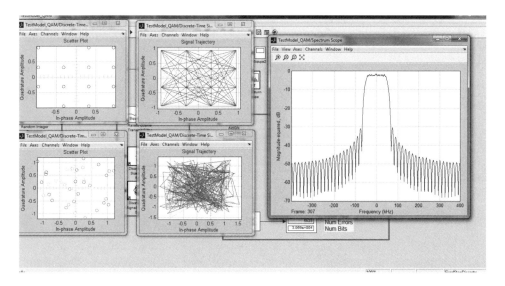

Figure 8.28   Various performance plots for the QAM modulation/demodulation in Example 1.

constellation, signal trajectory and BER of the received signal. Other channel impairments such as fading channels and nonlinear amplification can be added to the model.

The BER calculator compares the transmitted and received symbols and counts the number of errors during the simulation time. It calculates error rate, number of error events or total number of input events at the bit or symbol levels. The SNR can be varied from the AWGN channel block and a plot of BER vs. SNR can be developed.

8.13.2   Example 2: CDMA Transmitter

Example 2, shown in Figure 8.29 (File Name: `Example_CDMA.mdl`), simulates a CDMA transmitter and generates a baseband direct-sequence spread spectrum signal according to IS-95 mobile standard in Simulink®. Binary data at a sampling rate of 1/19200 Hz is generated using the "Bernoulli Random Binary Generator" block and fed into a "QPSK Modulator Block" that outputs complex modulation data. The modulated signal is then multiplied by a Walsh code generated from the "Hadamard Code Generator Block" that generates repeated Walsh code drawn from an orthogonal set of codes with length N (64 in this example). The sample time of the Walsh code (chip time) is $1/(19\,200*64)$ since every bit is multiplied by 64 chips. The spread data at the output of the multiplies is then fed to a SRRC pulse shaping filter with $\alpha = 0.3$, up-sampling of 8 samples and a group delay of 16 samples. The parameters of the various blocks of Example 2 are shown in Table 8.8.

Figure 8.30 shows the spectrum of the signal before spreading and after spreading where it is shown that the bandwidth of the spread signal is 64 times the bandwidth of the signal before spreading; that is, a processing gain of 64.

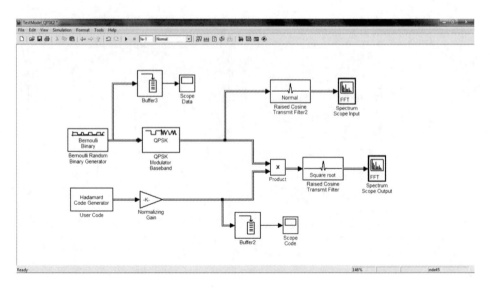

Figure 8.29   Example 2: CDMA transmitter model in Simulink®.

Table 8.8   Parameters of various blocks of Example 2

Block	Parameters
Random Binary Generator	Sample time $= 1/19\,200\,\text{Hz}$, Samples per frame $= 100$
QPSK Modulator	Phase offset $= \pi/4$, Contellation $=$ Binary
Hadamard Code Generator	Code length $= 64$, Sample time $= 1/(19\,200*64)$
RC Transmit Filter	Group delay $= 16$, $\alpha = 0.3$, Upsampling $= 8$

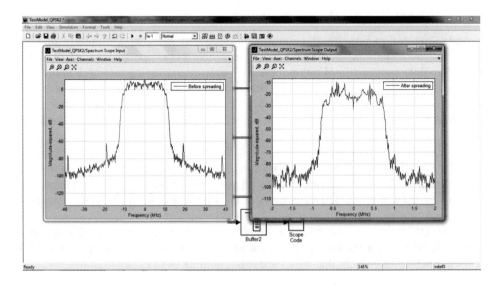

Figure 8.30   Input and output spectrum of CDMA transmitter in Example 2.

Table 8.9 Models of wireless standards in Simulink® (The Communication Blockset, 2009)

Model	Description
256-Channel ADSL	models part of the Asymmetric Digital Subscriber Line (ADSL) technology
Bluetooth® Frequency Hopping	simulates a simple Bluetooth® wireless data link
Bluetooth® Voice Transmission	models part of a Bluetooth® system
Bluetooth® Full Duplex Voice and Data Transmission	simulates the full duplex communication between two Bluetooth® devices
CDMA2000 Physical Layer	simulates part of the downlink physical layer of a wireless communication system according to the cdma2000 specification
Defense Communications: US MIL-STD-188-110B	implements an end-to-end baseband communications system compliant with the U. S. MIL-STD-188-110B military standard
Digital Video Broadcasting – Cable (DVB-C)	models part of the ETSI EN 300 429 standard for cable system transmission of digital television signals
Digital Video Broadcasting – Terrestrial	models part of the ETSI EN 300 744 standard for terrestrial transmission of digital television signals
DVB-S.2 Link, Including LDPC Coding	simulates the state-of-the-art channel coding scheme used in the second generation Digital Video Broadcasting Standard (DVB-S.2)
HIPERLAN/2	models part of HIPERLAN/2 (high-performance radio local area network), European (ETSI) Standard for high-rate wireless LANs
IEEE® 802.11a WLAN Physical Layer	represents an end-to-end baseband model of the physical layer of a Wireless Local Area Network (WLAN) according to the IEEE® 802.11a standard
IEEE® 802.11b WLAN Physical Layer	simulates an implementation of the Direct Sequence Spread Spectrum (DSSS) system.
IEEE® 802.16-2004 OFDM PHY Link, Including Space-Time Block Coding	simulates an end-to-end baseband model of the physical layer of a WMAN according to the IEEE® 802.16-2004 standard
WCDMA Coding and Multiplexing	simulation of the multiplexing and channel decoding structure for the Frequency Division Duplex (FDD) downlink as specified by 3GPP, Release 1999
WCDMA End-to-End Physical Layer	models part of the Frequency Division Duplex (FDD) downlink physical layer of 3G WCDMA.
WCDMA Spreading and Modulation	simulates spreading and modulation for an FDD downlink DPCH channels as specified by the 3GPP, Release 1999.

8.13.3 Simulation of Wireless Standards in Simulink®

The Communications Blockset contains demos of a number of wireless standards that are ready to use. These models simulate a number of common mobile communication and wireless networking standards and can be used to generate real-world communication signals. Table 8.9 lists the available models in Simulink® and their description.

Other wireless standard models such as IS-95 CDMA, GSM, EDGE systems are available at the MATLAB® Central website (MATLAB® Central, 2011).

8.14 Summary

In this chapter, the basics of communication system simulation in MATLAB® and Simulink® have been presented. The discussions on various elements of communication system simulations are supported by examples that can be used for evaluation of system performance for more complex systems. Various simulation issues such as sampling rate of simulations, measurement of system performance from simulated spectra and the choice of the parameters of MATLAB® functions and Simulink® blocks have also been discussed. This chapter serves as an introduction to the next chapter that discusses simulation of nonlinear systems where the objective is to assess the performance of a communication system under nonlinear channel impairments.

9

Simulation of Nonlinear Systems in MATLAB®

In the previous chapter, MATLAB® has been presented as a tool for the evaluation of the performance of communication systems. This chapter is dedicated to simulation of nonlinear system performance where tools for implementing nonlinear models in MATLAB® are presented. The simulation of performance metrics is based on the computation of the autocorrelation function and the power spectral density of the output of nonlinearity for different input signals. The autocorrelation functions are estimated from time autocorrelation functions assuming ergodicity, which is a valid assumption for most communication signals. In addition, simulation procedures for predicting in-band distortion using the orthogonal analysis of the nonlinear model, which was presented in Chapter 6, are presented. Programming issues related to the efficient simulation of nonlinear systems in MATLAB® and a comprehensive reference of simulation approaches of different types of nonlinearity are also provided.

9.1 Generation of Nonlinearity in MATLAB®

9.1.1 Memoryless Nonlinearity

A memoryless nonlinearity can be generated by simulation using one of the empirical nonlinear models discussed in Chapter 3 such as the Rapp and Saleh models. These models can be used to generate nonlinear characteristics over a range of input power by controlling the parameters of the nonlinear model such as the p parameter of the Rapp model and the α, β parameters of the Saleh model. The following script generates nonlinear characteristics using the Rapp and Saleh models.

Nonlinear Distortion in Wireless Systems: Modeling and Simulation with MATLAB®, First Edition.
Khaled M. Gharaibeh.
© 2012 John Wiley & Sons, Ltd. Published 2012 by John Wiley & Sons, Ltd.

```
% File Name:
% Rapp Model
p=1;
G=5;
Vsat=0.02;
N=5;
Pin=[-30:0.01:0]';
Vi=sqrt(.1*10.^(Pin/10));
Vo=G*Vi./((1+(abs(Vi)./Vsat).^(2*p)).^(1/(2*p)));
Po=10*log10((abs(Vo).^2)/.1);

% Saleh Model
alpha_a=5;
beta_a=.2;
alpha_p=10;
beta_p=10;
Pin=[-30:0.01:0]';
Vi=sqrt(.1*10.^(Pin/10));
A=alpha_a*abs(Vi)./(1+beta_a*(abs(Vi)).^2);
Phi=(alpha_p*(abs(Vi).^2)./(1+beta_p*(abs(Vi)).^2))+angle(Vi);
Vo=A.*(cos(Phi)+j*sin(Phi));
Po=10*log10((abs(Vo).^2)/.1);
```

Figure 9.1(a) shows nonlinear characteristics generated using Rapp model and Figure 9.1(b) shows the nonlinear characteristics generated using Saleh Model.

9.1.2 Nonlinearity with Memory

A nonlinearity with memory can be generated using the Wiener model that consists of a linear filter in cascade with a memoryless nonlinearity, as shown in Chapter 3. The memoryless nonlinearity block can be modeled using either Rapp, Saleh or other

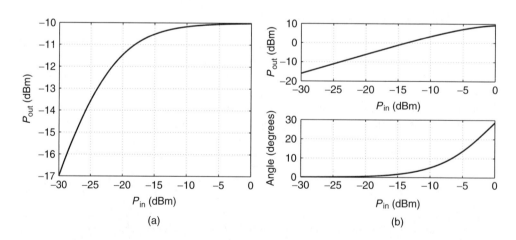

Figure 9.1   Nonlinear characteristics: (a) Rapp model and (b) Saleh model.

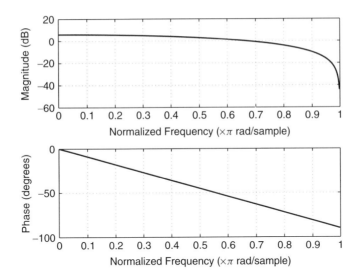

Figure 9.2 Linear filter frequency response.

nonlinear models. The transfer function of the linear filter used can be given in the z-domain by:

$$H(z) = \frac{a_0}{b_0 + b_1 z^{-1} + \ldots + b_n z^{-n}} \tag{9.1}$$

where the filter coefficients can be chosen to give the desired memory effects.

The following script generates a nonlinearity with memory and applies it to a WCDMA signal. The memoryless nonlinear block is the Saleh model used in the previous subsection and the linear filter has coefficients: $a_0 = 1$, $b_0 = 0.7692$, $b_1 = 0.1538$ and $b_2 = 0.0769$. Figure 9.2 shows the frequency response of the linear filter used in the simulation.

```
% File Name:
clear all
% Load a WCDMA signal
load WCDMA_UP_TM1_16DPCH.mat
% Specify the sampling time of the signal
XDelta=4.8828e-008; % Sampling time
x0=double(Y(1:256000));
% Normalize power to 0 dBm; Assume 50 ohm impedance;
P=10*log10(mean(abs(x0).^2)/(50*.001));
Pscale=10^(P/10);
x1=sqrt(1/Pscale)*(x0); % normalized signal to 0 dBm
% Set input power level (dBm)
Pin=-0; % dBm
% Scale the amplitude of the signal
ps =sqrt(10^(Pin/10)); % amplitude scaling
X=ps*(x1); % scaled signal
%%%%%%%%%%%%%%%%%%%%%%%%%%%%%%%%%%%%%%%%%%%%%%%%%%%%%%%%%%%%%
```

```
% Apply a memoryless nonlinearity to input signal
% Saleh Model parameters
u1=5;
v1=2.3;
u2=10;
v2=10;
% Linear filter coefficients
bb=[0.7692 0.1538 0.0769];
aa=1;
freqz(bb,aa)
% Apply memoryless nonlinearity to input signal
AM=u1*abs(X)./(1+v1*(abs(X)).^2);
Phi=(u2*(abs(X).^2)./(1+v2*(abs(X)).^2))+angle(X);
Z=AM.*(cos(Phi)+j*sin(Phi));
% Apply a linear filter to the nonlinearity output
Y =filter(bb,aa,Z);
```

9.2 Fitting a Nonlinear Model to Measured Data

9.2.1 Fitting a Memoryless Polynomial Model to Measured Data

To develop the model coefficients of an envelope power-series model of the form

$$y(n) = \sum_{k=1}^{N} b_k |x(n)|^{k-1} x(n) \tag{9.2}$$

the input power of the a single tone signal is swept in M power steps and the output of the amplifier is measured at each power step. Let x_i, where $i \leq M$, be the data points obtained by sweeping the input power and let y_i be the measured output of the amplifier at the ith power step, then the polynomial model can be found using least squares as shown in Chapter 3.

Measured amplifier characteristics can be obtained from S_{21} measurements of a non-linear amplifier using a Vector Network Analzser (VNA). The S_{21} vs. input power measurements are done by sweeping the input power of a sinusoid at a given frequency f_0 in steps and measuring the S_{21} parameter that represents the complex gain of the amplifier at each power step. Therefore, given an amplifier characterized by its AM–AM and AM–PM conversions as in Equation (4.42):

$$y(t) = F(|x(t)|)e^{j[\Theta(|x(t)|)+\theta(t)]} \tag{9.3}$$

where F and Θ represent the AM–AM and AM–PM conversions at a given fundamental frequency f_0, these characteristics can be written in terms of the complex gain of the amplifier as in Equation (4.39):

$$\tilde{y}(t) = \tilde{x}(t)G(\tilde{x}(t)) \tag{9.4}$$

where G is a complex gain function defined as

$$G(\tilde{x}(t)) = \frac{F(|\tilde{x}(t)|)}{|\tilde{x}(t)|}e^{j\Theta(|\tilde{x}(t)|)} = G_I(|\tilde{x}(t)|) + jG_Q(|\tilde{x}(t)|) \tag{9.5}$$

The complex gain of an amplifier G is the S_{21} parameter of the amplifier under test and can be measured versus either the input frequency or the input power level. For developing a nonlinear model, the input power is swept and the corresponding complex gain is measured at each power step. The measured complex gain is stored in a text file–a feature supported by most modern VNAs–and the file can be loaded from MATLAB® using the `load` command. Figure 9.3 shows the structure of the text file that contains the complex S_{21} measurements. Note that the measured characteristics are power measurements and not voltage measurements. To convert the power measurements to voltages, the amplifier characteristics are assumed to have odd symmetry ($V_o(-V_{in}) = -V_o(V_{in})$) and hence, only odd-order envelope coefficients can be obtained from measurements. Fortunately, only the odd-order components contribute to intermodulation and cross-modulation components that lie inside the bandwidth of the input signals.

The following MATLAB® script produces the complex coefficients of a 5th-order polynomial model using measured amplifier characteristics obtained from a VNA measurements of an amplifier. The amplifier considered here has a gain of 21 dB, an output 1-dB compression point of 11 dBm, and an Output 3rd-Order Intercept (OIP3) of 18 dBm all at 2 GHz. The coefficients of the envelope model of the power amplifier were obtained

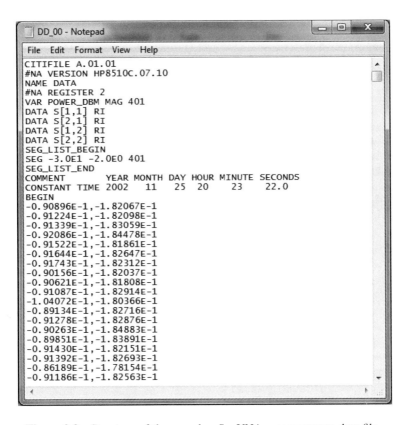

Figure 9.3 Structure of the complex S_{21} VNA measurements data file.

by fitting a fifth-order polynomial to the measured characteristics using classical least squares polynomial fitting.

```
% File Name:
clear all
% Specify the polynomial order
N=5;
% Load measured S21 measured data
load DD200.txt;
GainI=DD200(:,1);
GainQ=DD200(:,2);
% Create the complex gain vector
Pin=[-30:0.075:-0]';    % Input power vector
Gain=GainI+j*GainQ; % Complex gain
Po=Pin+20*log10(abs(Gain)); % Output power vector
% Generate input voltage vector
Vin=(sqrt(.1*10.^(Pin/10)));
V_neg=-Vin;
V_neg=flipud(V_neg);
X=[V_neg; 0; Vin];
% Generate output voltage vector
y=Gain.*Vin;
y_neg=-y;
y_neg=flipud(y_neg);
Y=[y_neg; 0; y];
% Generate the Vandermonde matrix (Phi)
Phi=[];
for k=1:length(X)
    B=[];
    for n=1:2:N
        B=[B X(k).*abs(X(k)).^(n-1)];
    end
    Phi=[Phi;B];
end
% Estimate the polynomial coefficients using LS
b=(inv(Phi'*Phi))*(Phi'*Y);

% Compute the model output
g=0;
Y_fit=zeros(size(X));
for n=1:2:N
    g=g+1;
    Y_fit=Y_fit+b(g)*X.*abs(X).^(n-1);
end
P_in=10*log10((abs(X).^2)/.1);
Po_fit=10*log10(abs(Y_fit).^2/.1);
ang_fit=(180/pi)*atan(imag(Y_fit)./real(Y_fit));
ang0=(180/pi)*atan(GainQ./GainI);

% Plot measured and fitted characteristics
figure
```

```
plot(Pin,Po,P_in,Po_fit,'k-.')
figure
plot(P_in,ang,Pin,ang0,'k-.')
```

Figure 9.4 shows measured and fitted AM−AM and AM−PM characteristics of a nonlinear amplifier and Table 9.1 shows the complex coefficients of the polynomial model.

In simulation of nonlinear systems, if measured data is not available, a simulated nonlinearity is usually obtained using any of the empirical nonlinear models described in Chapter 3. The same procedure in the previous subsection is followed except that complex gain data are generated from simulated nonlinearity. As an example where the amplifier characteristics are obtained using the Saleh model is shown in Figure 9.5. The parameters of the Saleh model were chosen to be $\alpha_a = 5$, $\beta_a = 5$, $\alpha_\theta = 12$, $\beta_\theta = 12$. A polynomial of order 5 was fitted to the complex data using the above formulation and a set of envelope coefficients (b_n) were obtained. Table 9.2 shows the envelope coefficients of the envelope power-series model developed from the nonlinear characteristics generated from Saleh model.

Figure 9.6 shows the condition number vs. the polynomial order of the polynomial model developed in the previous subsection. It is clear that increasing the polynomial order results in a higher condition number, which means lower stability of the model.

Figure 9.7 shows the NMSE versus the polynomial order of the polynomial model developed in the previous subsection, where it is clear that increasing the model order results in better accuracy of the model.

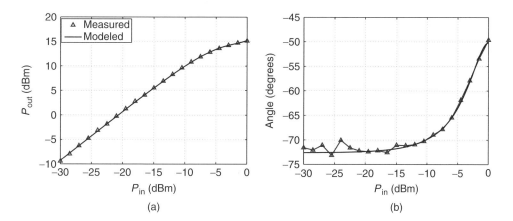

(a)   (b)

Figure 9.4 Polynomial fit to measured data: (a) AM−AM and (b) AM−PM.

Table 9.1 Complex polynomial coefficients

Coefficient	Value
b_1	$3.2607 - 10.4127i$
b_3	$1.8762e + 001 + 1.0952e + 002i$
b_3	$-1.4625e + 002 - 4.9562e + 002i$

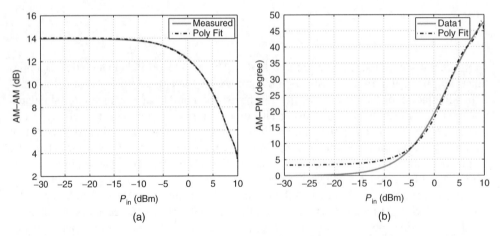

Figure 9.5 Polynomial fit to AM−AM and AM−PM characteristics generated using Saleh Model.

Table 9.2 Envelope power-series coefficients developed from the nonlinear characteristics generated from Saleh model

Coefficient	Value
b_1	$5.0235 + 0.2890i$
b_3	$-13.8939 + 12.9397i$
b_5	$22.8705 - 39.1527i$
b_7	$-19.6424 + 45.5835i$
b_9	$6.6778 - 18.6326i$

9.2.2 Fitting a Three-Box Model to Measured Data

Parameter extraction of the three-box model was discussed in Chapter 3 where the transfer functions of the linear filters can be obtained by measuring the gain characteristics at saturation $H_{\text{sat}}(f)$ (for example at the 1-dB compression point), and measuring the small signal linear frequency response $H_{\text{ss}}(f)$.

As an example, measured amplifier characteristics of the amplifier used in the previous subsection are used to develop the parameters of the three box model as shown in Gharaibeh and Steer (2005). The transfer functions $H_{\text{ss}}(f)$ and $H_{\text{sat}}(f)$ are obtained from S_{21} of the amplifier at a low power level (-20 dBm) and at the 1-dB compression point (-8 dBm input power approximately), see Figure 9.8. The transfer functions, of the linear filters, $H_1(f)$ and $H_2(f)$, can be computed from the measured filter transfer functions, as shown in Chapter 3. Figure 9.9 shows the measured frequency responses $H_{\text{ss}}(f)$ and $H_{\text{sat}}(f)$. One of the assumptions behind the model extraction procedure is that the phase-frequency response is linear. This response is shown in Figure 9.10 for the amplifier under test, where it is seen that it is indeed linear across the band.

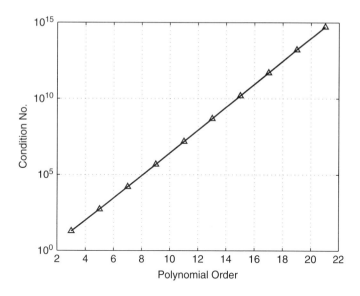

Figure 9.6 Condition number vs. polynomial order (N).

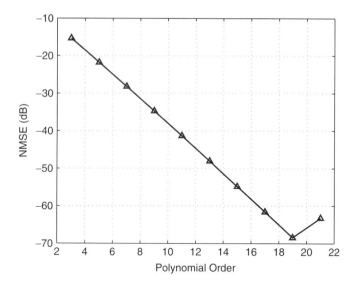

Figure 9.7 Modeling error (NMSE) vs. polynomial order (N).

The static nonlinearity, the middle block in Figure 3.4(b), is extracted as the measured single-tone AM–AM and AM–PM characteristics taken at a reference carrier frequency f_{ref}. The required input-output characteristics of the nonlinear block are the memoryless nonlinear gain characteristics (instantaneous response) given by Equation (4.39). The reference AM–AM and AM–PM characteristics are chosen to be measured at the middle

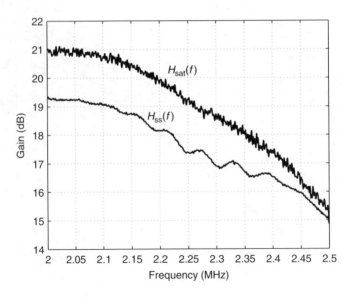

Figure 9.8 Three-box model filter frequency response.

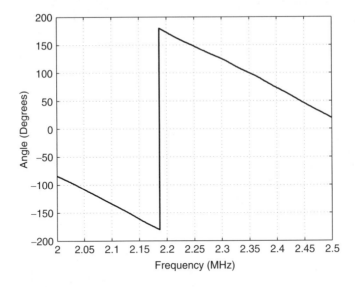

Figure 9.9 Phase response of the PA.

frequency of the band of interest. The choice of the reference is arbitrary when the phase-frequency response is linear. The coefficients of the reference static nonlinearity are obtained by measuring the AM–AM and AM–PM characteristics at the reference frequency. A polynomial of order 7 is then fitted to the complex data and a set of envelope coefficients (b_n) are obtained.

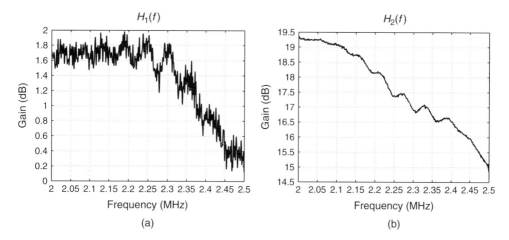

Figure 9.10 Frequency response of the linear filters in a three-box model; (a) input filter and (b) output filter.

The following code develops the parameters of the three-box model from measured data. Measured AM–AM and AM–PM characteristics of the amplifier were done at a reference frequency $f_{ref} = 2.25$ GHz. The frequency response of the amplifier was measured in the range 2–2.5 GHz and at two input power levels: -20 dBm (small-signal response) and –6 dBm (near saturation) as shown in Figure 9.8 and Figure 9.9, which shows the phase response of the amplifier. These measurements are used to develop the filter response of the input and output filters of the three-box model, as shown in Figure 9.10. The code also predicts single-tone characteristics of the amplifier at different frequencies other than the reference frequency as shown in Figure 9.11. The three-box model is seen to capture the single-tone behavior of the amplifier and should be entirely satisfactory when used to characterize the multichannel response of the amplifier.

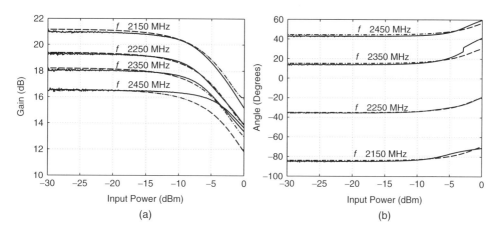

Figure 9.11 Predicted nonlinear characteristics at different frequencies from the three-box model; (a) AM–AM and (b) AM–PM.

```
% File Name: Behavmod_PolynomialFit_3box.m
clear all
% close all
%% Calculate the frequency response of the linear filters
% Load measured frequency response of the amplifier at
% saturation (Pin=-6 dBm)
load DD_F60.txt
Hsat=(DD_F60(:,1)+1i*DD_F60(:,2));
% Calculate the frequency response of the output filter
H2=interp(Hsat,32);
% Load measured frequency response of the amplifier at
% small signal (Pin=-20 dBm)
load DD_F200.txt
Hs=DD_F200(:,1)+1i*DD_F200(:,2);
Hss=interp(Hs,32);
% Calculate the frequency response of the input filter
H1=Hss./H2;
% Compute the frequency vector for piloting the frequency response
delf=500/length(H2);
f=2000:500/length(H2):2500-500/length(H2);
% Plot the frequency response of the input and output filters
figure
plot(f/1000,20*log10(abs(Hss)))
hold on
plot(f/1000,20*log10(abs(H2)),'r')
figure
plot(f/1000,20*log10(abs(H2)))
figure
plot(f/1000,20*log10(abs(H1)))
% Find the frequency response of the amplifier at the
% reference frequency (2.25GHz)
df=1;
f_ref=2250;
fs_ref=find(f>f_ref-df  &  f<f_ref+df);
Si=(mean(H1(fs_ref)));  % Input filter
So=(mean(H2(fs_ref)));  % Output filter
%%%%%%%%%%%%%%%%%%%%%%%%%%%%%%%%%%%%%%%%%%%%%%%%%%%%%%%%%%%
%% Polynomial fit at reference frequency (f=2.25 GHz)
% Specify the polynomial order
N=5;
% Load measured AM-AM and AM-PM data
load DD_225.txt
GainI=DD_225(:,1);
GainQ=DD_225(:,2);
Pin=[-30:0.075:-0]';   % Input Power vector
Gain=GainI+j*GainQ; % Complex gain
Po=Pin+20*log10(abs(Gain)); % Output power
% Generate input voltage vector
Vin=(sqrt(.1*10.^(Pin/10)));
```

```
V_neg=-Vin;
V_neg=flipud(V_neg);
X=[V_neg; 0; Vin];
% Generate output voltage vector
y=Gain.*Vin;
y_neg=-y;
y_neg=flipud(y_neg);
Y=[y_neg; 0; y];
% Generate the Vandermonde matrix Phi
Phi=[];
for k=1:length(X)
    B=[];
    for n=1:2:N
        B=[B X(k).*abs(X(k)).^(n-1)];
    end
    Phi=[Phi;B];
end
% Estimate the polynomial coefficients at the reference
% frequency using LS
b=(inv(Phi'*Phi))*(Phi'*Y);
%%%%%%%%%%%%%%%%%%%%%%%%%%%%%%%%%%%%%%%%%%%%%%%%%%%%%%%%%%%%%%%
%% Calculate the model output at another frequency (2GHz)
% Load measured AM-AM and AM-PM at another frequency (2GHz)
load DD_200.txt
GainI=DD_200(:,1);
GainQ=DD_200(:,2);
Pin=[-30:0.075:-0]';% Measured Input Power vector
Gain=GainI+j*GainQ;   % Measured Complex gain vector
Ang=(180/pi)*atan(GainQ./GainI); % Measured angle vector
Po=Pin+20*log10(abs(Gain)); % Measured output power vector
% Calculate the frequency response of the amplifier at the
% new frequency (2GHz)
df=1;
f1=2000;
fs1=find(f>f1-df  & f<f1+df);
Si1=(mean(H1(fs1)))/Si;
So1=(mean(H2(fs1)))/(So);
% Generate input voltage vector
Vin=(sqrt(.1*10.^(Pin/10)));
V_neg=-Vin;
V_neg=flipud(V_neg);
X=[V_neg; 0; Vin];
% Scale the input by the frequency response of the input filter
X1=X*Si1;
% Compute the model output
g=0;
Y_fit=zeros(size(X1));
for n=1:2:N
```

```
    g=g+1;
    Y_fit=Y_fit+b(g)*X1.*abs(X1).^(n-1);
end
% Scale the output by the frequency response of the output filter
Y_fit1=Y_fit*So1;
% Model Input power vector
Pin1=10*log10((abs(X).^2)/.1);
% Model output power vector
Po_fit1=10*log10(abs(Y_fit1).^2/.1);
% Model output angle vector
Ang_fit1=(180/pi)*atan(imag(Y_fit1)./real(Y_fit1));
% Compare measured and predicted characteristics
figure(1)
plot(Pin,Po-Pin,Pin1,Po_fit1-Pin1,'k-.')
hold on
figure(2)
plot(Pin,Ang,Pin1,Ang_fit1,'k-.')
hold on
```

9.2.3   Fitting a Memory Polynomial Model to a Simulated Nonlinearity

A memory polynomial model is represented as (Morgan *et al.*, 2006):

$$y(n) = \sum_{k=1}^{K} \sum_{m=0}^{P-1} a_{km} x(n-m)|x(n-m)|^{k-1}$$

$$= \sum_{k=1}^{K} \sum_{m=0}^{P-1} a_{km} u^{k}(n-m) \qquad (9.6)$$

where $u(k)(n) = x(n)|x(n)|^{k-1}$ and P is the memory depth.

With memory polynomials, single-tone signals cannot be used as test signals as they represent a single frequency while the model exhibits wideband effects. To develop the parameters of a memory polynomial, a WCDMA signal is used as a test signal where a nonlinearity with memory is applied and the extraction of the memory polynomial coefficients is based on fitting the model to time-domain data.

The following script uses the formulation in Chapter 3 to develop the memory polynomial coefficients using a simulated nonlinearity with memory based on a filter-nonlinearity model. The input/output data is generated at a given input power level and applying a linear filter succeeded by a memoryless nonlinearity (based on the Saleh model) to the input WCDMA signal. Figure 9.12 shows the predicted output of the nonlinearity and the predicted output using a memory polynomial with $N = 3$ and $P = 5$. Table 9.3 shows the coefficients of the memory polynomial.

```
% File Name:
clear all;
% Specify the memory order of the memory polynomial
P=5;
```

```
% Specify the polynomial order
N=5;
% Specify the Saleh model parameters
u1=5;
v1=2.3;
u2=20;
v2=20;
G=u1;
%%%%%%%%%%%%%%%%%%%%%%%%%%%%%%%%%%%%%%%%%%%%%%%%%%%%%%%%%%%%%%%%%
% Load and power scale a WCDMA signal
load  WCDMA_DN_5DPDCH.mat
x0=double(Y);
P1=10*log10(mean(abs(x0).^2)/(50*.001));
Pscale=10^(P1/10);
x1=sqrt(1/Pscale)*(x0);
Pin=0; %dBm
ps =sqrt(10^(Pin/10));
x=ps*(x1(1:64000));
% Specify the linear filter coefficients
bb = [0 1 0 0];
aa = [1 -0.6 0.08];
% Compute the output signal of Wiener-Hammerstein model
y0 =filter(bb,aa,x);
y = Saleh(y0,u1,v1,u2,v2);
% Compute the regressor_matrix of the memory polynomial model
X = zeros(length(x),P*(N+1)/2);
idx = 1;
for k = 0:P-1
    % Delay signal with k samples
    x_kp = [zeros(k,1); x(1:end-k)];
    for n = 1:2:N
        X(:,idx) = x_kp.*abs(x_kp).^(n-1);
        idx = idx+1;
    end
end
% Find the coefficients of the memory polynomial
a_kp = pinv(X)*y;
% Compute the model output
y_model = X*a_kp;
% Plot the true and the modeled output
t=[0:length(x)-1]*XDelta;
plot(t,G*x,t,y,t,y_model)
```

9.3   Autocorrelation and Spectrum Estimation

9.3.1   Estimation of the Autocorrelation Function

As discussed in Appendix B, the power spectral density of random signals is computed from the Fourier transform of the autocorrelation function per the Wiener–Khinchin Theorem (Papoulis, 1994). The autocorrelation function of stationary random processes

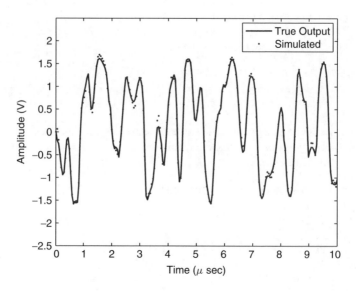

Figure 9.12 Predicted output signal of a memory polynomial.

Table 9.3 Memory polynomial coefficients

Coefficient	Value
b_{11}	$-0.28 - 0.14i$
b_{31}	$4.19 + 0.98i$
b_{51}	$-12.03 + 3.46i$
b_{12}	$2.96 + 2.05i$
b_{32}	$-21.77 - 3.11i$
b_{52}	$57.09 - 13.92i$
b_{13}	$5.53 + 2.97i$
b_{33}	$-51.82 + 6.45i$
b_{53}	$185.95 - 58.16i$
b_{14}	$-0.94 - 0.124i$
b_{34}	$-2.15 + 4.92i$
b_{54}	$1.12 + 1.35i$
b_{15}	$0.92 + 1.55i$
b_{35}	$0.85 - 1.86i$
b_{55}	$22.8705 - 39.1527i$

can be obtained from their time averages assuming ergodicity (Gard *et al.*, 2005). There-fore, the autocorrelation function can be evaluated from the second-order time average of the signal as

$$R_{xx}(\tau) = E[x(t)x^*(t+\tau)] = \lim_{T \to \infty} \frac{1}{2T} \int_{-T}^{T} x(t)x^*(t+\tau)dt. \qquad (9.7)$$

In discrete time, the autocorrelation function can be written as:

$$R_{xx}(m) = E[x(n+m)x^*(n)] = E[x(n)x^*x(n-m)] \tag{9.8}$$

The estimation of the autocorrelation in MATLAB® is done under the assumption that most communication signals are ergodic, and hence the autocorrelation function can be estimated from a single signal realization.

The function XCORR produces an estimate of the autocorrelation of a random (stationary) sequence over a range of lags:

```
[Rxx,LAGS] = XCORR(x,x,SCALEOPT,MAXLAG)
```

where

```
SCALEOPT:
      'biased'    - scales the raw cross-correlation by 1/M.
      'unbiased' - scales the raw correlation by 1/(M-abs(lags)).
      'coeff'     - normalizes the sequence so that the
                    auto-correlations at zero lag are identically 1.0.
      'none'      - no scaling (this is the default).
```

The range of lags is: -MAXLAG to MAXLAG, that is, 2*MAXLAG+1 lags. If missing, the default is MAXLAG = M-1.

9.3.2 Plotting the Signal Spectrum

The PSD of the input and output signals of a nonlinearity can be computed from the Fourier transform of the autocorrelation function as shown in Chapter 6. The following script computes the PSD of a WCDMA signal and plots it versus frequency. The PSD is computed from the FFT of the computed autocorrelation functions. A Hanning window is used to remove spectral leakage. The signal is normalized to 0 dBm in order to allow scaling to any power level and then the autocorrelation estimate is calculated using the xcorr function.

```
% File Name:
clear all
close all
% Load a WCDMA signal
load WCDMA_UP_TM1_16DPCH.mat
% Specify the sampling time of the signal
XDelta=4.8828e-008; % Sampling time
x0=double(Y(1:256000));
% Normalize power to 0 dBm; Assume 50 ohm impedance;
P=10*log10(mean(abs(x0).^2)/(50*.001));
Pscale=10^(P/10);
x1=sqrt(1/Pscale)*(x0); % normalized signal to 0 dBm
% Set input power level (dBm)
Pin=-10; % dBm
% Scale the amplitude of the signal
```

```
ps =sqrt(10^(Pin/10)); % amplitude scaling
x=ps*(x1); % scaled signal
% Compute the ACF of the signal
Rx=xcorr(x,x,2^10,'biased'); % Autocorrelation function
% Compute the PSD of the signal
len=length(Rx);
Sx=fftshift(fft(hanning(len).*Rx/len)); % Power spectral density
SxdB=10*log10(Sx/0.05);  % Power spectral density in dBm
len=length(Rx);
m=len/2;
% Create a frequency vector
Fs=1/XDelta;
delf=Fs*1e-6/len;% Frequency resolution normalized to 1 MHz
xx=-(m):(m-1);
f=delf*xx; % Frequency vector
% Plot the PSD of the signal
figure
plot(f,(SxdB))
axis([-5 5 -70 -30])
```

MATLAB® has also built-in functions for computing the PSD of communication signals. The following script plots the Welch PSD of a WCDMA signal. Figure 9.13 shows the resulting PSD of a WCDMA signal.

```
hPsd = spectrum.welch('Hamming',1024);
hopts = psdopts(hPsd);
set(hopts,'SpectrumType','twosided','NFFT',1024,'Fs',Fs,...
,'CenterDC',true)
PSD1 = psd(hPsd,Signal,hopts);
data = dspdata.psd([PSD1.Data],PSD1.Frequencies,'Fs',Fs);
plot(data)


hPsd =
    EstimationMethod: 'Welch'
       SegmentLength: 1024
      OverlapPercent: 50
          WindowName: 'Hamming'
        SamplingFlag: 'symmetric'


hopts =
             FreqPoints: 'All'
                   NFFT: 1024
    NormalizedFrequency: false
                     Fs: 2.0480e+007
           SpectrumType: 'Twosided'
              CenterDC: true
              ConfLevel: 'Not Specified'
           ConfInterval: []


data =
```

```
             Name: 'Power Spectral Density'
             Data: [1024x1 double]
     SpectrumType: 'Onesided'
NormalizedFrequency: false
               Fs: 2.0480e+007
      Frequencies: [1024x1 double]
        ConfLevel: 'Not Specified'
     ConfInterval: []
```

9.3.3   Power Measurements from a PSD

To compute the power within a certain bandwidth of the power spectrum, the indices of the frequency components corresponding to that bandwidth are found within a frequency vector using the `find` command. The power within that bandwidth is simply the sum of absolute values of the PSD points corresponding those indexes. The following script calculates the power in the main channel and the power in the adjacent channel of the spectrum generated in the previous subsection.

```
% File Name:
clear all
close all
% Load a WCDMA signal
load WCDMA_UP_TM1_16DPCH.mat
% Specify the sampling time of the signal
XDelta=4.8828e-008; % Sampling time
x0=double(Y(1:256000));
% Normalize power to 0 dBm; Assume 50 ohm impedance;
P=10*log10(mean(abs(x0).^2)/(50*.001));
Pscale=10^(P/10);
x1=sqrt(1/Pscale)*(x0); % normalized signal to 0 dBm
% Set input power level (dBm)
Pin=-10; % dBm
% Scale the amplitude of the signal
ps =sqrt(10^(Pin/10)); % amplitude scaling
x=ps*(x1); % scaled signal
Fs=1/XDelta;
% Compute the ACF of the signal
Rx=xcorr(x,x,2^10,'biased'); % Autocorrelation function
% Compute the PSD of the signal
len=length(Rx);
Sx=fftshift(fft(hanning(len).*Rx/len)); % Power spectral density
SxdB=10*log10(Sx/0.05);  % Power spectral density in dBm

% Create a frequency vector
len=length(Rx);
m=len/2;
Fs=1/XDelta;
delf=Fs*1e-6/len;% Frequency resolution normalized to 1 MHz
xx=-(m):(m-1);
f=delf*xx; % Frequency vector
```

```
% Compute output power within a bandwidth specified
% by two frequencies f1 and f2
freq_bw=find(f>f1  & f<f2);
Po_main=10*log10(sum(abs(Sx(freq_bw))/.05))
```

9.4 Spectrum of the Output of a Memoryless Nonlinearity

9.4.1 Single Channel

The analysis presented in Chapter 6 requires the estimation of higher-order autocorrelation functions and spectra for the effective evaluation of nonlinear distortion. Therefore, the autocorrelation functions in Equation (5.14) can be evaluated from the time average of powers of the input signal as

$$R_{x_n x_m}(\tau) = \lim_{T\to\infty} \frac{1}{2T} \int_{-T}^{T} x_1^{\frac{(n+1)}{2}} x_1^{*\frac{(n-1)}{2}} x_2^{\frac{(n-1)}{2}} x_2^{*\frac{(n+1)}{2}} \, dt. \tag{9.9}$$

where $x_1 = x(t)$ and $x_2 = x(t + \tau)$. To evaluate the higher-order ACFs, a similar approach to the one presented in the previous subsection can be used. Therefore, using the function xcorr, the ACF in Equation (9.9) can be evaluated as

```
[Rnm,LAGS] = XCORR((x^((n+1)/2)).*conj(x)^((n-1)/2),...
conj((x^((m-1)/2)).*conj(x)^((m+1)/2)),SCALEOPT,MAXLAG)
```

where $n = 1 : N$.

The higher-order PSDs can be evaluated from the Fourier transform of the ACF as

```
% Compute higher order PSD's
Snm=fftshift(fft(hanning(len).*Rnm/len));
```

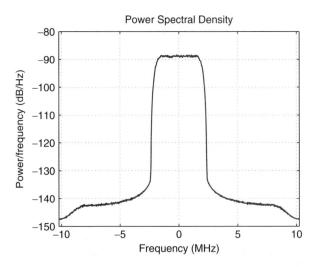

Figure 9.13 PSD of a WCDMA signal.

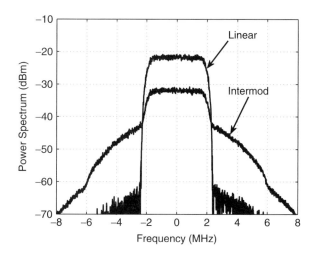

Figure 9.14 Output spectrum partitioned into linear and intermodulation components.

The following script computes and plots the power spectrum of the output of a nonlinearity by computing the higher-order spectra. The output power spectrum consists of linear, gain compression and intermodulation distortion spectra as shown in Figure 9.14.

```
% File Name:
clear all
close all
% Load a WCDMA signal
load WCDMA_UP_TM1_16DPCH.mat
% Specify the sampling time of the signal
XDelta=4.8828e-008; % Sampling time
x0=double(Y(1:256000));
% Normalize power to 0 dBm; Assume 50 ohm impedance;
P=10*log10(mean(abs(x0).^2)/(50*.001));
Pscale=10^(P/10);
x1=sqrt(1/Pscale)*(x0); % normalized signal to 0 dBm
z=x1(1:128000);
cz=conj(z);
% Specify the bandwidth of the input signal
wcdma_bw=3.84;
% Set the nonlinear order
N=5;
% Load the envelope coefficients
load envcoef
b1(1)=b(1);
b1(3)=b(2);
b1(5)=b(3);
% Set the width of autocorrelation window length
W=2^11;
% Compute the powers of the signal to be used in computation of
```

```
% the higher order ACF's
t0=clock;
A=[];
for n=1:2:N
    for m=1:2:N
        for l=0:(n-1)/2
            for k=0:(m-1)/2
                A=[A;((n+1)/2) ((n-1)/2) ((m-1)/2) ((m+1)/2)];
            end
        end
    end
end
% Compute the higher order ACF's
for k=1:size(A,1);
    R(k,:)=xcorr((z.^A(k,1)).*(cz.^A(k,2)),...
    conj((z.^A(k,3)).*(cz.^A(k,4))),W,'biased');
end
% Compute the higher order PSDs
len=size(R,2);
for k=1:size(A,1);
    S(k,:)=fftshift(fft(hanning(len)'.*R(k,:)/len));
    k;
end
% Sweep the power of the signal and compute the PSD
    Pin=-5:2:-5;
    for w=1:length(Pin)
        ps1=sqrt(10^((Pin(w))/10));
        h3=0;
        h2=0;
        h1=0;
        for n=1:2:N
            for m=1:2:N
                        h1=h1+1;
                        C=(ps1^(n+m))*b1(n)*conj(b1(m));
                        % Total spectrum
                        Sz(h1,:)=C*S(h1,:);
                        % PSD of linear term with gain compression
                        if (n==1 | m==1)
                            h2=h2+1;
                            Sz1(h2,:)=C*S(h1,:);
                        end
                        % PSD of intermodulation
                        if n~=1  & k~=1
                            h3=h3+1;
                            Sz2(h3,:)=C*S(h1,:);
                        end
                        h1;
            end
        end
        Sy=sum(Sz,1);   % The total spectrum
```

```
              Sl=sum(Sz1,1); % Linear and gain compression spectrum
              Sd=sum(Sz2,1); % Intermodulation distortion spectrum
              % Compute the spectrum in dBm
              Syd(:,w)=10*log10(((Sd))/(50*.001));
              Syl(:,w)=10*log10(((Sl))/(50*.001));
              Syy(:,w)=10*log10(((Sy))/(50*.001));
              % Create a frequency vector
              len=length(Szt);
              m=len/2;
              Fs=1/XDelta;
              delf=Fs*1e-6/len;% Frequency resolution normalized to 1 MHz
              xx=-(m):(m-1);
              f=delf*xx; % Frequency vector
              % Compute the inband frequency vector
              freq_bw=find(f>-wcdma_bw/2  & f<+wcdma_bw/2);
              % Compute the inband power
              % Total power
              Po_tot(w)=10*log10(sum(abs(Sy(freq_bw))))/.05);
              % Linear output power
              Po_lin(w)=10*log10(sum(abs(Sl(freq_bw))))/.05);
              % Intermodulation power
              Po_imod(w)=10*log10(sum(abs(Sd(freq_bw))))/.05);
         end
% Plot the PSD of the signal
figure
plot(f,Syd,'r',f,Syl,'k')
axis([-5 5 -90 -10])
```

9.4.2 *Two Channels*

For the case of multichannel systems, the following script computes the higher-order spectra and evaluates the PSD of the output of a nonlinearity at the center frequency of one of the input signals. The output spectrum in this case consists of linear, gain compression, intermodulation and cross-modulation spectral components.

```
% File Name:
load WCDMA_UP_TM1_16DPCH.mat
% Specify the sampling time of the signal
XDelta=4.8828e-008; % Sampling time
x0=double(Y(1:256000));
% Normalize power to 0 dBm; Assume 50 ohm impedance;
P=10*log10(mean(abs(x0).^2)/(50*.001));
Pscale=10^(P/10);
x1=sqrt(1/Pscale)*(x0); % normalized signal to 0 dBm
z=x1(1:128000); % The first WCDMA signal
u=z;            % The second WCDMA signal
cz=conj(z);
cu=conj(u);
```

```
wcdma_bw=3.84;
% Set the nonlinear order
N=5;
% Load the envelope coefficients
load envcoef
b1(1)=b(1);
b1(3)=b(2);
b1(5)=b(3);
% Find the cross envelope coefficients
for n=1:2:N
    h=0;
    for l=0:(n-1)/2
        h=h+1;
        c(n,h)=b1(n)*factorial(n)/...
        (factorial(((n-1)/2)-l)*factorial(((n+1)/2)-l)...
        *factorial(l)*factorial(l));
    end
end
% Set the width of autocorrelation window length
W=2^11;
% Compute the powers of the signal to be used in computation
% of the higher order ACF's
t0=clock;
A=[];
for n=1:2:N
    for m=1:2:N
        for l=0:(n-1)/2
            for k=0:(m-1)/2
                A=[A;((n+1)/2)-l ((n-1)/2)-l ((m-1)/2)-k...
                  ((m+1)/2)-k l l k k];
            end
        end
    end
end
% Compute the higher order ACF's
for k=1:size(A,1);
    R(k,:)=xcorr((z.^A(k,1)).*(cz.^A(k,2)).*(u.*cu).^A(k,5),...
    conj((z.^A(k,3)).*(cz.^A(k,4)).*(u.*cu).^A(k,7)),W,'biased');
end
% Compute the higher order PSDs
len=size(R,2);
for k=1:size(A,1);
    S(k,:)=fftshift(fft(hanning(len)'.*R(k,:)/len));
    k
end
% Sweep the power of the two signals and compute the PSD
for Pin2=-150:5:-150;
    Pin1=-15:2:-15;
    for w=1:length(Pin1)
        ps2=sqrt(10^((Pin2)/10));
```

```
ps1=sqrt(10^((Pin1(w))/10));
h3=0;
h2=0;
h1=0;
h=0;
for n=1:2:N
    for m=1:2:N
        for l=0:(n-1)/2
            for k=0:(m-1)/2
                n1=((n+1)/2)-l;
                m1=((m+1)/2)-k;
                h=h+1;
                B=c(n,l+1)*conj(c(m,k+1));
                C=(ps1^(n+m-2*l-2*k))*(ps2^(2*l))...
                *(ps2^(2*k))*B;
                % Total spectrum
                Sz(h,:)=C*S(h,:);
                % PSD of linear term with gain compression
                if l==0  & k==0  & (n==1 | m==1)
                    h1=h1+1;
                    Sz1(h1,:)=C*S(h,:);
                end
                % PSD of linear term with intermodulation
                if l==0  & k==0
                    h2=h2+1;
                    Sz2(h2,:)=C*S(h,:);
                end
                h;
            end
        end
    end
end
Szt=sum(Sz,1);   % The total spectrum
Szt1=sum(Sz1,1); % Linear and gain compression spectrum
Szt2=sum(Sz2,1);
Sdt=Szt2-Szt1;   % Distortion spectrum
Sct=Szt-Szt2;    % Cross modulation spectrum
% Compute the spectrum in dBm
Syd(:,w)=10*log10(((Szt2-Szt1))/(50*.001));
Syc(:,w)=10*log10(((Szt-Szt2))/(50*.001));
Syl(:,w)=10*log10(((Szt1))/(50*.001));
Syy(:,w)=10*log10(((Szt))/(50*.001));
% Create a frequency vector
len=length(Szt);
m=len/2;
Fs=1/XDelta;
delf=Fs*1e-6/len;% Frequency resolution normalized to 1 MHz
xx=-(m):(m-1);
f=delf*xx; % Frequency vector
% Compute the inband frequency vector
freq_bw=find(f>-wcdma_bw/2  & f<+wcdma_bw/2);
```

```
% Compute the inband power
 % total output power
Po_tot(w)=10*log10(sum(abs(Szt(freq_bw)))/.05);
% linear power
Po_lin(w)=10*log10(sum(abs(Szt1(freq_bw)))/.05);
% intermodulation power
Po_imod(w)=10*log10(sum(abs(Sdt(freq_bw)))/.05);
% crossmodulation power
Po_cross(w)=10*log10(sum(abs(Sct(freq_bw)))/.05);

    end
end
% Plot the PSD of the signal
figure
plot(f,Syl,f,Syd,'r',f,Syc,'k')
```

Figure 9.15 shows the output spectrum at one of the carriers divided into linear with gain compression, intermodulation and cross-modulation components. The two signals are set to an input power of $-10\,$dBm.

9.5 Spectrum of the Output of a Nonlinearity with Memory

9.5.1 Three-Box Model

The spectrum of the output of the three-box model can be developed in a similar way to memoryless nonlinearity except that the output spectrum needs to be multiplied by the squared magnitude of the frequency response of the linear filers incorporated in the nonlinear model. In the following code, the output spectrum of a three-box model is generated using the derivations of the output spectrum presented in Chapter 6 and measured

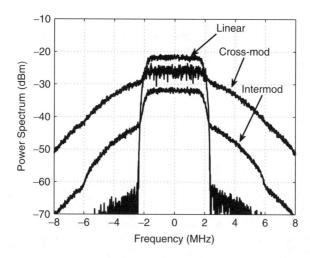

Figure 9.15 Output spectrum partitioned into linear, intermodulation and cross-modulation components.

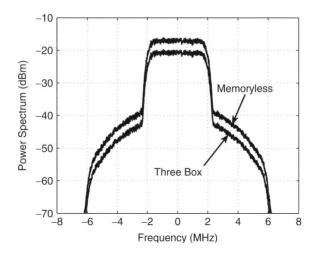

Figure 9.16   Output spectrum of a three-box model.

filter frequency responses. Figure 9.16 shows the output spectrum of a three-box model for a single WCDMA signal input.

```
% File Name:
clear all
% close all
% Load a WCDMA signal
load WCDMA_UP_TM1_16DPCH.mat
% Specify the sampling time of the signal
XDelta=4.8828e-008;      % Sampling time
x0=double(Y(1:256000));
% Normalize power to 0 dBm; Assume 50 ohm impedance;
P=10*log10(mean(abs(x0).^2)/(50*.001));
Pscale=10^(P/10);
x1=sqrt(1/Pscale)*(x0); % normalized signal to 0 dBm
x=x1(1:256000);
cx=conj(x);
% Specify the bandwidth of the input signal
wcdma_bw=3.84;
% Set the nonlinear order
N=5;
% Load the envelope coefficients at reference frequency
load envcoef_ref
b1(1)=b(1);
b1(3)=b(2);
b1(5)=b(3);
%%%%%%%%%%%%%%%%%%%%%%%%%%%%%%%%%%%%%%%%%%%%%%%%%%%%%%%%%%%%%%%%
% Load input and output filters frequency responses
load H1
load H2
fh=f;
```

```
delf=delf;
% Set the carrier frequency of the WCDMA signal
fc=2446;
% Bandwidth corresponding to the WCDMA signal bandwidth
bw=find(fh>fc-wcdma_bw/2  & fh<fc+wcdma_bw/2);
% Small bandwidth around reference frequency
df=find(fh>f_ref-10*delf  & fh<f_ref+10*delf);
% Calculate the frequency response the input and output
% filter within signal bandwidth
J1=mean(abs(H1(bw)))./abs(mean(H1(df)));
J2=mean(abs(H2(bw)))./abs(mean(H2(df)));
% Set the width of autocorrelation window length
W=2^11;
% Compute the powers of the signal to be used in computation
% of the higher order ACF's
t0=clock;
A=[];
for n=1:2:N
    for m=1:2:N
        for l=0:(n-1)/2
            for k=0:(m-1)/2
                A=[A;((n+1)/2) ((n-1)/2) ((m-1)/2) ((m+1)/2)];
            end
        end
    end
end
% Compute the higher order ACF's
for k=1:size(A,1);
    R(k,:)=xcorr((x.^A(k,1)).*(cx.^A(k,2)),...
    conj((x.^A(k,3)).*(cx.^A(k,4))),W,'biased');
    k;
end
% Compute the higher order PSDs
len=size(R,2);
for k=1:size(A,1);
    S(k,:)=fftshift(fft(hanning(len)'.*R(k,:)/len));
    k;
end
% Sweep the power of the signal and compute the PSD
Pin=-7:2:-7;
for w=1:length(Pin)
    ps1=sqrt(10^((Pin(w))/10));
    h3=0;
    h2=0;
    h1=0;
    for n=1:2:N
        for m=1:2:N
            h1=h1+1;
            C=(J1*ps1^(n+m))*b1(n)*conj(b1(m));
            % Total spectrum
            Sx(h1,:)=C*S(h1,:);
```

```
        % PSD of linear term with gain compression
        if (n==1 | m==1)
            h2=h2+1;
            Sx1(h2,:)=C*S(h1,:);
        end
        % PSD of intermodulation
        if n~=1  & k~=1
            h3=h3+1;
            Sx2(h3,:)=C*S(h1,:);
        end
        h1;
    end
end
Sy=sum(Sx,1).*(J2).^2; % Total
Sl=sum(Sx1,1).*(J2).^2;% Linear and gain compression
Sd=sum(Sx2,1).*(J2).^2;% Intermodulation distortion
% Compute the spectrum in dBm
Syd(:,w)=10*log10(((Sd))/(50*.001));
Syl(:,w)=10*log10(((Sl))/(50*.001));
Syy(:,w)=10*log10(((Sy))/(50*.001));
% Create a frequency vector
len=length(Sy);
m=len/2;
Fs=1/XDelta;
delf=Fs*1e-6/len;% Frequency resolution normalized to 1 MHz
xx=-(m):(m-1);
f=delf*xx;        % Frequency vector
end
% figure
plot(f,Syy,'r')
axis([-8 8 -70 -10])
hold on
```

9.5.2 *Memory Polynomial Model*

The spectrum of the output of the memory polynomial model is developed using the coefficients of the memory polynomial developed in Section 9.3.3. In the following script, the memory polynomial model is applied to a WCDMA signal and the spectrum of the output of nonlinear model (filter-nonlinearity) and the modeled output using memory polynomial are plotted. Figure 9.17 shows the spectrum of the output of the nonlinear model, the spectrum of the output of the memory polynomial model and the spectrum of the output of a memoryless power-series model. It is clear that memory polynomials give more accurate model than the memoryless polynomial when the system exhibits memory effects.

```
% File Name:
clear all;
% Specify the memory order of the memory polynomial
P=5;
```

```
% Specify the polynomial order
N=5;
% Specify the Saleh model parameters
u1=5;
v1=2.3;
u2=10;
v2=10;
G=u1;
%%%%%%%%%%%%%%%%%%%%%%%%%%%%%%%%%%%%%%%%%%%%%%%%%%%%%%%%%%%%%%%%%%%%
% Load and power scale a WCDMA signal
load  WCDMA_DN_5DPDCH.mat
x0=double(Y);
P1=10*log10(mean(abs(x0).^2)/(50*.001));
Pscale=10^(P1/10);
x1=sqrt(1/Pscale)*(x0);
Pin=0; % dBm
ps =sqrt(10^(Pin/10));
x=ps*(x1(1:64000));
% Specify the linear filter coefficients
bb = [0 1 0 0];
aa = [1 -0.6 0.08];
% Compute the output signal of a Filter-Nonlinearity model
y0 =filter(bb,aa,x);
y = Saleh(y0,u1,v1,u2,v2);
% Compute the output of the memory polynomial model
% Load the memory polynomial coefficients
load MemoryCoeffs

% Compute the model output
X = zeros(length(x),P*(N+1)/2);
idx = 1;
for k = 0:P-1
    % Delay signal with k samples
    x_kp = [zeros(k,1); x(1:end-k)];
    for n = 1:2:N
        X(:,idx) = x_kp.*abs(x_kp).^(n-1);
        idx = idx+1;
    end
end
y_model = X*a_kp;
% Plot the true and the modeled output
t=[0:length(x)-1]*XDelta;
plot(t,y,t,y_model)

% Calculate the spectrum of the true and the modeled output
Fs=1/XDelta;
hPsd = spectrum.welch('Hamming',1024);
```

```
hopts = psdopts(hPsd);
set(hopts,'SpectrumType','twosided','NFFT',1024,...
'Fs',Fs,'CenterDC',true)
PSD1 = psd(hPsd,y,hopts);
PSD2 = psd(hPsd,y_model,hopts);
data1 = dspdata.psd([PSD1.Data],PSD1.Frequencies,'Fs',Fs);
data2 = dspdata.psd([PSD2.Data],PSD2.Frequencies,'Fs',Fs);
% Plot Spectrum
figure
plot(data1)
hold on
plot(data2)
```

9.6   Spectrum of Orthogonalized Nonlinear Model

As discussed in Chapter 6, the orthogonalization of the nonlinear model consists of two main steps: the first is the computation of the orthogonal model coefficients from envelope coefficients and the second is the computation of the PSDs of the linear and distortion components. Therefore, given a set of envelope coefficients, the orthogonal model coefficients are computed by finding the correlation coefficients α_{nm} between nth- and mth-model branches using Equation (6.39). This requires the computation of the higher-order spectra as shown in the previous subsections. Figure 9.18 shows a flow chart for the computation of the orthogonal model spectra.

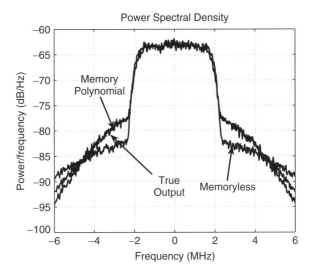

Figure 9.17   Output spectrum of a memory polynomial model.

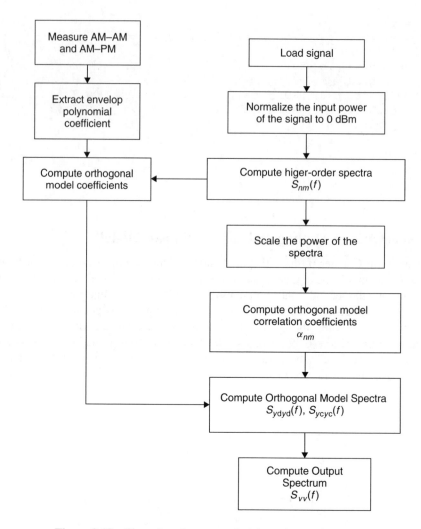

Figure 9.18   Flow chart for computing the orthogonal spectra.

The following script computes the orthogonal model coefficients of a power-series model for a WCDMA signal input.

```
% File Name: SpectrumPlot_Orthogonal.m
clear all;
% close all
load WCDMA_UP_3DPCH_SCH.mat
% load WCDMA_UP_TM1_16DPCH.mat
% Specify the sampling time of the signal
XDelta=4.8828e-008; % Sampling time
x0=double(Y(1:256000));
% Normalize power to 0 dBm; Assume 50 ohm impedance;
```

```
P=10*log10(mean(abs(x0).^2)/(50*.001));
Pscale=10^(P/10);
x1=sqrt(1/Pscale)*(x0); % normalized signal to 0 dBm
CDMA=x1(1:128000)';
% Specify the bandwidth of the input signal
wcdma_bw=3.84;
% Set the nonlinear order
N=5;
% Load envelope coefficients
load coeff.mat
coeff=b;
% Set the input power of the signal
Pin=-5;
% Compute higher order spectra
[Sz1,Sz13,Sz31,Sz33,Sz15,Sz51,Sz35,Sz53,Sz55]=SpecGen(CDMA);
% Compute orthogonal spectra
[Scc,Sdd,Syy]=OrthSpec(Pin,coeff,Sz1,Sz13,Sz31,Sz33,...
Sz15,Sz51,Sz35,Sz53,Sz55);
% Create a frequency vector
len=length(Sz1);
m=len/2;
Fs=1/XDelta;
delf=Fs*1e-6/len;% Frequency resolution normalized to 1 MHz
xx=-(m):(m-1);
f=delf*xx; % Frequency vector
% Compute the spectrum in dBm
Sy=10*log10(Syy/.05);      % total
Sd=10*log10(Sdd/.05);      % in_band
Sc=10*log10(Scc/.05);      % correlated
% Plot the correlated and uncorrelated spectra
plot(f,Sc,'r',f,Sd)
axis([-8 8 -70 0])
xlabel('Frequency (MHz)')
ylabel('Power Spectrum (dBm)')
grid
```

The functions SpecGen and OrthSpec are used to generate the higher-order spectra and the orthogonal spectra as shown in the following script.

```
% File Name: SpecGen.m
function [Sz1,Sz13,Sz31,Sz33,Sz15,Sz51,Sz35,Sz53,Sz55]...
=SpecGen(Signal);
% scale the input power of the signal to 0 dBm
Pin=10*log10(mean(abs(Signal).^2)/(50*.001));
Pscale=10^(Pin/10);
Signal=sqrt(1/Pscale)*(Signal);
% Set the width of autocorrelation window length
W=2^11;
% Compute the autocorrelation estimate
cx=transpose(Signal);
cjx=conj(cx);
```

```
r11=xcorr(cx,cx,W,'biased');
r13=xcorr(cx,conj(cx.*cjx.^2),W,'biased');
r31=xcorr(cx.^2.*cjx,conj(cjx),W,'biased');
r33=xcorr(cx.^2.*cjx,conj(cx.*cjx.^2),W,'biased');
r15=xcorr(cx,conj(cx.^2.*cjx.^3),W,'biased');
r51=xcorr(cx.^3.*cjx.^2,conj(cjx),W,'biased');
r35=xcorr(cx.^2.*cjx,conj(cx.^2.*cjx.^3),W,'biased');
r53=xcorr(cx.^3.*cjx.^2,conj(cx.*cjx.^2),W,'biased');
r55=xcorr(cx.^3.*cjx.^2,conj(cx.^2.*cjx.^3),W,'biased');
% Compute the nth order odd power spectrums
% Hanning window is used to surpress FFT spectral leakage
len=length(r11);
Sz1=fftshift(fft(hanning(len).*r11/len));
Sz13=fftshift(fft(hanning(len).*r13/len));
Sz31=fftshift(fft(hanning(len).*r31/len));
Sz33=fftshift(fft(hanning(len).*r33/len));
Sz15=fftshift(fft(hanning(len).*r15/len));
Sz51=fftshift(fft(hanning(len).*r51/len));
Sz35=fftshift(fft(hanning(len).*r35/len));
Sz53=fftshift(fft(hanning(len).*r53/len));
Sz55=fftshift(fft(hanning(len).*r55/len));


File Name: OrthSpec.m
function [Scc,Sdd,Syy]=OrthSpec(Pin,coeff,Sz1,Sz13,...
Sz31,Sz33,Sz15,Sz51,Sz35,Sz53,Sz55)
ps=sqrt(10^(Pin/10));
b1=coeff(1);
b3=coeff(2);
b5=coeff(3);
% Scale the power of the higher order spectra
Sx1x1=ps^2*Sz1;
Sx1x3=ps^4*Sz13;
Sx1x5=ps^6*Sz15;
Sx3x1=ps^4*Sz31;
Sx5x1=ps^6*Sz51;
Sx3x3=ps^6*Sz33;
Sx3x5=ps^8*Sz35;
Sx5x3=ps^8*Sz53;
Sx5x5=ps^10*Sz55;
% Compute the correlation coeff's
alpha13=sum(abs(Sx1x3))./sum(abs(Sx1x1));
alpha31=sum(abs(Sx3x1))./sum(abs(Sx1x1));
alpha15=sum(abs(Sx1x5))./sum(abs(Sx1x1));
alpha51=sum(abs(Sx5x1))./sum(abs(Sx1x1));
S33=Sx3x3-conj(alpha31).*Sx3x1...
    -alpha31.*Sx1x3+(alpha31.*conj(alpha31)).*Sx1x1;
S53=Sx5x3-conj(alpha31).*Sx5x1;
alpha53=sum(abs(S53))./sum(abs(S33));
```

```
% Compute the orthogonal model coefficients
a1=b1+alpha31*b3+alpha51*b5;
a3=b3+alpha53*b5;
a5=b5;
% Compute the spectra of the orthogonal branches
Sd3=(abs(a3).^2).*(Sx3x3+(abs(alpha31).^2).*Sx1x1...
    -alpha13.*Sx1x3-conj(alpha31).*Sx3x1);
Sd5=(abs(a5).^2).*(Sx5x5-alpha53.*Sx3x5-(alpha51...
    -alpha53.*alpha31).*Sx1x5...
    -conj(alpha53).*Sx5x3+(abs(alpha53).^2).*Sx3x3...
    +alpha53.*conj(alpha51-alpha53.*alpha31).*Sx3x1...
    -(alpha51-alpha53.*alpha31).*Sx1x5...
    +(alpha51-alpha53.*alpha31).*conj(alpha53).*Sx1x3...
    +(abs(alpha51-alpha53.*alpha31).^2).*Sx1x1);
Sdd=Sd3+Sd5; % Distortion spectrum
Scc=(abs(a1).^2).*Sx1x1; % Correlated spectrum
Syy=Scc+Sdd; % Total spectrum
```

Table 9.4 shows the envelope coefficients and the corresponding orthogonal model coefficients for a WCDMA signal and at an input power of -5 dBm. Figure 9.19 shows the total output spectrum and the uncorrelated distortion spectrum of three different customized forward-link WCDMA signals; forward-link 3 DPCH and 16 DPCH and a reverse-link 5DPCH signal all simulated at $P_{in} = -5$ dBm, which is the 1-dB compression point of the PA. Note that the shape of the uncorrelated components depends on the signal and its statistics. Figure 9.20 shows the probability density functions of the three signals. In the case of 3DPCH and 5 DPCH cases shown in Figures 9.19(a) and (b), the uncorrelated distortion inside the main channel is below spectral regrowth in the adjacent channel. The case of 16 DPCH shown in Figure 9.19 (c) represents the worst case since it exhibits the highest Peak-to Average Ratio (PAR). The shape of the uncorrelated spectrum resembles that of Gaussian signals that is flat over the signal bandwidth because its distribution can be approximated by a Gaussian distribution by the central limit theorem.

The same approach can be used to develop the uncorrelated spectrum of multisine signal by replacing the input signal by a multisine signal that can be generated as shown in the previous chapter. Figures 9.21(a)–(f) show the output spectrum and the uncorrelated distortion spectrum of multisine input signals with uniformly distributed random phases and Figures 9.22(a)–(f) show the output spectrum and the uncorrelated distortion spectrum of multisine input signals with zero initial phases.

Table 9.4 Envelope polynomial coefficients and corresponding orthogonal model coefficients

Envelope Coeff.	Value	Orth. Model. Coeff.	Value
b_1	$3.2607 - 10.4127i$	a_1	$3.6344 - 7.6754i$
b_3	$18.762 + 109.52i$	a_3	$4.7582 + 62.0682i$
b_3	$-146.25 - 495.62i$	a_3	$-146.25 - 495.62i$

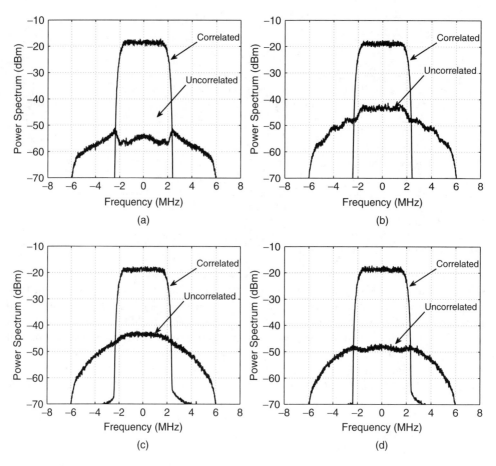

Figure 9.19 Output spectrum of a nonlinearity partitioned into correlated and uncorrelated components; (a) forward link 1 DPCH, (b) forward link 3 DPCH, (c) forward link 16 DPCH and (d) reverse link 5 DPCH.

9.7 Estimation of System Metrics from Simulated Spectra

Using the orthogonalized model, the output spectrum is represented as the sum of the spectra of the uncorrelated signal components. This partition is useful for the separation of the uncorrelated output distortion from the useful or undistorted component and hence to estimate the effective in-band distortion and all system-performance metrics. The effective in-band distortion can be expressed in terms of the PSD of the uncorrelated output components as

$$P_{\text{In-band}} = \int_{-B/2}^{B/2} S_{\tilde{y}_d \tilde{y}_d}(f) df \tag{9.10}$$

where B is the bandwidth of the input signal.

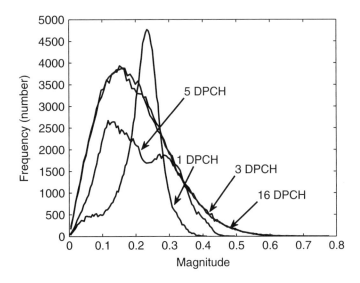

Figure 9.20   PDF of different WCDMA signals.

9.7.1 Signal-to-Noise and Distortion Ratio (SNDR)

The effective in-band distortion can be found from power measurements of the uncorrelated spectrum found in the previous section and can be used to estimate all system metrics such as SNDR, EVM and rho. The following script estimates the effective inband distortion and all system metrics from the uncorrelated distortion spectrum using the formulation of system metrics in Chapter 7.

```
File Name: SNDR.m
% Load spectra computed from the script SpectrumPlot_Orthogonal.m
% Scc(f), Sdd(f), Syy(f) and the frequency vector f
load SpecComp
% Specify the bandwidth of the input signal
wcdma_bw=3.84;
% Specify the signal to AWGN power (Eb/N0) ratio
EbN0dB=30;
EbN0=(10^((EbN0dB)/10));
% Specify the symbol rate
T=XDelta;
% Compute the inband frequency vector
freq_bw=find(f>-wcdma_bw/2  & f<+wcdma_bw/2);
% Compute output power
Po=sum(abs(Syy(freq_bw)));
% Compute the AWGN power
N0=T*Po/(EbN0);
% Compute SNDR
SNDR=10*log10(sum(abs(Scc(freq_bw)))/(sum(abs(Sdd(freq_bw)))+N0/T));
```

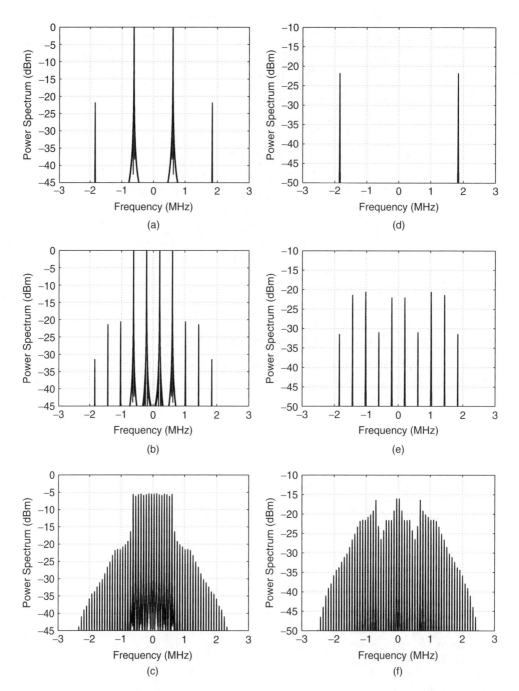

Figure 9.21 Spectra of the response of a nonlinearity to multisines with random phases; (a,b,c) output spectrum and (d,e,f) uncorrelated distortion spectrum.

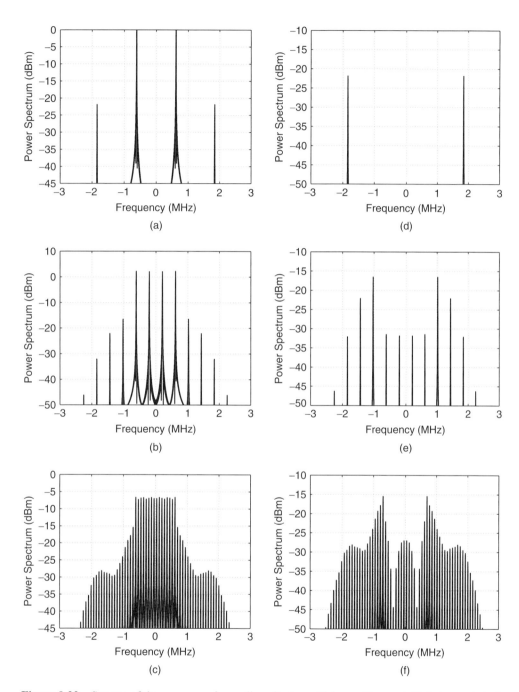

Figure 9.22 Spectra of the response of a nonlinearity to multisines with zero initial phases; (a,b,c) output spectrum and (d,e,f) uncorrelated distortion spectrum.

9.7.2 EVM

Simulation of EVM can be done in two ways; the first is using the approach described in Chapter 7 where EVM is estimated from the uncorrelated signal spectrum and the second is by using the EVM measurement object which is a built-in function in MATLAB®.

The following script estimates EVM from orthogonal spectra using the script developed in Section 9.7.

```
% File Name: EVM_spec.m
clear all;
% Load and power scale a WCDMA signal
load  WCDMA_DN_5DPDCH.mat
Fs=1/XDelta;
x0=double(Y);
P1=10*log10(mean(abs(x0).^2)/(50*.001));
Pscale=10^(P1/10);
x1=sqrt(1/Pscale)*(x0);
Pin=-10; % dBm
ps =sqrt(10^(Pin/10));
x=ps*(x1(1:128000));
% Compute the output of the memory polynomial model
load coeff.mat
coeff=b;
y=polyeval(coeff,x,5);
% Calculate linear gain
G= mean(y./x);
% Create EVM measurement object
h = commmeasure.EVM
update(h, y, G*x)


EVM =

    4.5239
```

EVM can be estimated using the EVM measurement object:

```
h = commmeasure.EVM
h = commmeasure.EVM('PropertyName',PropertyValue,...)
```

which constructs a default EVM object, h. The EVM object measures RMS EVM, maximum EVM, and percentile EVM. The unit for each measurement is a percentage. The command

```
h = commmeasure.EVM('PropertyName',PropertyValue,...)
```

constructs an EVM object, h, with property values set to PropertyValues as shown in Table 9.5

The following script estimates EVM using the commeaure.EVM command after applying a memoryless polynomial to an input WCDMA signal with the same parameters as in the above script EVM_Spec.m.

Table 9.5 Parameters of EVM measurements

Property	Description
Type	`'EVM Measurements'`. This property is read only.
NormalizationOption	Specify EVM normalization method. Possible values are: `'AveragePower'`, `'PeakPower'`, and `'RMSEVM'`.
AveragePower	Average constellation power. Relevant when `NormalizationOption` is `'Average constellation power'`.
PeakPower	Peak constellation power. Relevant when `NormalizationOption` is `'Peak constellation power'`.
RMSEVM RMS	EVM measurement result. Relevant when `NormalizationOption` is `'RMSEVM'`. This property is read only.
MaximumEVM	Maximum EVM measurement result. This property is read only.
Percentile	Percentile value to calculate `PercentileEVM`.
PercentileEVM	Percentile EVM measurement result. This property is read only.
NumberOfSymbols	Number of processed symbols.

```
% File Name:
% File Name: EVM_MATLAB\circledRObj.m
clear all;
% Load and power scale a WCDMA signal
load  WCDMA_DN_5DPDCH.mat
Fs=1/XDelta;
x0=double(Y);
P1=10*log10(mean(abs(x0).^2)/(50*.001));
Pscale=10^(P1/10);
x1=sqrt(1/Pscale)*(x0);
Pin=-10; % dBm
ps =sqrt(10^(Pin/10));
x=ps*(x1(1:128000));
% Compute the output of the memory polynomial model
load coeff.mat
coeff=b;
y=polyeval(coeff,x,5);
% Calculate linear gain
G= mean(y./x);
% Create EVM measurement object
h = commmeasure.EVM
update(h, y, G*x)
```

The resulting EVM measurements are:

```
h =

            Type: 'EVM Measurements'
          RMSEVM: 5.5501
      MaximumEVM: 40.7064
      Percentile: 95
   PercentileEVM: 13.1100
 NumberOfSymbols: 128000
```

9.7.3 ACPR

ACPR can be computed from power measurements in the ACPR bandwidth as specified by the standards of different wireless communication signals. Most standards define ACPR as the ratio of the power in the main channel to the power in the adjacent channel measured in a given bandwidth and a given offset from the carrier. The following script computes the ACPR of a WCDMA signal using the spectra of the output of a nonlinearity computed in Section 9.5.1. ACPR is measured in a 3.84-MHz bandwidth at a 5-MHz offset from the carrier.

```
% File Name: ACPR_Spec.m
clear all
% Load spectra computed from the script SpectrumPlot_Orthogonal.m
% Scc(f), Sdd(f), Syy(f) and the frequency vector f
load SpecComp
% Specify the bandwidth of the input signal
wcdma_bw=3.84;
% Specify the ACPR bandwidth and offset from carrier
acp_bw=3.84;
offset=5;
% Compute the inband frequency vector
freq_bw=find(f>-wcdma_bw/2  & f<+wcdma_bw/2);
% Compute output power within the signal bandwidth
freq_main=find(f>-wcdma_bw/2  & f<wcdma_bw/2);
Po_main=10*log10(sum(abs(Syy(freq_main))/.05))
% Compute adjacent channel power
freq_acp=find(f>offset-acp_bw/2  & f<offset+acp_bw/2);
ACP=10*log10(sum(abs(Syy(freq_acp))/.05))
% ACPR=Adjacent Channel Power/Main Channel Power
ACPR=ACP-Po_main
```

```
Po_main =
     9.9399
ACP =
   -26.7683
ACPR =
   -36.7082
```

MATLAB® has an ACPR Object for calculating ACPR at the output of a nonlinearity (The Communication Toolbox, 2009). The command:

```
hACPR = commmeasure.ACPR('Fs',Fs)
```

creates an ACPR object with a sampling frequency Fs. The ACPR object has the following attributes:

```
hACPR =
Type: 'ACPR Measurement'
NormalizedFrequency
Fs
MainChannelFrequency
```

```
MainChannelMeasBW
AdjacentChannelOffset
AdjacentChannelMeasBW
MeasurementFilter
SpectralEstimatorOption
FrequencyResolutionOption
FFTLengthOption
MaxHold
PowerUnits
FrameCount
```

For example, to calculate ACPR of a WCDMA signal after nonlinear amplification, the attributes of the ACPR object are defined according to the WCDMA standard as

```
hACPR.Fs=30.72
hACPR.MainChannelFrequency = 0;
hACPR.MainChannelMeasBW = 3.84e6;
hACPR.AdjacentChannelOffset = 5e6;
hACPR.AdjacentChannelMeasBW = 3.84e6;
```

ACPR measurement can be obtained by using the following command:

```
[ACPR mainChannelPower adjChannelPower] =run(hACPR,SignalAfterAmp)
```

where `SignalAfterAmp` is the amplifier output signal.

9.8 Simulation of Probability of Error

The BER of a communication system with amplifier nonlinearity can be simulated using the BERtool, as shown in Chapter 8. In the following example, the BERtool is used to simulate the BER of a communication system with amplifier nonlinearity where the simulation function models a communication system that uses QPSK modulation/demodulation with raised cosine pulse shaping. Nonlinearity is applied to the modulated signal using the Saleh model before the AWGN channel. The received signal is then demodulated and the number of errors are counted. The simulated BER is compared to the theoretical BER of QPSK modulation without pulse shaping and amplifier nonlinearity. The following code is the simulation function used in the Monte Carlo simulations in the BERtool. Figure 9.23 shows BER vs. E_bN_o for different values of the Saleh model parameters simulated using MC simulations and compared to theoretical BER without nonlinear distortion.

```
% File Name:
function [ber, numBits] = bertool_simfcn_nonlinear(EbNo,...
 maxNumErrs, maxNumBits)
% Import Java class for BERTool.
import com.mathworks.toolbox.comm.BERTool;
%% Initialize variables related to exit criteria.
totErr = 0;  % Number of errors observed
numBits = 0; % Number of bits processed
%%--- Set up parameters. ---
% Set up initial parameters.
siglen = 1000; % Number of bits in each trial
```

```
M = 4; % QPSK is binary.
k=log2(M);
% Create a QPSK modulation/demodulation object
hMod = modem.pskmod('M', M);
hDemod = modem.pskdemod(hMod);
% Create pulse shaping filter object
rolloff=0.3;
Fd=1;
Fs=4;
Delay=4;
rrcfilter = rcosine(Fd,Fs,'fir/sqrt',rolloff,Delay);
ntrials = 0; % Number of passes through the loop

%%%%%%%%%%%%%%%%%%%%%%%%%%%%%%%%%%%%%%%%%%%%%%%%%%%%%%%%%%%%%%%%%%%
%% Simulate until number of errors exceeds maxNumErrs
% or number of bits processed exceeds maxNumBits.
while((totErr < maxNumErrs)  && (numBits < maxNumBits))
    % Check if the user clicked the Stop button of BERTool.
    if (BERTool.getSimulationStop)
        break;
    end
%%%%%%%%%%%%%%%%%%%%%%%%%%%%%%%%%%%%%%%%%%%%%%%%%%%%%%%%%%%%%%%%%%%
%%--- Proceed with simulation.
msg = randint(siglen, 1, M); % Generate message sequence.
% Modulate the message signal
txsig = modulate(hMod, msg); % Modulate.
% Apply a raised cosine pulse shaping transmitter filter
txsigpulse=rcosflt(txsig,Fd,Fs,'filter',rrcfilter);
% Apply a nonlinearity
txSaleh=Saleh(txsigpulse,1,0,0,0);
% Apply AWGN
snr = EbNo + 10*log10(k) - 10*log10(Fs/Fd);
rxsignoisy = awgn(txSaleh, snr, 'measured'); % Add noise.
% Apply a raised cosine pulse shaping receiver filter
rxsigpulse=rcosflt(rxsignoisy,Fd,Fs,'Fs/filter',rrcfilter);
rxsig=downsample(rxsigpulse,(Fs/Fd));
rxsig = rxsig(2*Delay+1:end-2*Delay);
% Demodulate the received signal
decodmsg = demodulate(hDemod, rxsig); % Demodulate.
%% Count Errors
newerrs = biterr(msg,decodmsg); % Errors in this trial
ntrials = ntrials + 1; % Update trial index.
% Update the total number of errors.
totErr = totErr + newerrs;
% Update the total number of bits processed.
numBits = ntrials * siglen;
%%%%%%%%%%%%%%%%%%%%%%%%%%%%%%%%%%%%%%%%%%%%%%%%%%%%%%%%%%%%%%%%%%%
    %--- Be sure to update totErr and numBits.
    %--- INSERT YOUR CODE HERE.
end % End of loop
% Compute the BER.
ber = totErr/numBits;
```

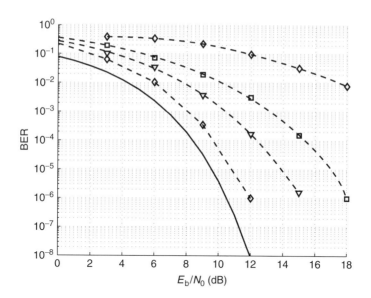

Figure 9.23 BER vs. $E_b N_o$ for different values of the Saleh model parameters: Solid: Theoretical (no nonlinearity) and dashed: Monte Carlo Simulations.

As another example, the following script (BERSWP) performs Monte Carlo simulation of BER of an IS-95 system based on the standard transmitter/receiver model of IS-95 systems (IS95 Standard, 1993) and considering nonlinear distortion generated by a power-series model. The script generates a CDMA signal, applies it to a nonlinearity and AWGN, demodulates the signal and then counts the number of bits in error. BER is then computed versus swept E_b/N_o and for different input power levels.

```
File Name: BERSWP.m
clear all;
% close all
WalshSize=4; % Walsh code size (Max no. of users)
NumUser=4;   % Number of active users
EbN0=0:4:18; % Signal-to-AWGN ratio vector
Pin=-4:.5:-4;% Specify the input power vector for power sweep
OverSamp=4; % Oversampling ratio
% Load baseband filter coefficients
load cdma100.txt
FiltCoef=cdma100;
% Set nonlinear amplifier coefficients
AmpCoef=[11.3963 -122.4691];
nBits=2^10; % Number of generated bits
Bits=nBits*ones(size(Pin)); % Input bit stream
% Generate PN codes for the I and Q channels
PNI=2*randerr(1,WalshSize*nBits,WalshSize*nBits/2)-1;
PNQ=2*randerr(1,WalshSize*nBits,WalshSize*nBits/2)-1;
% Set the number of loops used to calculate the number of error bits
loops=2^8*ones(size(EbN0));
```

```
% Sweep the input power
for w=1:length(Pin)
    for u=1:length(EbN0)
        BitsInError1=0;
        % Simulate the BER and count the number of bits in error
        for k=0:loops(u);
            [BitsInError,TotBits,TimeSignal]=...
            BERSim(Bits(w),Pin(w),EbN0(u),OverSamp,NumUser,...
            WalshSize,AmpCoef,FiltCoef,PNI,PNQ);
            BitsInError1=BitsInError1+BitsInError;
        end
        BER_s(w,u)=BitsInError1/(TotBits)/(k+1);
    end
    w;
    clear BitsInError BitsInError1 TimeSignal TotBits
end
% Plot simulated and analytical BER
figure(1)
figure(1)
semilogy(EbN0,BER_s,'ko-')
hold on
semilogy(EbN0,BER_a,'r^-')
axis([0 18 10^-5 1])
```

The function CDMAGen generates a CDMA signal for a given number of bits, Walsh code size and number of active users. Four bits streams of 2^{12} bits are generated using a Gaussian random number generator in MATLAB®. Each bit stream is multiplied by a Walsh code drawn from a set of 4 codes and then filtered by a standard baseband filter of IS-95 CDMA system. The resulting signal is then applied to a nonlinearity that is modeled as a third-order polynomial and then AWGN is added to the resulting signal.

```
% File Name: CDMAGen.m
function [TimeSignal,Data,Huser]=...
CDMAGen(nBits,FiltCoef,OverSamp,NumUser,WalshSize,PNI,PNQ)
% Generate a Hadamard matrix (Walsh codes)
H=hadamard(WalshSize);
r=zeros(1,WalshSize);
% Select the number of active users in the composite CDMA signal
users=randperm(WalshSize);
act_user=users(1:NumUser);
% Select the user data for which BER is to be simulated
r(act_user)=ones(1,NumUser);
User=act_user(1);
% Initialize the coded bit stream
c=1:WalshSize*nBits;
Huser=H(User,rem(c+WalshSize-1,WalshSize)+1);
Ichan=zeros(1,WalshSize*nBits*OverSamp);
Qchan=Ichan;
% Generate the composite CDMA signal of no. of users =act_user
for d=1:length(act_user)
    % Generate random bit stream
```

```
      bits=2*randerr(1,nBits,nBits/2)-1;
      if act_user(d)==User
          Data=bits;
      end
      i=1:WalshSize*nBits;
      % Generate the I and Q bit streams
      yI=bits(ceil(i/WalshSize));
      yQ=bits(ceil(i/WalshSize));
      % Encode the I and Q bit streams with Walsh codes
      walshbitsI=r(act_user(d))*H(act_user(d),...
      rem(i+WalshSize-1,WalshSize)+1).*yI;
      walshbitsQ=r(act_user(d))*H(act_user(d),...
      rem(i+WalshSize-1,WalshSize)+1).*yQ;
      % Encode the I and Q Walsh coded bits with PN codes
      pnbitsI=walshbitsI.*PNI;
      pnbitsQ=walshbitsQ.*PNQ;
      % 4x Oversampling for BB filtering purposes
      chipbitsI=zeros(1,OverSamp*WalshSize*nBits);
      chipbitsQ=chipbitsI;
      chip=1:WalshSize*nBits;
      chipbitsI((chip-1)*OverSamp+1)=pnbitsI(chip);
      chipbitsQ((chip-1)*OverSamp+1)=pnbitsQ(chip);
      % Filter the I and Q data with an FIR filter
      Ichanf=filter(FiltCoef,1,chipbitsI);
      Qchanf=filter(FiltCoef,1,chipbitsQ);
      % Generate the composite CDMA signal with all user data
      Ichan=Ichan+Ichanf;
      Qchan=Qchan+Qchanf;
end
% Complex CDMA signal
CDMA=(Ichan+1i*Qchan);
% Normalize power to 0 dBm
pin=10*log10(mean(abs(CDMA).^2)/(.05));
pscale=10^(pin/10);
TimeSignal=CDMA/sqrt(pscale);
```

The function `BERSim` performs the simulation of BER by generating random data and then counting the number of bits in error after applying the transmitted signal to the receiver model. The input power of the signal is swept (in $[-10:-4]$ dBm) and the number of erroneous bits are counted at each power level and signal-to-noise ratio (E_b/N_o) which is swept in [0:20] db. The simulations are repeated to account for the different combinations of consecutive bits where a loop is used to repeat the simulations and then the average BER is computed.

```
File Name: BERSim.m
function [BitsInError,TotBits,TimeSignal,Data,Huser]=...
BERSim(nBits,Pin,EbN0,OverSamp,NumUser,WalshSize,AmpCoef,...
FiltCoef,PNI,PNQ)
% Generate a CDMA signal
[TimeSignal,Data,Huser]=...
CDMAGen(nBits,FiltCoef,OverSamp,NumUser,WalshSize,PNI,PNQ);
```

```
% Amplifier model coefficients
b1=AmpCoef(1);
b3=AmpCoef(2);
% Generate the output of the nonlinearity
ps=sqrt((10^((Pin)/10)));
V=b1*ps*TimeSignal+b3*ps^3*(abs(TimeSignal).^2).*TimeSignal;
% Add AWGN for a given signal to noise ration (EbNo)
TxSignal=awgn(V,EbN0-10*log10(OverSamp)+...
10*log10(NumUser/WalshSize),'measured',1234,'dB');
% Receiver operation
rI=real(TxSignal);
rQ=imag(TxSignal);
% Apply a Baseband filter
RI=filter(FiltCoef,1,rI);
RQ=filter(FiltCoef,1,rQ);
L=length(FiltCoef);
% Detection of Tx chips
drecI=(RI(L:OverSamp:length(RI)));
drecQ=(RQ(L:OverSamp:length(RQ)));
% Despread the signal using PN sequence identical to
% the Tx PN codes and detect the transmitted bits
drec=drecI.*PNI(1:length(drecI))+drecQ.*PNQ(1:length(drecI));
for i=0:1:(length(drec)-WalshSize)/WalshSize;
    S(i+1)=sum(drec(i*WalshSize+1:(i+1)*WalshSize).*...
    Huser(i*WalshSize+1:(i+1)*WalshSize));
end
E=sign(S)+Data(1:length(S));
% Count the bit errors by comparing the detected
% bits with the transmitted bits
Error=length(find(E==0));
BitsInError=Error; % Bits in error
TotBits=length(S); % Total number of bits
```

Figure 9.24 shows the probability of error versus the signal to noise ratio (E_b/N_o) with nonlinear amplification and at different input power levels. The figure shows how the BER is increased as the amplifier is driven close to its nonlinear region where nonlinear distortion has more impact on BER than AWGN.

9.9   Simulation of Noise-to-Power Ratio

Simulation of NPR is done using the concepts discussed in Chapter 7 where NPR is calculated by measuring the power within a notch in the output spectrum of nonlinear amplifier. NPR simulations are performed using WCDMA signals that has a bandwidth of 3.84 MHz as follows: a notch filter of around 20 kHz bandwidth is applied to the input signals using a band stop FIR filter in order to create a notch at the center of the signal frequency band. The resulting signal is then applied to the nonlinear model and higher-order spectra are computed. The correlated and uncorrelated spectra and the output spectrum of the output of the notch filter are developed using the procedure in the Section 9.5. NPR is measured from the output spectrum as the power within the notch

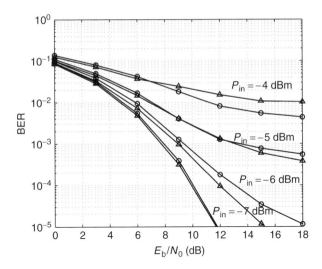

Figure 9.24 BER vs. E_b/N_o at different input power levels; $\triangle$: simulated and $\circ$: analytical (Gharaibeh *et al.*, 2004).

bandwidth and is compared to the in-band distortion measured within the notch bandwidth on the output spectrum without the notch applied to the input signal.

The following script calculates the NPR and compares it to the effective in-band distortion.

```
% File Name:
clear all
% load WCDMA_UP_3DPCH_SCH.mat
% load WCDMA_DN_5DPDCH.mat
load WCDMA_UP_TM1_16DPCH.mat
% load WCDMA_UP_1DPCH_small1.mat
CDMA=transpose(double(Y(1:256000)));
% Specify the bandwidth of the input signal
wcdma_bw=3.84;
% Load polynomial coefficients
load coeff.mat
coeff=b;
% Scale the input power of the input signal
Pin=10*log10(mean(abs(CDMA).^2)/(50*.001));
Pscale=10^(Pin/10);
CDMA_O=sqrt(1/Pscale)*(CDMA);
% Apply a notch of 10 kHz bandwidth to the input signal
N     = 842;    % Order
N     = 242;    % Order
Fstop = 0.001;  % Stopband Frequency
Fpass = 0.01;   % Passband Frequency
Wstop = 1;      % Stopband Weight
Wpass = 1;      % Passband Weight
```

```
% Calculate the coefficients using the FIRLS function.
FiltCoeff   = firls(N, [0 Fstop Fpass 1], [0 0 1 1], [Wstop Wpass]);
yx=filter(FiltCoeff,1,CDMA);
CDMA_notch=yx(N:length(yx));
% Compute higher order spectra
[Sz1,Sz13,Sz31,Sz33,Sz15,Sz51,Sz35,Sz53,Sz55]=SpecGen(CDMA_notch);
% Sweep the input power of the signal and calculate NPR
Pin=-15:-5;
for k=1:length(Pin)
    % Compute orthogonal spectra
    [Scc,Sdd,Syy]=OrthSpec(Pin(k),coeff,Sz1,Sz13,Sz31,...
    Sz33,Sz15,Sz51,Sz35,Sz53,Sz55);
    % Compute the spectrum in dBm
    Sy=10*log10(Syy/.05);      % total
    Sd=10*log10(Sdd/.05);      % uncorrelated
    Sc=10*log10(Scc/.05);      % correlated
    % Create a frequency vector
    len=length(Sz1);
    m=len/2;
    Fs=1/XDelta;
    delf=Fs*1e-6/len;% Frequency resolution normalized to 1 MHz
    xx=-(m):(m-1);
    f=delf*xx; % Frequency vector
    freq_bw=find(f>=-wcdma_bw/2  & f<+wcdma_bw/2);
    freq_bw_npr=find(f>=-.01  & f<.01);
    % Power measurements
    % Total output power
    Pout(k)=10*log10(sum(abs(Syy(freq_bw))))/0.05);
    % Power within notch BW
    NPR(k)=10*log10(sum(abs(Syy(freq_bw_npr))))/.05);
    % In-band dist within notch BW
    In_band(k)=10*log10(sum(abs(Sdd(freq_bw_npr))))/.05);
end
% Plot the correlated and uncorrelated spectra
figure
plot(f,Sy,'r')
axis([-8 8 -70 0])
xlabel('Frequency (MHz)')
ylabel('Power Spectrum (dBm)')
grid
hold on
```

Figure 9.25 shows the frequency response of the notch filter. Figure 9.26 shows simulated $S_{\hat{y}_c\hat{y}_c}(f)$ and $S_{\tilde{y}_d\tilde{y}_d}(f)$ of three different WCDMA signals. The figure shows that the power level within the notch is above the level of the effective in-band distortion for some signals, while it matches the uncorrelated in-band distortion for other signals.

It is worth noting that the power difference between $S_{\hat{y}_c\hat{y}_c}(f)$ and $S_{\tilde{y}_d\tilde{y}_d}(f)$ within the notch bandwidth depends on the statistical properties of the input signal, as is evident from Figures 9.26(a)–(c) where it is clear that the power within the notch matches the effective

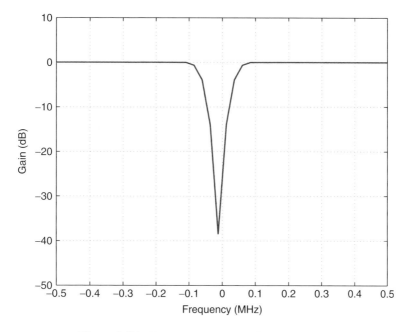

Figure 9.25 Frequency response of a notch filter.

in-band distortion and hence the NPR test represents a good measure of the effective in-band distortion in this case. In WCDMA systems the selection of the channelization codes (Orthogonal Variable Spreading Factor (OVSF) codes for example) is an important factor in determining the level of the spike that appears in the distortion spectrum and hence the validity of NPR as a measure of in-band distortion. In the case of heavily loaded WCDMA signal (big number of active code channels) the signal approaches the Gaussian distribution and the equality $S_{\hat{y}_d \hat{y}_d}(0) = S_{\tilde{y}_d \tilde{y}_d}(0)$ holds as shown in the previous subsection (Gharaibeh, 2009).

9.10 Simulation of Nonlinear Noise Figure

Figure 9.27 shows a flow chart for the procedure used in the simulation of the NF of a nonlinear amplifier using the concepts developed in Chapter 7. A WCDMA signal and a NBGN signal with bandwidths B_s and B_n, respectively are first generated and then their individual power levels are set by adjusting the total input power and the input SNR. The NBGN signal is designed to have twice the bandwidth of WCDMA signal ($B_n = 4B_s$). The two signals are then summed to produce the input signal $\hat{x}(t)$. In the next step, the individual autocorrelation functions of $s(t)$, $n(t)$ and $x(t)$ and the higher-order autocorrelation functions of $x(t)$ are computed. Once the orthogonal model coefficients are available, the correlated output signal spectrum ($|c_1|^2 S_{\tilde{s}\tilde{s}}(f)$), the correlated output noise spectrum ($|c_1|^2 S_{\tilde{n}\tilde{n}}(f)$) and the uncorrelated distortion spectrum ($S_{\tilde{y}_d \tilde{y}_d}(f)$) can be computed, and hence the output SNR can be evaluated using Equation (7.46). System NF is then evaluated by comparing the input and output SNRs using Equation (7.47) (Gharaibeh, 2010).

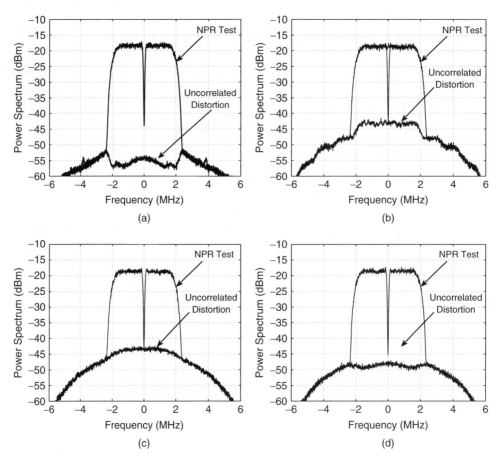

Figure 9.26 Output PSDs $S_{\hat{y}_c \hat{y}_c}(f)$ and $S_{\tilde{y}_d \tilde{y}_d}(f)$; (a) forward-link 1 DPCH, (b) forward-link 3 DPCH, (c) forward-link 16 DPCH and (d) reverse-link 5 DPCH.

The following script calculates the nonlinear NF using the above procedure.

```
% File Name:
clear all;
% Set the input SNR
SNR=30;
% Load amplifier polynomial coefficients
load coeff.mat
Ampcoeff=b;
% Load WCDMA signal
load WCDMA_DN_5DPDCH.mat
CDMA=transpose(double(Y));
WcdmaBW=3.84;  % bandwidth of the input signal
% Specify the sampling time of the signal
XDelta=4.8828e-008; % Sampling time
```

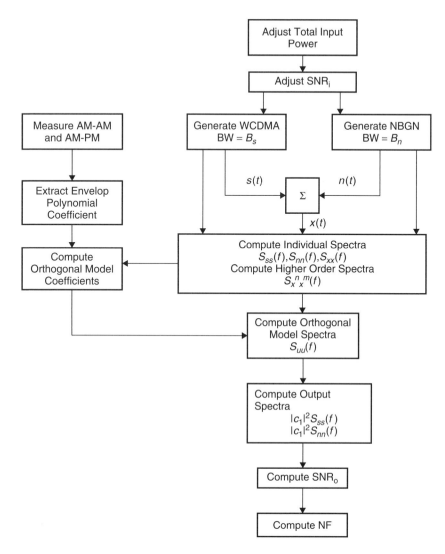

Figure 9.27 Flow chart for simulation of nonlinear NF (Gharaibeh, 2009).

```
N=length(CDMA);
% Scale the input power of the input signal
P=10*log10(mean(abs(CDMA).^2)/(50*.001));
Pscale=10^(P/10);
Sig=sqrt(1/Pscale)*(CDMA);
% Generate NBGN signal of the same length
% and with Bandwidth=NoiseBW
NoiseBW=2*WcdmaBW;
Noise=NBGNGen(N,NoiseBW,XDelta);
% Sweep the input power of the sum signal
Pin=-10:4:-10;
```

```
for k=1:length(Pin)
    ps=sqrt(10^(Pin(k)/10));
    % Compute orthogonal spectra
    % Sc: total correlated output spectrum
    % Sd: total distortion output spectrum
    % Sy: total output spectrum
    % Ssi: spectrum of the input signal component
    % Sni: spectrum of the input noise component
    % Sso: spectrum of the correlated output signal
    % Sno: spectrum of the correlated output noise
    [Sc,Sd,Sy,Ssi,Sni,Sso,Sno]=OrthSpecNF(Pin(k),...
    Ampcoeff,SNR,NoiseBW,Sig,Noise);
    Sndo=Sd+Sno; % Noise and distortion spectrum
    Ssumi=Ssi+Sni;
    % Create a frequency vector
    len=length(Sc);
    m=len/2;
    delf=1e-6/(XDelta*len);
    x=-(m):(m-1);
    f=delf*x;
    freq_bw=find(f>=-WcdmaBW/2 & f<+WcdmaBW/2);
    % Power measurements
    % Output signal power
    Pso(k)=10*log10(sum(abs(Sso(freq_bw)))/.05);
    % Output noise and distortion power
    Pndo(k)=10*log10(sum(abs(Sndo(freq_bw)))/.05);
    % Input signal power
    Psi(k)=10*log10(sum(abs(Ssi(freq_bw)))/.05);
    % Input noise power
    Pni(k)=10*log10(sum(abs(Sni(freq_bw)))/.05);
    SNRi(k)=Psi(k)-Pni(k); % Input SNR
    SNRo(k)=Pso(k)-Pndo(k);% Output SNR
    NF(k)=SNRi(k)-SNRo(k); % Noise Figure
    % Calculate spectra in dBm
    SsumidB(:,k)=10*log10((Ssumi)/.05);
    SydB(:,k)=10*log10(Sy/.05);
    SddB(:,k)=10*log10(Sd/.05);
    ScdB(:,k)=10*log10(Sc/.05);
    SndodB(:,k)=10*log10((Sndo)/.05);
    SnodB(:,k)=10*log10((Sno)/.05);
    SsodB(:,k)=10*log10((Sso)/.05);
    SnidB(:,k)=10*log10((Sni)/.05);
    SsidB(:,k)=10*log10((Ssi)/.05);
end
% Plot spectra
figure
plot(f,SndodB)
axis([-10 10 -90 -10])
```

The function OrthSpecNF computes the orthogonal spectra and the output correlated noise spectrum:

```
File Name: OrthSpecNF.m
function [Sc,Sd,Sy,Ssi,Sni,Sso,Sno]=OrthSpecNF(Pin,Ampcoeff,...
SNR,NoiseBW,Sig,Noise)
ps=sqrt(10^(Pin/10));
% Amplifier coefficients
b1=Ampcoeff(1);
b3=Ampcoeff(2);
b5=Ampcoeff(3);
WcdmaBW=3.84; % BW of input signal
% Scale the input signal and noise for a given SNR
x=10.^((SNR-10*log10(NoiseBW/WcdmaBW))/10);
p=1./sqrt(1+x);
q=sqrt(1-p^2);
Sum=p*Noise+q*Sig; % Sum signal
% Normalize the power of the sum signal to 0 dB
P=10*log10(mean(abs(Sum).^2)/(50*.001));
Pscale=10^(P/10);
SUM=sqrt(1/Pscale)*(Sum);
% Compute the individual ACF and PSD of the
% input signal and noise
W=2^11;
Rs=xcorr(Sig',Sig',W,'biased');
Rn=xcorr(Noise',Noise',W,'biased');
len=length(Rs);
Ss=fftshift(fft(hanning(len).*Rs/len));
Sn=fftshift(fft(hanning(len).*Rn/len));
Ssi=ps^2*q^2*Ss; % Input noise spectrum
Sni=ps^2*p^2*Sn; % Input signal spectrum
% Calculate higher order spectra of the sum signal
[Sz1,Sz13,Sz31,Sz33,Sz15,Sz51,Sz35,Sz53,Sz55]=SpecGen(SUM);
% Scale the power of the higher order spectra
Sx1x1=ps^2*Sz1;
Sx1x3=ps^4*Sz13;
Sx1x5=ps^6*Sz15;
Sx3x1=ps^4*Sz31;
Sx5x1=ps^6*Sz51;
Sx3x3=ps^6*Sz33;
Sx3x5=ps^8*Sz35;
Sx5x3=ps^8*Sz53;
Sx5x5=ps^10*Sz55;
% Compute the correlation coeff's
alpha13=sum(abs(Sx1x3))./sum(abs(Sx1x1));
alpha31=sum(abs(Sx3x1))./sum(abs(Sx1x1));
alpha15=sum(abs(Sx1x5))./sum(abs(Sx1x1));
alpha51=sum(abs(Sx5x1))./sum(abs(Sx1x1));
S33=Sx3x3-conj(alpha31).*Sx3x1...
    -alpha31.*Sx1x3+(alpha31.*conj(alpha31)).*Sx1x1;
S53=Sx5x3-conj(alpha31).*Sx5x1;
alpha53=sum(abs(S53))./sum(abs(S33));
% Compute the orthogonal model coefficients
```

```
a1=b1+alpha31*b3+alpha51*b5;
a3=b3+alpha53*b5;
a5=b5;
% Compute the spectra of the orthogonal branches
% of the nonlinear model
Sd3=(abs(a3).^2).*(Sx3x3+(abs(alpha31).^2).*Sx1x1...
    -alpha13.*Sx1x3-conj(alpha31).*Sx3x1);
Sd5=(abs(a5).^2).*(Sx5x5-alpha53.*Sx3x5...
    -(alpha51-alpha53.*alpha31).*Sx1x5...
    -conj(alpha53).*Sx5x3+(abs(alpha53).^2).*Sx3x3...
    +alpha53.*conj(alpha51-alpha53.*alpha31).*Sx3x1...
    -(alpha51-alpha53.*alpha31).*Sx1x5...
    +(alpha51-alpha53.*alpha31).*conj(alpha53).*Sx1x3...
    +(abs(alpha51-alpha53.*alpha31).^2).*Sx1x1);
Sd=Sd3+Sd5; % Distortion spectrum of the sum signal
Sc=(abs(a1).^2).*Sx1x1; % Correlated spectrum of the sum signal
Sy=Sc+Sd; % Total spectrum of the sum signal
% Compute the spectra of the correlated output
% signal and output noise
Sso=Ssi*(abs(a1).^2); % Correlated output signal spectrum
Sno=Sni*(abs(a1).^2); % Correlated output noise spectrum
```

Figures 9.28 show the input and output spectra when the input consists of the sum of a WCDMA signal and NBGN with the total input power held at -10 dBm and SNR_i at 30 dB with NBGN signal having twice the bandwidth of the WCDMA signal. Figures 9.29(a) and 9.29(b) show the spectra of the correlated signal component and the spectrum of the output correlated noise plus the uncorrelated distortion that are used

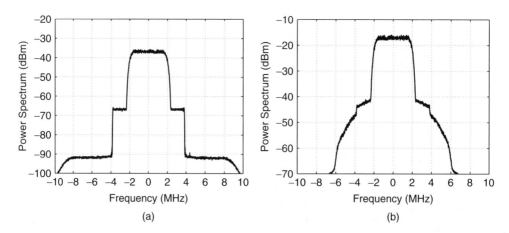

(a)   (b)

Figure 9.28 The input and output spectra of a nonlinearity for an input that consists of the sum of a WCDMA signal and NBGN.

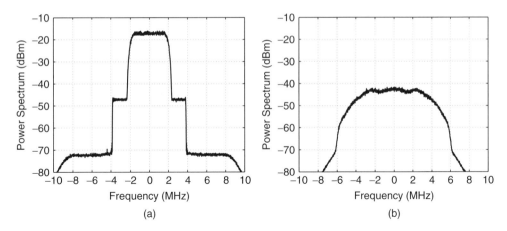

Figure 9.29 Output spectrum of a nonlinearity partitioned into (a) correlated and (b) uncorrelated components for an input which consists of the sum of a WCDMA signal and NBGN.

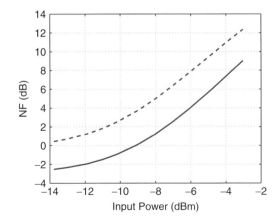

Figure 9.30 NF vs. input power; solid: input=WCDMA+NBGN and dashed: input=NBGN+NBGN.

to compute the NF. The total noise and distortion power is measured within the signal bandwidth from the noise and distortion spectrum. Figure 9.30 shows the simulated NF versus input power level when the amplifier input consists of a WCDMA signal plus NBGN signal and when the input consists of the sum of two NBGN signals. The figure shows the NBGN assumption overestimates NF by about 3 dB.

9.11 Summary

In this chapter, MATLAB® has been used to simulate nonlinear models and nonlinear distortion and their impact on communication-system performance. Matlab codes that can be used for the development of various nonlinear models from measured data have been presented. Then scripts that generate different communication signals were given. These simulated signals along with nonlinear models are then used in the simulation of nonlinear distortion and the various parameters that quantify its effect on the performance of a communication system. Therefore, this chapter consists of a complete set of MATLAB tools that enables nonlinearity in wireless communication systems to be modeled and simulated. The next chapter will use the same approach with Simulink®.

10

Simulation of Nonlinear Systems in **Simulink**®

Simulink® is an environment for model-based design of a wide variety of systems including communications, controls, signal processing, video processing, and many other electrical systems. The Simulink® environment has many features that make it an efficient tool for the modeling and simulations of electrical systems. First, it consists of block libraries and an interactive graphical environment that is easy to use in the design and simulation of systems. Secondly, Simulink® is integrated with MATLAB® where all MATLAB® tools are accessible in the Simulink® environments, which means that a customizable and expandable libraries of predefined blocks can be developed, and thirdly, Simulink® provides an ability to manage complex designs by segmenting models into hierarchies of design components.

Through The Communications Blockset and The RF Blockset, Simulink® provides an efficient tool for the modeling and simulation of nonlinear systems in wireless communications. Simulink® has a variety of models for nonlinearity that can be used to simulate nonlinear distortion in wireless systems. One of the most efficient models is the General Amplifier Model that is available in The RF Blockset and simulates a real amplifier environment. The amplifier model can be fed with measured amplifier characteristics such as AM–AM and AM–PM conversions or S-parameters as a function of frequency. Simulation of system performance under nonlinearity can be done by special setups and using the proper signal sources.

This chapter explains how to use Simulink® to analyze and evaluate the performance of wireless communication systems under nonlinearity using the Simulink® library. We present simulation experiments in Simulink® that lead to the prediction of nonlinear distortion and nonlinear system metrics. The various amplifier models available in The RF Blockset are presented as simulation test vehicle when fed with measured amplifier characteristics to simulate a real amplifier. Measurement setups used to simulate the concepts presented in the previous chapters are introduced including signal generation, methods for measuring in-band and out-of-band distortion, and methods for measuring nonlinear system metrics such as SNR, NPR, EVM, NF, etc.

Nonlinear Distortion in Wireless Systems: Modeling and Simulation with MATLAB®, First Edition.
Khaled M. Gharaibeh.
© 2012 John Wiley & Sons, Ltd. Published 2012 by John Wiley & Sons, Ltd.

10.1   RF Impairments in Simulink®

RF impairments in communications systems such as nonlinearity, phase noise, etc. are
included as blocks under both The Communications Blockset and The RF Blockset (The
RF Blockset, 2009; The Communications Blockset, 2009). In the following subsections,
these blocks are presented and explained.

10.1.1   Communications Blockset

The Communications Blockset contains a number of models for RF impairments in a
communication system. These include free-space path loss, I/Q imbalance, phase noise,
phase/frequency offset, receiver thermal noise and a memoryless nonlinearity (The
Communications Blockset, 2009). Figure 10.1 shows the RF Impairment library in
The Communications Blockset. These blocks apply to baseband communication signals
where a number of parameters for each block can be specified.

The "RF Impairment" library includes the following blocks (The Communications
Blockset, 2009):

- The Free-Space Path Loss block: models the power loss that results from signal propagation in space. The free-space path loss is mainly dependent on the distance the signal
  travels and on its frequency. In this block, only the value of the loss can be specified
  (in dB units).
- The I/Q Imbalance block: models the imbalance between the I and Q channels of
  the transmitted signal that is caused by different channel conditions on the I and Q
  channels. The I/Q imbalance is manifested as a change in the shape of the constellation
  diagram of the signal. In this model, amplitude imbalance (in dB) and phase imbalance
  (in degrees) can be specified as fixed values.
- The Phase Noise block: models phase noise in receivers that alters only the phase and
  frequency of the signal. The phase noise is modeled as a noise signal whose spectrum
  has 1/f slope. The block allows the specification of the phase noise spectrum level in
  dBc/Hz measured at a given frequency offset from the carrier.
- The Receiver Thermal Noise block: models the effects of thermal noise on a complex
  baseband signal. This block applies thermal noise to a complex baseband signal and
  allows thermal noise to be specified by either noise temperature, noise figure or noise
  factor.
- The Phase/Frequency Offset block: applies fixed phase and frequency offsets to the
  input complex baseband signal.
- The Memoryless Nonlinearity block models the AM-to-AM and AM-to-PM distortion
  in nonlinear amplifiers. A detailed description of this block is given in the next section.

10.1.2   The RF Blockset

The RF Blockset contains a number of libraries for modeling RF systems and components.
These include RF amplifiers, RF filters, RF mixers and transmission lines. In addition,
this library contains models for general RF network that are modeled by their S, Y or
Z parameters.

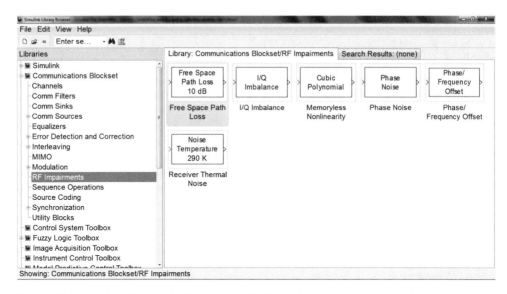

Figure 10.1 RF impairments in the "Communications Blockset" (The Communications Blockset, 2009).

The RF Blockset consists of two types of models: the first type is mathematical models and the second is physical models. Figure 10.2 shows the RF blocks available from the RF Mathematical library. Mathematical RF blocks model an RF component using a mathematical expression that represents the transformation that the signal undergoes when applied to the RF component.

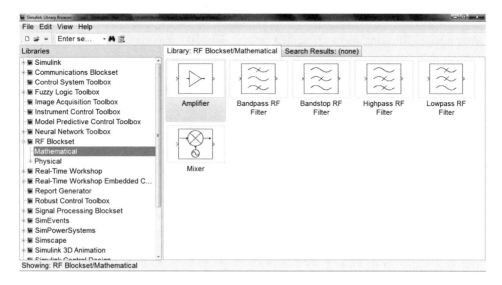

Figure 10.2 Mathematical RF blocks in The RF Blockset (The RF Blockset, 2009).

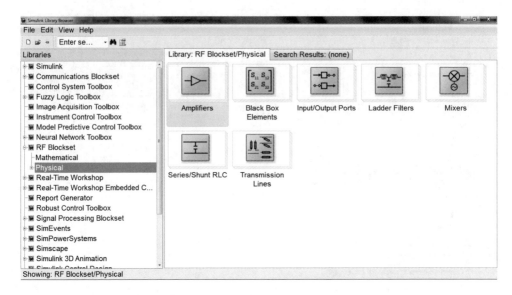

Figure 10.3 Physical RF blocks in The RF Blockset (The RF Blockset, 2009).

The RF Mathematical block uses a unidirectional power flow where perfect impedance matching (no loading effects) at a nominal input impedance of 1 ohm is assumed. Therefore, mathematical blocks can be connected directly to a signal source without the need for any impedance matching. On the other hand, the Physical block consists of a number of physical RF blocks where a block models an RF system by specifying physical properties of that component or by importing measured data of a real device. Figure 10.3 shows the RF blocks available from the RF Physical library. The Physical library includes the following sub libraries (The RF Blockset, 2009):

- Amplifiers: RF amplifiers, specified using network parameters, noise data, and nonlinearity data, or a data file containing these parameters.
- Ladder Filters: RF filters, specified using LC parameters. The software calculates the network parameters and noise data of the blocks from the topology of the filter and the LC values.
- Series/Shunt RLC: Series and shunt RLC components for designing lumped-element cascades, specified using RLC parameters. The software calculates the network parameters and noise data of the blocks from the topology of the components and the RLC values.
- Mixers: RF mixers that contain local oscillators, specified using network parameters, noise data, and nonlinearity data, or a data file containing these parameters.
- Transmission Lines: RF filters specified using physical dimensions and electrical characteristics. The software calculates the network parameters and noise data of the blocks from the specified data.
- Black Box Elements: Passive RF components specified using network parameters, or a data file containing these parameters. The software calculates the network parameters and noise data of the blocks from the specified data.

- Input/Output Ports: Blocks for specifying simulation information that pertains to all blocks in a physical subsystem, such as center frequency and sample time.

10.2 Nonlinear Amplifier Mathematical Models in Simulink®

10.2.1 The "Memoryless Nonlinearity" Block-Communications Blockset

The Memoryless Nonlinearity block is available from the RF Impairments sub library in The Communications Blockset and models nonlinear characteristics of nonlinear amplifiers. The block takes a complex baseband input and models AM–AM and AM–PM using one of the following memoryless or quasi-memoryless nonlinear models (The Communications Blockset, 2009):

- Cubic polynomial;
- Hyperbolic tangent;
- Saleh model;
- Ghorbani model;
- Rapp model.

Figure 10.4 shows the internal structure of the Memoryless Nonlinearity block where AM–AM and AM–PM conversions are modeled.

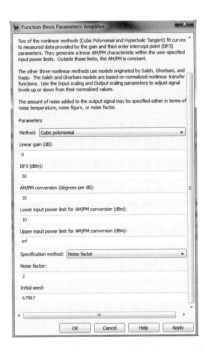

Figure 10.4 Internal structure of the "Memoryless Nonlinearity" block (The Communications Blockset, 2009).

10.2.2 Cubic Polynomial Model

The cubic polynomial models the AM–AM conversion by applying a gain factor to the magnitude of the input signal according to the following cubic polynomial equation (The Communications Blockset, 2009):

$$F_{AM-AM} = u - \frac{u^3}{3} \tag{10.1}$$

where u is the magnitude of the scaled input signal (normalized input voltage).

In this model, the AM–AM conversion is applied after scaling the input signal by a scaling factor that corresponds to the third-order input intercept point IIP3 (dBm) of the amplifier and limiting the scaled input signal to a maximum value of 1. The scaling factor is given by the factor, f, that scales the input signal before the Amplifier block applies the nonlinearity (The Communications Blockset, 2009):

$$f = \sqrt{\frac{3}{\text{IIP3(Watts)}}} \tag{10.2}$$

$$= \sqrt{\frac{3}{10^{(\text{IIP3(dBm)}-30)/10}}} \tag{10.3}$$

hence, the input signal to the AM–AM conversion block is given by $u = f|x_{in}|$.

The AM–PM conversion is specified by the AM–PM conversion parameter (degrees per dB) that specifies the phase change in the input signal by nonlinearity. The model adds a constant phase change to the input signal that depends on the input power level of the signal linearly within a range of input powers specified by the `upper input power limit` and the `lower input power limit` in dBm. Outside those limits, the phase change is constant at the values corresponding to the lower and upper input power limits, which are zero and (AM–PM conversion) (upper input power limit-lower input power limit), respectively.

$$F_{AM-PM} = \begin{cases} G_{PM}(P_U - P_L) & u > P_U \\ G_{PM}u & P_L \leq u \leq P_U \\ 0 & u < P_L \end{cases} \tag{10.4}$$

The linear gain (dB) parameter scales the output signal by the linear gain of the amplifier after the applying the nonlinear model to the normalized input.

10.2.3 Hyperbolic Tangent Model

The hyperbolic tangent model uses the same approach used in the cubic nonlinearity except for using the `tanh` function instead of the cubic polynomial to model the AM–AM conversion (The Communications Blockset, 2009):

$$F_{AM-AM} = \tanh(u) \tag{10.5}$$

where u is the input signal magnitude normalized and scaled in a similar way to the cubic polynomial model. For the AM–PM conversion, the Hyperbolic Tangent model applies a linear phase change to the normalized input in a similar way to the cubic nonlinearity model.

10.2.4 Saleh Model

In the Saleh model, the model parameters $(\alpha_a, \beta_a, \alpha_\theta, \beta_\theta)$ are used to specify the AM–AM and AM–PM conversions according to Saleh model as (The Communications Blockset, 2009):

$$F_{AM-AM}(u) = \frac{\alpha_a u}{1 + \beta_a u^2} \tag{10.6}$$

and

$$F_{AM-PM}(u) = \frac{\alpha_\theta u^2}{1 + \beta_\theta u^2} \tag{10.7}$$

where $u = f|x_{in}|$ and f is an input scaling parameter. The input scaling is specified in (dB) units where $f = 10^{\text{InputScaling}/10}$ and scales the input signal before the nonlinearity is applied. If the input scaling parameter is set to be the inverse of the input signal amplitude, the scaled signal has amplitude normalized to 1. The output scaling (dB) parameter scales the output signal in a similar way to the input scaling. The model parameters are specified in vectors: AM–AM parameters $= [\alpha_a \ \beta_a]$ and AM–PM parameters $= [\alpha_\theta \ \beta_\theta]$.

10.2.5 Ghorbani Model

In the Ghorbani model, the AM–AM conversion is specified by the parameters $[x_1, x_2, x_3, x_4]$ and the AM–PM conversion is specified by the parameters $[y_1, y_2, y_3, y_4]$ according to the following equations (The Communications Blockset, 2009):

$$F_{AM-AM}(u) = \frac{x_1 u^{x_2}}{1 + x_3 u^{x_2}} + x_4 u \tag{10.8}$$

and

$$F_{AM-PM}(u) = \frac{y_1 u^{y_2}}{1 + y_3 u^{y_2}} + y_4 u \tag{10.9}$$

where $u = f|x_{in}|$ and f is an input scaling parameter that is equal to $f = 10^{\text{InputScaling}/10}$. The output scaling (dB) parameter scales the output signal in a similar way to the input scaling.

10.2.6 Rapp Model

As discussed in Chapter 3, the Rapp model considers only the AM–AM conversion and is defined by a smoothness factor and output saturation level parameters. The AM–AM conversion is computed according to the following equation (The Communications Blockset, 2009):

$$F_{AM-AM}(u) = \frac{Gu}{\left(1 + \left(\frac{u}{V_{sat}}\right)^{2p}\right)^{1/2p}}$$

(10.10)

where u is the magnitude of the scaled signal, p is the smoothness factor and V_{sat} is the output saturation level. The output saturation level parameter limits the output signal level while the smoothness factor parameter controls the amplitude–gain characteristics.

10.2.7 Example

Figure 10.5 shows a basic model (File Name: Amp_Comm.mdl) where a sine wave is input to the "Memoryless Nonlinearity" block and the output is simulated using these nonlinear models. Since the model accepts only complex input, a constant input (zero) is added as an imaginary part to the sine wave to produce a complex input. Figure 10.6 shows the mask of the "Amplifier" block where the block parameters for all the nonlinear models shown in Table 10.1 can be specified. Figures 10.7 (b)–(f) show the output of the Amplifier block for different choices of the nonlinear model.

10.2.8 The "Amplifier" Block–The RF Blockset

The Amplifier block available from the Mathematical RF sublibrary generates a complex baseband model of a nonlinear amplifier with additive noise. The block models nonlinearity using the same nonlinear models used in the Memoryless Nonlinearity block in The Communications Blockset. It also models additive thermal white noise by specifying Noise Figure, Noise Factor or Noise Temperature (The RF Blockset, 2009). Figure 10.8 shows the mask of the Amplifier block in The RF Blockset.

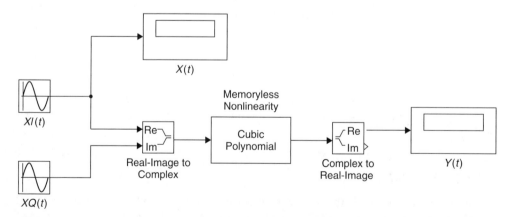

Figure 10.5 A basic model involving a "Memoryless Nonlinearity" block (The Communications Blockset, 2009).

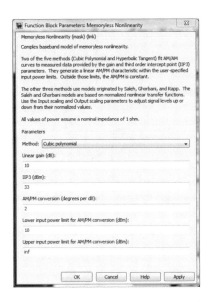

Figure 10.6 Mask of the "Memoryless Nonlinearity" block (The Communications Blockset, 2009).

Table 10.1 Parameters of the "Memoryless Nonlinearity" block (The Communications Blockset, 2009)

Model	Parameters
Cubic Polynomial	Linear Gain $= 10$ dB, IIP3 $= 33$ dBm, AM−PM $= 2$ deg/dB
Saleh	AM−AM paramters $=$ [2.1587 1.1517], AM−PM paramters $=$ [4.0033 9.1040]
Hyperbolic Tangent	Linear Gain $= 10$ dB, IIP3 $= 33$ dBm, AM−PM $= 2$ deg/dB
Ghorbani	AM−AM paramters $=$ [8.1081 1.5413 6.5202−0.0718], AM−PM paramters $=$ [4.6645 2.0965 10.88−0.003]
Rapp	Linear Gain $= 10$ dB, Smoothing Factor $= 0.5$

Thermal noise in the "Amplifier" block can be specified in three ways:

- Noise Temperature: Specifies the noise in kelvin.
- Noise Factor: Specifies the noise by the following equation: Noise Factor $= 1 + \dfrac{\text{Noise Temperature}}{290}$.
- Noise figure: Specifies the noise in decibels relative to a noise temperature of 290 kelvin. In terms of Noise Factor, Noise Figure $= 10 \log(\text{Noise Factor})$.

The block assumes a nominal impedance of 1 ohm (The RF Blockset, 2009).

Figure 10.9 shows a basic model (File Name: `Amp_RF_Math.mdl`) where a complex WCDMA signal is input to an amplifier block and the output is simulated using different nonlinear models with parameters as specified in Table 10.2. Figures 10.10 (b)–(f) show the output spectrum of the amplifier for different choices of the nonlinear models.

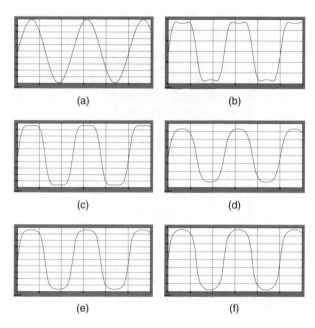

Figure 10.7 Output of "Memoryless Nonlinearity" block; (a) Input, (b) cubic polynomial model output, (c) Saleh model output, (d) hyperbolic tangent model output (e) Ghorbani model output and (f) Rapp model output.

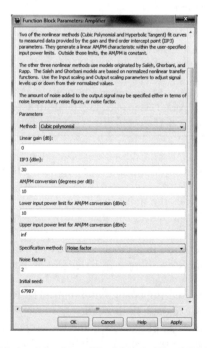

Figure 10.8 Mask of the "Amplifier" block (The RF Blockset, 2009).

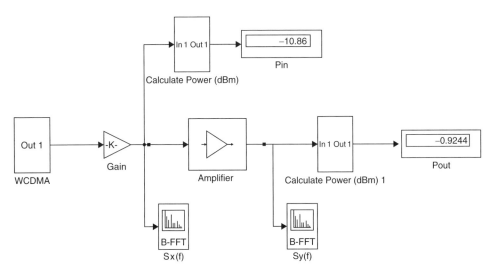

Figure 10.9 A basic model that uses the "Amplifier" block (The RF Blockset, 2009).

Table 10.2 Parameters of the "Amplifier" block (The RF Blockset, 2009)

Model	Parameters
Cubic Polynomial	Linear Gain = 10 dB, IIP3 = 33 dBm, AM–PM = 2 deg/dB
Saleh	AM–AM paramters = [2.1587 1.1517], AM–PM paramters = [4.0033 9.1040]
Hyperbolic Tangent	Linear Gain = 10 dB, IIP3 = 33 dBm, AM–PM = 2 deg/dB
Ghorbani	AM–AM paramters = [8.1081 1.5413 6.5202−0.0718], AM–PM paramters = [4.6645 2.0965 10.88−0.003]
Rapp	Linear Gain = 10 dB, Smoothing Factor = 0.5

10.3   Nonlinear Amplifier Physical Models in **Simulink**®

The Physical Amplifier sublibrary, available from the physical RF library in The RF Blockset, contains blocks that model an RF amplifier using either power data such as input/output power sweep, S-parameter, Y-parameter, Z-parameter data, or by using third-order intercept data. The RF Physical Amplifier model sublibrary consists of the following amplifier models (see Figure 10.11):

- General Amplifier;
- S-Parameters Amplifier;
- Y-Parameters Amplifier;
- Z-Parameters Amplifier.

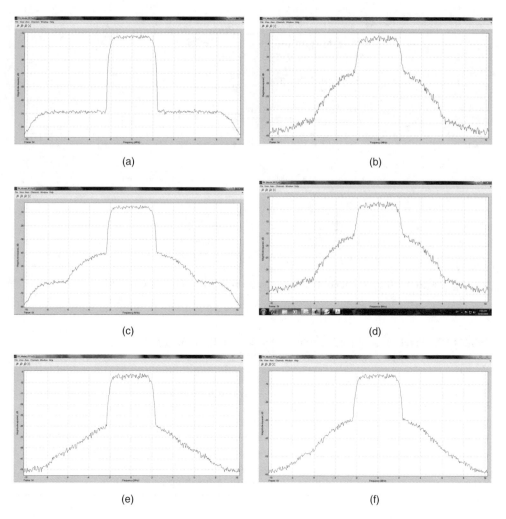

(a)   (b)

(c)   (d)

(e)   (f)

Figure 10.10 Spectrum at the output of a "Memoryless Nonlinearity" block; (a) Input, (b) cubic polynomial model output, (c) Saleh model output, (d) hyperbolic tangent model output (e) Ghorbani model output and (f) Rapp model output.

10.3.1 "General Amplifier" Block

In a General Amplifier block, one of the following specifications can be chosen:

- Power data (using a P2D, S2D, or AMP data file).
- Third-order intercept data or one or more power parameters in the block dialog box.

Figure 10.12 shows the mask of the General Amplifier block. The mask contains a number of tabs where amplifier specifications can be fed in.

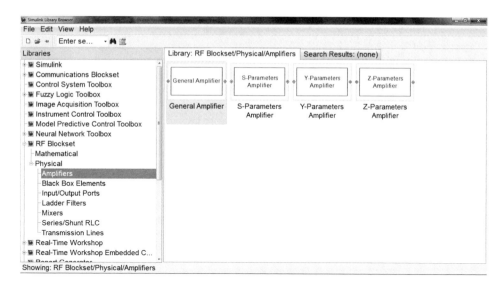

Figure 10.11 "Physical Amplifier" models in The RF Blockset (The RF Blockset, 2009).

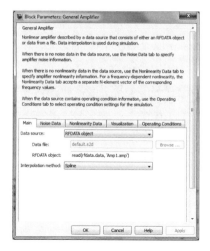

Figure 10.12 Mask of the "General Amplifier" block (The RF Blockset, 2009).

10.3.1.1 Data Source

The Data Source field consists of either an RF Blockset data (rfdata.data) object or data from a file. The Data File option enables a data file, which contains amplifier data with extensions ".s2d" or ".s2p" to be chosen. The second option is the RFDATA object that enables a data file with extension".amp" to be chosen using the command `read(rfdata. data, 'Amp1.amp')`. The structures of these data files are shown in Figures 10.13 to 10.15.

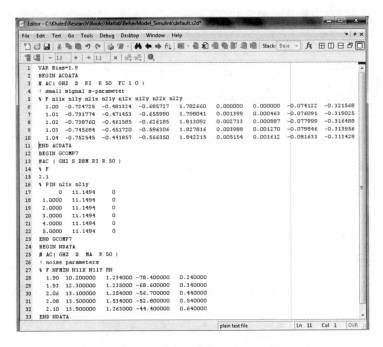

Figure 10.13   Structure of the "Amplifier" data files (The RF Blockset, 2009).

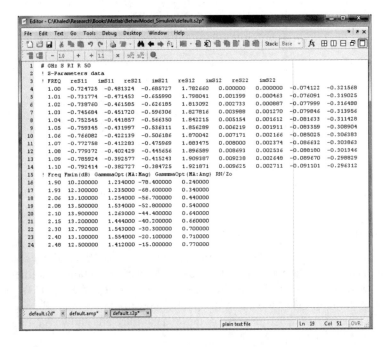

Figure 10.14   Structure of the "S-parameter Amplifier" data file (The RF Blockset, 2009).

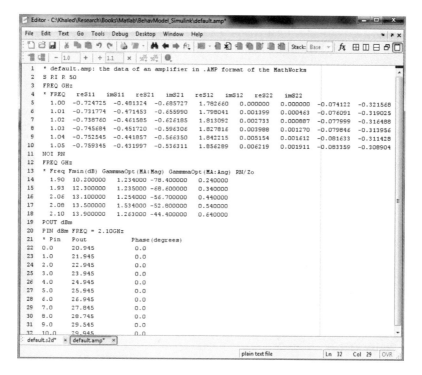

Figure 10.15 Structure of the amplifier data files (The RF Blockset, 2009).

The interpolation method is the method used to determine the values of S-parameters given in the data source (which correspond to certain frequencies) at the modeling frequency. Three interpolation methods can be used: linear, spline and cubic (Arabi, 2008; The RF Blockset, 2009).

10.3.1.2 Nonlinearity Data

Nonlinearity data can be specified if the data source does not contain power data. In this case, the following power parameters can be specified:

- IP3 type parameter: IIP3 or OIP3.
- IP3(dBm) value: Third-order intercept values at one or more frequency points.
- 1-dB gain compression power (dBm): Gain compression power in the 1-dB gain compression power (dBm) parameter.
- Output saturation power (dBm): The saturation power of the amplifier.
- Frequency (Hz): Corresponding frequency values.

Depending on which of these parameters is specified, the block computes the AM–AM and AM–PM conversions using the available data. For example, if the 1-dB compression power is not specified, the AM–AM conversion is computed as (The RF Blockset, 2009):

$$F_{\text{AM–AM}} = u - \frac{u^3}{3} \tag{10.11}$$

On the other hand, if the 1-dB gain compression power is specified, the block ignores the output saturation power specification and computes and adds the nonlinearity from the OIP3 or IIP3 value for each specified frequency point. Nonlinearity is modeled as a polynomial model where the AM–AM conversion is given by (The RF Blockset, 2009):

$$F_{\text{AM–AM}} = c_1 s + c_3 |s|^2 s + c_5 |s|^4 s + c_7 |s|^6 s \tag{10.12}$$

where the coefficients c_1, c_3, c_5, and c_7 are computed from IP3 data (The RF Blockset, 2009). The coefficients c_3, c_5, and c_7 are formulated as the solutions to a system of one, two, or three linear equations, depending on the number of provided parameters.

If the data source contains power data, the block extracts the AM-AM and AM-PM nonlinearities from the power data and all the fields in the "Nonlinearity Data" tap will be disabled.

10.3.1.3 Noise Data

If noise data does not exist in the data source, block noise can be specified in the block dialog box using one of the following noise parameters (The RF Blockset, 2009):

- Spot noise data (rfdata.noise class) object: the block uses this data to calculate noise figure.
- Spot noise data: minimum noise figure, optimal reflection coefficient and noise resistance must be specified.
- Noise figure, noise factor, or noise temperature value.
- Frequency-dependent noise figure data (rfdata.nf) object: calculates noise figure at modeling frequencies using the specified interpolation method.

10.3.1.4 Visualization

This tab provides visualization of nonlinearity data where operating frequencies and power data are either entered manually or extracted from power data file. It provides a number of plot types such as composite data, $X-Y$-plane, polar plane and Smith chart plots of the S-parameter (The RF Blockset, 2009).

10.3.1.5 Operating Conditions

Operating conditions of the General Amplifier model are specified in the data file that contains the amplifier data. Data files with extensions".p2d" and".psd" contain operating conditions at the beginning of the file. If other operating conditions need to be used, these

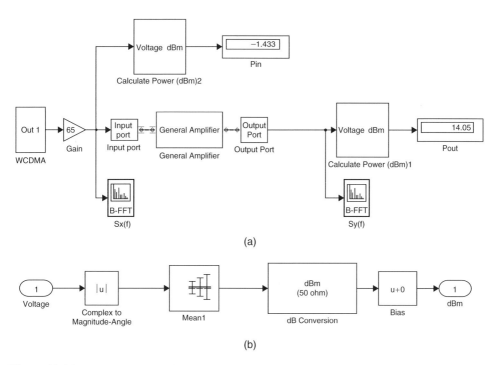

Figure 10.16 (a) Example that involves the "General Amplifier" block and (b) power calculator block.

operating conditions can be entered in the dialogue box of the amplifier block (The RF Blockset, 2009).

10.3.1.6 Example

Figure 10.16 shows a setup for simulation of a nonlinear amplifier using the General Amplifier model (File Name: `Amp_RF_GeneralPA.mdl`). The model uses data from a".amp" data file that contains power data of real amplifier obtained from measured AM–AM and AM–PM data. The amplifier has a gain of 21 dB, an output 1-dB compression point of 11 dBm, and an Output 3rd-Order Intercept (OIP3) of 18 dBm all at 2 GHz.

The input signal is a standard WCDMA signal with 5 active channels (5 DPCH) and is measured using a Vector Signal Generator (VSG) and then saved as ".mat" file. The signal is fed to the amplifier through an input port and its power is measured in dBm using the Power Calculator subsystem shown in Figure 10.16. The Input Port block enables the parameter data needed to calculate the modeling frequencies and the baseband-equivalent impulse response of the amplifier to be provided. The output of the amplifier model is fed to an output port and then the output power is calculated in dBm using the Power Calculator subsystem. The Output Port block produces the baseband-equivalent time-domain response of an input signal traveling through the amplifier model. The spectrum scopes measure the input and output spectra of the amplifier model using an FFT length

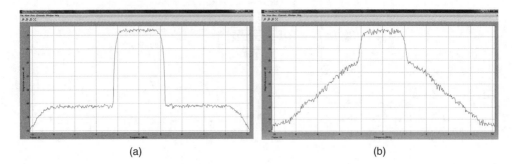

(a)   (b)

Figure 10.17 Spectra at the input and the output of the amplifier model; (a) input spectrum and (b) output spectrum.

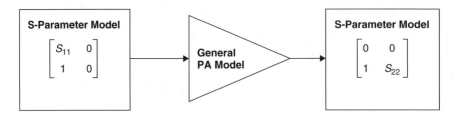

Figure 10.18 "S-Parameter Amplifier" model (Arabi, 2008).

of 5096 and a hanning window. Figures 10.17(a) and (b) show the spectra at the input and the output of the amplifier model.

10.3.2 "S-Parameter Amplifier" Block

Figure 10.18 shows the internal structure of the S-Parameter Amplifier model and Figure 10.19 shows the mask of the S-Parameter Amplifier block. The parameters of the model are (The RF Blockset, 2009):

- S-Parameters: The S-parameter vector can be specified if the Data Source field is set to "S-parameters". S-parameters are inserted as a 2-by-2-by-M array; $[S_{11}, S_{12}; S_{21}, S_{22}]$. S-parameters at different frequencies can be inserted using the cat function to create a 2-by-2-by-M array where M is the number of frequencies.
- Frequency (Hz): The frequency vector contains the frequencies of the S-parameters as an M-element vector where the order of the frequencies corresponds to the order of the S-parameters in "S-Parameters" vector. All frequencies must be positive.
- Reference Impedance (ohms): The reference real or complex impedance of the S-parameters as a scalar or a vector of length M.
- Interpolation Method: The method used to interpolate the network parameters.

Noise data and nonlinearity data can be specified in a similar way to the General Amplifier block.

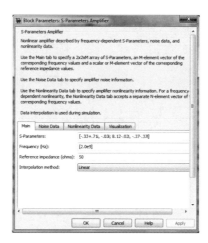

Figure 10.19 Mask of the "S-Parameter Amplifier" block (The RF Blockset, 2009).

10.4 Measurements of Distortion and System Metrics

In the following subsections, Simulink® is used to simulate a real amplifier. An amplifier model is created using the General Amplifier model available in The RF Blockset. Measured AM–AM and AM–PM characteristics are plugged in into the amplifier model. Different blocks are added to fit the purpose of simulating different figures of merit as will be shown next.

10.4.1 Adjacent-Channel Distortion

In Simulink®, there is no specific block that can be used to measure ACPR. Instead, and depending on the wireless standard, RF bandpass filters can be used to select the adjacent-channel and the main-channel bands. Figure 10.20 shows a Simulink® setup (File Name: `ACPR.mdl`) used for measurement of ACP of WCDMA signals. ACP is measured using an RF bandpass filter centered at 5 MHz offset from the carrier and with a bandwidth of 3.84 MHz. Another filter is used to measure the power in the main channel with a bandwidth of 3.84 MHz. ACPR is measured as the ratio of the power in the main channel (output power of filter 2) to the ACP power (output power of filter 1). Figure 10.21 shows the output spectra of the PA, filter 1 and filter 2.

10.4.2 In-Band Distortion

To measure in-band distortion, a feed-forward (FF) cancelation loop is used where the correlated output ($\tilde{y}_c(t)$) is subtracted from the total output of nonlinearity to produce the uncorrelated distortion component ($\tilde{y}_d(t)$) as shown in Chapter 6. Figure 10.22 shows a Simulink® model (File Name: `InBand.mdl`) for measuring uncorrelated distortion using a FF cancelation loop. Figures 10.23(a)–(c) show the spectra of the nonlinear output, the correlated spectra and the uncorrelated distortion spectrum for a WCDMA input signal.

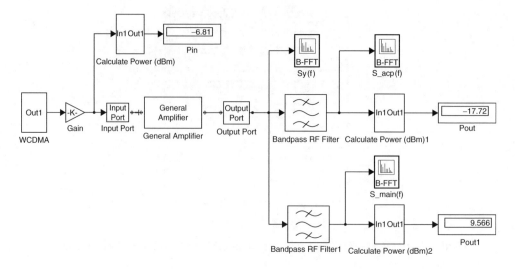

Figure 10.20 A Simulink® model for ACPR measurements.

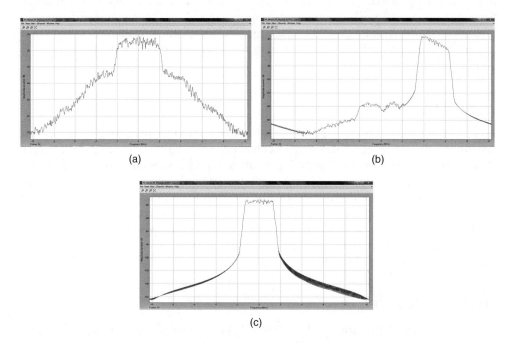

Figure 10.21 Simulation of ACPR in Simulink® using the model in Figure 10.20; (a) output spectrum, (b) ACP spectrum and (c) main-channel spectrum.

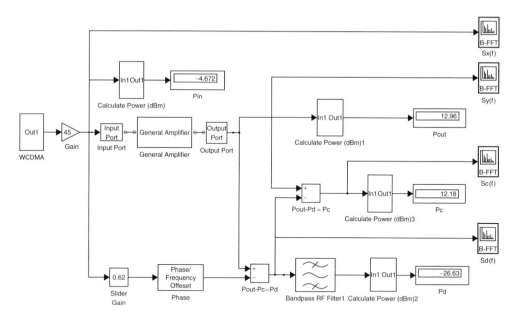

Figure 10.22 A Simulink® model for simulation of in band distortion.

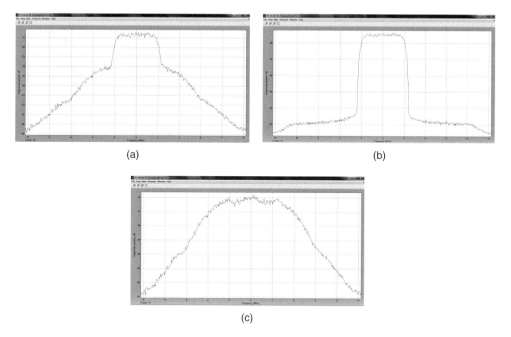

(a)

(b)

(c)

Figure 10.23 Simulated output spectrum from the model in Figure 10.22; (a) total spectrum, (b) correlated spectrum(c) uncorrelated (distortion) spectrum.

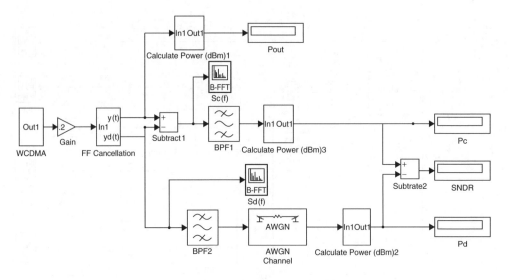

Figure 10.24   Simulink® model for simulation of SNDR.

In-band distortion is measured within the main channel bandwidth of the distortion spectrum using a bandpass filter of bandwidth of 3.84 MHz. The input power level is swept by adjusting the input signal level using a gain block to measure the distortion spectrum at different power levels. The gain and phase of the cancelation loop are adjusted at each input power level using gain and phase blocks to obtain a perfect cancelation of the correlated output.

10.4.3   Signal-to-Noise and Distortion Ratio

SNDR can be measured using the in-band distortion measurement setup used in the previous subsection. Figure 10.24 shows a model (File Name: SNDR.mdl) for measurement of SNDR in Simulink®. SNDR is obtained by subtracting the power of the uncorrelated output (in dBm) from the power of the correlated output (in dBm) within the main channel bandwidth.

10.4.4   Error Vector Magnitude

As shown in Chapter 7, EVM is defined as the normalized magnitude of the difference between the input and output of a nonlinearity. Figure 10.25 (File Name: EVM_Sim.mdl) shows a Simulink® measurement setup for EVM where nonlinear distortion and AWGN are considered and using the normalized magnitude of the difference between the input and output of a nonlinearity.

EVM can also be calculated from the ratio of the distortion and noise powers to the correlated output power, as discussed in Chapter 7. Figure 10.26 shows a measurement

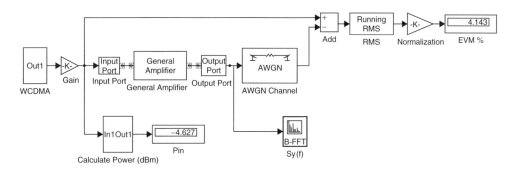

Figure 10.25 Simulink® model for EVM simulation using time-domain measurements.

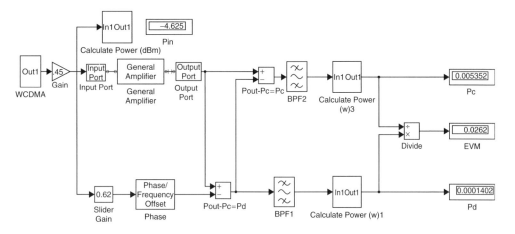

Figure 10.26 Simulink® model for EVM simulation using uncorrelated distortion spectrum.

setup for EVM (File Name: EVM_Spec.mdl) using the setup used for measurement of in-band distortion where EVM is calculated as the ratio of the output distortion power to the correlated output power (in Watts).

10.5 Example: Performance of Digital Modulation with Nonlinearity

This example is the same example presented in Chapter 8 but with a nonlinear amplifier added to the signal path after the AWGN channel, as shown in Figure 10.27 (File Name: QAM_Sim_PA.mdl). Figure 10.28 shows various plots generated by various performance blocks where it is shown how nonlinear distortion affects the constellation, signal trajectory and BER of the received signal.

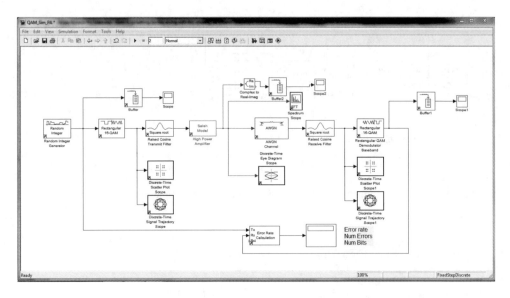

Figure 10.27 Example: QAM modulation/demodulation with amplifier nonlinearity.

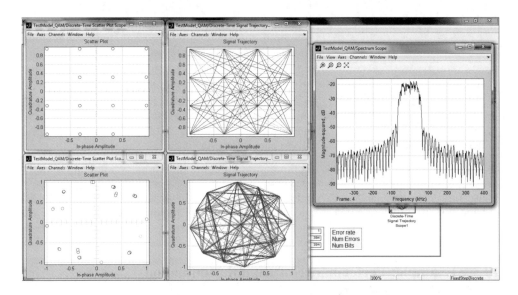

Figure 10.28 Various performance plots for the model in Figure 10.27.

10.6 Simulation of Noise-to-Power Ratio

NPR measurements are simulated in Simulink® using the setup shown in Figure 10.29 (File Name: NPR_Sim.mdl). A lightly loaded WCDMA signal (1 DPCH) signal is first applied to a band stop filter model (20 kHz stop band bandwidth) to create a notch in the

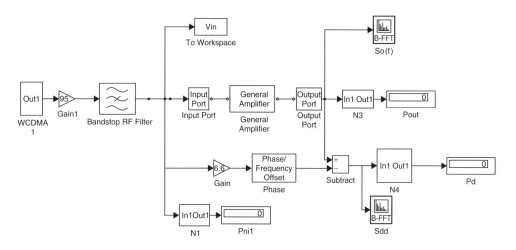

Figure 10.29 Simulink® model for simulation of NPR.

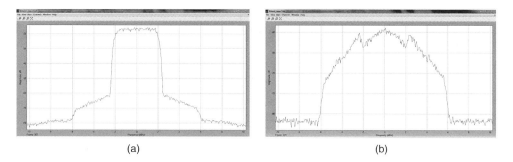

Figure 10.30 Output spectrum without the notch filter in Figure 10.29 (a) total output spectrum and (b) uncorrelated distortion spectrum.

middle of its band. The resulting signal is then applied to the amplifier model and the in-band distortion power of the output signal is measured from the output signal within the notch bandwidth. The uncorrelated distortion spectrum when the notch exists ($S_{\hat{y}_d \hat{y}_d}(f)$) is measured using the same amplifier model and using a FF cancelation loop as shown in Figure 10.22. The uncorrelated distortion spectrum when the notch filter is removed ($S_{y_d y_d}(f)$) is also measured in order to compare the effective in-band distortion with the in-band distortion obtained from NPR test.

Figures 10.30(a) and 10.30(b) show the output spectrum and the uncorrelated distortion spectrum when the notch filter is removed while Figures 10.31(a) and 10.31(b) show the output spectrum and the uncorrelated distortion spectrum when the notch filter exists. It is clear that NPR overestimates the effective in-band distortion by 8 to 10 dB for the lightly loaded signal (1 DPCH) while it represents a good estimate of the effective in-band distortion for the heavily loaded signal (16 DPCH).

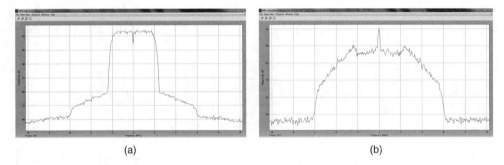

(a)   (b)

Figure 10.31 Output spectrum with the notch filter in Figure 10.29; (a) total output spectrum and (b) uncorrelated distortion spectrum.

10.7   Simulation of Noise Figure in Nonlinear Systems

The noise figure of linear amplifiers can be simulated in Simulink® using noise-modeling approaches in The RF Blockset. Noise can be included in the simulation of RF systems by enabling the noise option "Add Noise" in the Input Port block. When noise information is included in a model, the blockset simulates the noise of the physical system by combining the noise contributions from each individual block. Noise data of active RF blocks such as amplifiers and mixers can be specified in the RF data files of a certain block. If noise data is not included in the RF data file, it can be specified using the Noise Data tab in the dialog box of an RF block. On the other hand, noise data of passive devices/networks can either be automatically determined from their network parameters or from· RF data files.

Simulation of NF of nonlinear amplifiers in Simulink® cannot be done by specifying noise data as in linear amplifiers. When the amplifier is nonlinear, nonlinear distortion must be taken into account in the calculation of the noise figure since nonlinear distortion adds to system output noise. The model shown in Figure 10.32 (File Name: NF_Nonlinear.mdl) is used to simulate NF of a nonlinear amplifier. The figure shows a similar procedure presented in Chapter 9 for simulation of nonlinear noise figure in MATLAB®. The NF is calculated by measuring noise power and SNR at the input and output of a General Amplifier block when the input consists of a communication signal plus a NBGN signal. The NBGN signal represents the input noise and has twice the bandwidth of the communication signal ($B_n = 2B_s$). Nonlinear distortion that represents distortion created by both signal and noise components is measured using a feed forward cancelation loop.

In order to separate the output noise from the output signal and distortion, the WCDMA signal is fed to a replica of the first amplifier ("General Amplifier 2"). In each amplifier a FF cancelation loop is used to separate nonlinear distortion from the correlated signal component. In the first amplifier, the cancelation loop subtracts a scaled version of the WCDMA signal from the output of the amplifier, hence, the output of the FF cancelation loop consists of output noise and distortion ($P_{no} + P_d$). The output of the FF cancelation

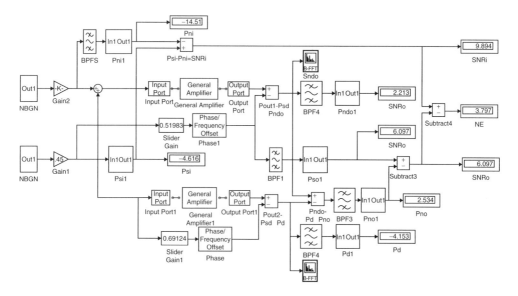

Figure 10.32   Simulink® model for simulation of NF.

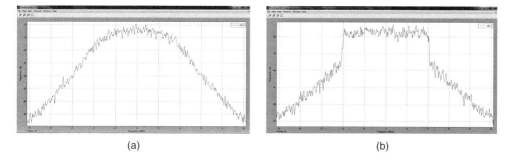

         (a)   (b)

Figure 10.33   Output spectrum of the NF simulation setup; (a) first FF cancelation loop and (b) second FF cancelation loop.

loop in the second amplifier consists of nonlinear distortion only (P_d). The effective output noise power (P_{no}) is then measured by subtracting the output of the cancelation loop of the second amplifier ($P_{no} + P_d$) from the output of the cancelation loop of the first amplifier (P_d). The output signal power P_{so} is measured from the output of the cancelation branch of the first amplifier. The output SNR is then measured as $P_{so} - P_{no}$ (in dB scale). The input power of WCDMA and noise signals and hence, the input SNR, are adjusted using a "Gain" blocks to the desired total input power and input SNR. The noise figure is then calculated as NF $= SNR_i - SNR_o$. Figure 10.33 shows the simulated spectrum at the output of each of the cancelation loops.

10.8   Summary

The simulation models presented in this chapter represent the basics of nonlinear simulations in Simulink®. Various models have been used for the simulation of nonlinear systems/circuits in Simulink® including models from The Communications Blockset and The RF Blockset. The examples presented in this chapter can be used directly and can be modified to fit different applications.

Appendix A

Basics of Signal and System Analysis

Modern wireless communication signals are usually produced as a baseband digital signal that modulates a carrier. A baseband digital signal results from either a digital data source or from an analogue data source after ADC. The modulation process is needed for the purpose of transmission over a wireless channel and for frequency division multiplexing.

Signals are represented either in time or frequency domains. The frequency-domain representation is developed using the Fourier series and the Fourier transform. The Fourier theory originated from the approximation theory where a time function can be represented by a series of weighted basis functions of known forms. The Fourier representation provides an insight into the frequency contents of the signal and hence, signals are usually characterized by the shape and width of their frequency spectrum. On the other hand, the Fourier representation is useful in characterizing the effect of passing a signal into linear or nonlinear systems where the system frequency response transforms the input signal spectrum into a new shape that may include the enhancement or attenuation of certain frequency components as in linear systems or the addition of new frequency components as in nonlinear systems.

Real-world signals are usually continuous, however, when these signals are to be generated, simulated or analyzed by computers, these signals are approximated by their samples where the sampling process result in discrete time signals. For modulated signals, the sampling frequency is usually high, which makes simulation of signals inefficient. Since the modulation process only translates the frequency band of the baseband signal to a higher frequency, useful information in a digitally modulated signal is contained in its low pass equivalent, also known as the "complex envelope". In system-level simulations, the complex envelope form of the signal is used since envelope simulations offer an advantage over simulations of passband signals; as the sampling frequency is commensurate with the bandwidth of the baseband signal and not with the carrier frequency. This makes the simulation of modulated signals more practical and more numerically efficient.

Nonlinear Distortion in Wireless Systems: Modeling and Simulation with MATLAB®, First Edition.
Khaled M. Gharaibeh.
© 2012 John Wiley & Sons, Ltd. Published 2012 by John Wiley & Sons, Ltd.

In this chapter we provide a concise treatment of signal and system representation in order to establish the theory needed to model and simulate nonlinear distortion in wireless systems.

A.1   Signals

Signals can be classified into two types: continuous and discrete. A continuous signal is defined as a real or complex function of time $x(t)$, where the independent variable t is continuous. Continuous signals constitute the basic building blocks of the construction of any other signal. Thus, continuous signals are used to define models for systems (Jeruchem *et al.*, 2000). Discrete-time signals arise from either a discrete signal source of from sampling of continuous signals at discrete times nT_s, where T_s is the sampling frequency and n is a positive integer. Discrete-time signals are therefore, real or complex functions of a discrete time variable $x(nT_s)$ or sometimes functions of integers $x(n)$ where the sampling time T_s is dropped.

On the other hand, signals can also be described by their frequency spectrum that defines the frequency contents of the signal. The frequency spectrum of a continuous signal can be developed from the Fourier transform, which is defined as:

$$X(f) = \int_{k=-\infty}^{\infty} x(t)e^{-j2\pi ft}dt \tag{A.1}$$

The Fourier transform is useful in many applications in engineering especially in the simulation of signals and systems. The Fourier transform of discrete signals is called the Discrete Fourier Transform (DFT) and is defined as:

$$X(n\Delta f) = \frac{1}{N}\sum_{k=0}^{N-1} x(kT_s)e^{-j2\pi n\Delta fkT_s} = \frac{1}{N}\sum_{k=0}^{N-1} x(kT_s)e^{-j2\pi nk/N} \tag{A.2}$$

where $x(n)$ is a discrete signal with length equal N.

Another form of DFT is the Fast Fourier Transform (FFT) that plays an important role in the simulation of signals and systems. The FFT is defined in a similar way to the DFT, however, it involves signals of length N that is a power of 2, which means that FFT can be implemented in computer simulations using time-efficient algorithms.

A.2   Systems

A system is defined as a transformation of a time function $x(t)$ (or $x(n)$ in the discrete case) into a new time function $y(t)$ (or $y(n)$ in the discrete case):

$$y(t) = G[x(t)] \tag{A.3}$$

where $G[\cdot]$ represents the transformation of the input signal by the system.

Systems are classified according to their properties such as linearity, memory, causality, stability and time invariance.

- Linear Systems: A system is linear if it enjoys the properties of superposition and homogeneity. In other words, for $y_1(t) = G[x_1(t)]$ and $y_2(t) = G[x_2(t)]$ then

$$G[\alpha x_1(t) + \beta x_2(t)] = \alpha y_1(t) + \beta y_2(t) \tag{A.4}$$

  where α and β are scaling parameters. For example: the system $y(t) = ax(t - t_0)$ is linear, while the system $y(t) = a\log(x(t))$ is nonlinear.

- Time-Invariant Systems: A system is time invariant if a time shift in the input (t_0) results in a similar time shift in the output, that is,

$$G[x(t - t_0)] = y(t - t_0) \tag{A.5}$$

  For example: the system $y(t) = tx(t)$ is time varying.

- Memoryless Systems and Systems with Memory: A memoryless system has an output at time t_0 that depends only on the input at t_0 and hence, memoryless systems are often called instantaneous. Conversely, the system is said to have memory if its output at a certain time instant depends not only at the input at that time instant but also on previous time instants (sometimes called a dynamic system). For example, the system $y(t) = ax^2(t)$ is a memoryless system, while $y(t) = ax(t - 5)$ is a system with memory.

- Causal Systems: A causal system is the system whose output at a certain time instant does not depend on future values of the input signal. For example, the system $y(t) = x(t + 1)$ is a noncausal system.

Of particular interest in most communication system simulations are linear time invariant systems that model filters, wireless channels, etc. When a system is considered as linear and time invariant, it can be modeled using the concept of the impulse response $h(t)$. The impulse response of an LTI system is defined as the response of the system to an impulse function $\delta(t)$

$$h(t) = G[\delta(t)] \tag{A.6}$$

Therefore, using the properties of linearity and time invariance, the response of an LTI system to an input signal $x(t)$ can be written as:

$$y(t) = \int_{-\infty}^{\infty} x(\tau)h(t - \tau)d\tau) = x(t) * h(t) \tag{A.7}$$

which is called the convolution operation indicated by $*$.

On the other hand, nonlinear-time invariant systems can be represented by a Volterra series that is an extension to the linear system case. Therefore, a nonlinear time invariant system can be written as:

$$y(t) = \sum_{n=1}^{\infty} \int_{-\infty}^{\infty} \ldots \int_{-\infty}^{\infty} h_n(t - \lambda_1, \ldots, t - \lambda_n) \prod_{i=1}^{n} x(\lambda_i)d\lambda_i \tag{A.8}$$

Appendix B

Random Signal Analysis

Digital communication signals exhibit a random nature where the signal envelope at any point in time does not have a deterministic value that can be predicted from past values. The spectrum of random signals cannot be characterized by the direct voltage Fourier transform because the voltage Fourier transform may not exist for most sample functions of the process. Even if it exists for a realization of the process, it may not exist for other realizations and hence the Fourier transform of a single realization does not represent the whole process.

Although random signals do not have defined amplitude, frequency or phase values, they do have some known features such as the probability distribution of amplitude, phase or frequency values that can be developed from experience gained by observing the signal over a long period of time. This rationale leads to the concept of ergodicity, which states that the statistics of a random signal can be obtained from a single realization of the signal. This concept is central to the simulation of communication signals where a long sequence of a signal is used to develop the signal statistical properties and its power spectrum.

Systems response to a random signal is also an important concept that needs to be understood when dealing with system simulations as the transformations of random signals usually entails transformation of their statistics. In linear systems, the effect of the linear transformation is manifested as a transformation to the shape of the signal spectrum. Nonlinear systems, however, produce more complex effects on the signal spectrum as well as on signal statistics, such as the probability distribution of signal parameters.

In this appendix we present the underlying theory of the probabilistic model that will be used to characterize communication signals and their nonlinear distortion. We present the theory of random variables and random processes and define their basic statistical measures such as the probability density function, mean, variance, moments, autocorrelation function, power spectrum, etc. In addition, we show how these statistics are transformed by linear and nonlinear systems.

Nonlinear Distortion in Wireless Systems: Modeling and Simulation with MATLAB®, First Edition.
Khaled M. Gharaibeh.
© 2012 John Wiley & Sons, Ltd. Published 2012 by John Wiley & Sons, Ltd.

B.1   Random Variables

In probability theory, a random variable is defined through the assignment of a real number $\mathbf{x}$ to every outcome ζ of an experiment S. A probability is then assigned to events generated by a random variable. Therefore, an event $\{\mathbf{x} < x\}$ represents a subset of S consisting of all outcomes such that $\{\mathbf{x}(\zeta) < x\}$. A probability measure P can be defined on these events by assigning a real number to each event of the form $\{\mathbf{x} < x\}$ and this number is called the distribution function $F_x(x)$ or the Cumulative Distribution Function (CDF). The distribution function is thus defined as:

$$F_x(x) = P(\mathbf{x} < x) \tag{B.1}$$

The distribution function $F_x(x)$ which can be continuous or discrete, is an increasing function with maximum value of unity since it is a probability.

A Probability Density Function (PDF) is defined as the derivative of the distribution function:

$$f_x(x) = \frac{dF_x(x)}{dx} \tag{B.2}$$

and hence the distribution function can be derived from the density function as:

$$F_x(x) = \int_{-\infty}^{x} f_x(\zeta)d\zeta \tag{B.3}$$

B.1.1   Examples of Random Variables

A random variable is uniformly distributed if it has a density function of the form:

$$f_x(x) = \left\{ \begin{array}{ll} \frac{1}{b-a}, & x \in [a, b]; \\ 0, & \text{otherwise.} \end{array} \right\}. \tag{B.4}$$

An exponentially distributed random variable has a density function:

$$f_x(x) = \left\{ \begin{array}{ll} \lambda e^{-\lambda x}, & x > 0; \\ 0, & \text{otherwise.} \end{array} \right\}. \tag{B.5}$$

A Rayleigh distributed random variable is a very common distribution because it models the envelope of a received communication signal. It is characterized by a density function of the form:

$$f_x(x) = \left\{ \begin{array}{ll} \frac{x}{\sigma^2} e^{-x^2/2\sigma^2}, & x \geq 0; \\ 0, & \text{otherwise.} \end{array} \right\}. \tag{B.6}$$

Other examples of random variables include Gaussian, Gamma, Chi-square, Nakagami-m, Beta, Cauchy, etc.

B.1.2   Functions of Random Variables

It is a well-known problem in engineering that given a random variable with a known distribution we wish to find the distribution of the random variable generated by a transformation (which can be linear or nonlinear, memoryless or with memory) of the original random variable. Therefore, let $\mathbf{x}$ be a random variable with known distribution $F_x(x)$ and let $g(x)$ be a function, we wish to find the distribution of the new random variable $\mathbf{y} = g(\mathbf{x})$. The distribution of y is then defined as:

$$F_y(y) = P(\mathbf{y} \le y) = P(g(\mathbf{x}) \le y) \tag{B.7}$$

A fundamental theorem for finding the probability density function of y in terms of the density function of x states that given $f_x(x)$ the density function of y is given by:

$$f_y(y) = \frac{f_x(x_1)}{|\acute{g}(x_1)|} + \ldots + \frac{f_x(x_n)}{|\acute{g}(x_n)|} \tag{B.8}$$

where x_n are the roots of the equation $y = g(x)$ and $\acute{g}(x)$ is the derivative of $g(x)$. For example, if $y = ax + b$; then

$$f_y(y) = \frac{1}{|a|} f_x(\frac{y-b}{a}) \tag{B.9}$$

The inverse problem in engineering is the problem of finding the transformation $g(x)$ to create a random variable with certain distribution from another random variable with a known distribution. Therefore, given a random variable $\mathbf{x}$ with distribution $F_x(x)$ we wish to find a function $g(x)$ so that the distribution of $\mathbf{y} = g(\mathbf{x})$ is $F_y(y)$. This problem is often used with random number generators where new distributions are usually generated from the transformation of uniform random variables.

B.1.3 Expectation

Let us assume that we play a game of chance in which we win the amount x_i with probability p_i. If the experiment is conducted K times we expect that x_n would be won Kp_n times. The average value of the amount won would then be

$$\frac{\sum_i K x_i p_i}{K} = \sum_i x_i p_i \tag{B.10}$$

For example, the average value of rolling a fair die is:

$$\sum_i x_i p_i = \sum_{i=1}^{6} i(\frac{1}{6}) = 3.5 \tag{B.11}$$

That is the average value of the outcomes is 3.5, which is not one of the outcomes.

If we assign a random variable to an experiment then the average value of a random variable taking values x_i with probability p_i is called the expected value (or mean value) $E(\mathbf{x}) = \sum_i x_i p_i$. For a continuous random variable the expected value is defined in terms of the probability density function as:

$$E[\mathbf{x}] = \int_{-\infty}^{\infty} x f_x(x) dx \tag{B.12}$$

The mean value of a random variable is a good parameter that describes a random variable, however, it is not always sufficient to represent its PDF.

B.1.4   Moments

The expected value of a function of a random variable is defined as:

$$E[g(\mathbf{x})] = \int_{-\infty}^{\infty} g(x) f_x(x) dx \tag{B.13}$$

Therefore, we can define the nth moment of a random variable by setting $g(x) = x^n$. Therefore, the nth moment denoted as m_n is defined as:

$$m_n = E[\mathbf{x}^n] = \int_{-\infty}^{\infty} x^n f_x(x) dx \tag{B.14}$$

where the first moment is the mean. The moment $E[(\mathbf{x} - \eta)^n]$, where $\eta = E[\mathbf{x}]$ is called the central moment and denoted μ_n:

$$\mu_n = E[(\mathbf{x} - \eta)^n] = \int_{-\infty}^{\infty} (x - \eta)^n f_x(x) dx \tag{B.15}$$

The first central moment is zero, the second central moment is called the variance, which is a measure of the width of the PDF. The third central moment is called the skew, which is a measure of the symmetry of the PDF. Higher-order moments are usually important in nonlinear system analysis. Note that the variance denoted by σ^2 is usually expressed as:

$$\sigma^2 = E[(\mathbf{x} - \eta)^2] = E[\mathbf{x}^2] - \eta^2 \tag{B.16}$$

The variance measures the spread of the distribution around the mean. A random variable with a zero variance is constant. If the random variable represents a random voltage or current, then the first moment $m_1 = \eta$ represents the DC component, the second moment m_2 represents the total average power, the second central moment (the variance σ^2 represents the average power in the AC component and the mean squared η^2 represents the power in the DC component.

B.2   Two Random Variables

Let us assume that we have two random variables $\mathbf{x}$ and $\mathbf{y}$ with distribution functions $F_x(x)$ and $F_y(y)$. The joint distribution function $F_{xy}(x, y)$ defines the probability of the event $\{\mathbf{x} \le x\} \cap \{\mathbf{y} \le y\} = \{\mathbf{x} \le x, \mathbf{y} \le y\}$, therefore:

$$F_{xy}(x, y) = P\{\mathbf{x} \le x, \mathbf{y} \le y\} \tag{B.17}$$

The marginal distribution functions $F_x(x)$ and $F_y(y)$ determine the separate distributions of the two random variables but not their joint distribution function, which is the function $F_{xy}(x, y)$. The joint distribution provides more information than the two individual distributions since it defines the dependency of $\mathbf{x}$ and $\mathbf{y}$. The joint density function is defined as:

$$f_{xy}(x, y) = \frac{\partial^2 F_{xy}(x, y)}{\partial x \partial y} \tag{B.18}$$

and it follows that the joint distribution function can be written in terms of the joint density function $f_{xy}(x, y)$ as

$$F_{xy}(x, y) = \int_{-\infty}^{x} \int_{-\infty}^{y} f_{xy}(u, v) du dv \tag{B.19}$$

The marginal distribution functions are defined in terms of the joint distribution function as

$$F_x(x) = F_{xy}(x, \infty)$$
$$F_y(y) = F_{xy}(\infty, y) \tag{B.20}$$

and hence the marginal density functions are:

$$f_x(x) = \int_{-\infty}^{-\infty} f_{x,y}(x, y) dy$$

$$f_y(y) = \int_{-\infty}^{-\infty} f_{x,y}(x, y) dx \tag{B.21}$$

B.2.1 Independence

Two random variables are said to be independent if their joint statistics can be written as:

$$F_{xy}(x, y) = F_x(x) F_y(y) \tag{B.22}$$
$$f_{xy}(x, y) = f_x(x) f_y(y) \tag{B.23}$$

Note that the independence of the random variables implies that their joint density and distribution functions do not provide additional information about the two random variables. If $\mathbf{x}$ and $\mathbf{y}$ are continuous and independent, then

$$f_{x+y}(x) = \int_{-\infty}^{-\infty} f_x(u) f_y(x - u) du = f_x(x) * f_y(y) \tag{B.24}$$

Hence, the PDF of the sum of two independent random variables is the convolution of their PDFs.

B.2.2 Joint Statistics

The expected value of a function of two random variables is given by: The expected value of a function of a random variable defined as:

$$E[g(\mathbf{x}, \mathbf{y})] = \int_{-\infty}^{\infty} \int_{-\infty}^{\infty} g(x, y) f_{xy}(x, y) dx dy \tag{B.25}$$

Therefore, we can define the nkth moment of a random variable by setting $g(x) = x^n y^k$. Therefore, the nkth moment denoted as m_{nk} is defined as:

$$m_{nk} = E[\mathbf{x}^n \mathbf{y}^k] = \int_{-\infty}^{\infty} x^n y^k f_{xy}(x, y) dx dy \tag{B.26}$$

The moments m_{01} and m_{10} are the means η_x and η_y, respectively, and the moment m_{11} is called the correlation: $m_{11} = E[\mathbf{x}\mathbf{y}] = R_{xy}$.

The nk central moments can be defined similarly as

$$\mu_{nk} = E[(\mathbf{x} - \eta_x)^n(\mathbf{y} - \eta_y)^k] = \int_{-\infty}^{\infty} \int_{-\infty}^{\infty} (x - \eta_x)^n(y - \eta_y)^k f_{xy}(x, y)dxdy \quad \text{(B.27)}$$

therefore, $\mu_{10} = \eta_x$, $\mu_{01} = \eta_y$. The moment $\mu_{11} = C_{xy}$ is called the covariance:

$$\mu_{11} = C_{xy} = E[(\mathbf{x} - \eta_x)(\mathbf{y} - \eta_y)] = E[xy] - E[x]E[y] = R_{xy} - \eta_x\eta_y \quad \text{(B.28)}$$

Also, $\mu_{20} = \sigma_x^2$ and $\mu_{02} = \sigma_y^2$, which are the variances of x and y.

The correlation coefficient of two random variables is defined as:

$$\rho_{xy} = \frac{C_{xy}}{\sigma_x\sigma_y} \leq 1 \quad \text{(B.29)}$$

The random variables are positively correlated if $C_{xy} > 0$ while they are uncorrelated if $C_{xy} = 0$. In this case $R_{xy} = E[\mathbf{x}]E[\mathbf{y}]$. The covariance is a measure of the dependence of the two random variables.

The idea is that if $E[\mathbf{x}\mathbf{y}]$ is larger than $E[\mathbf{x}]E[\mathbf{y}]$, then $\mathbf{x}$ and $\mathbf{y}$ tend to be large or small together more than if they were independent.

If the two random variables $\mathbf{x}$ and $\mathbf{y}$ are independent and continuous, then $f_{xy}(x, y) = f_x(x)f_y(y)$, so that $E[\mathbf{x}\mathbf{y}] = E[\mathbf{x}]E[\mathbf{x}]$. The covariance then equals zero. However, uncorrelatedness does not guarantee that the two random variables are independent; i.e. the equality $E[\mathbf{x}\mathbf{y}] = E[\mathbf{x}]E[\mathbf{x}]$ does guarantee that $f_{xy}(x, y) = f_x(x)f_y(y)$ unless the two random variables are jointly Gaussian, as will be shown in the following section. If $R_{xy} = E[\mathbf{x}\mathbf{y}] = 0$ then the two random variables are called *orthogonal*.

B.3 Multiple Random Variables

We define a random vector as the vector $X = [\mathbf{x}_1, \mathbf{x}_2, \ldots, \mathbf{x}_n]$ with $\mathbf{x}_i$ being random variables. The joint distribution and density functions are defined as

$$F(x_1, x_2, \ldots x_n) = P\{\mathbf{x}_1 \leq x_1, \ldots, \mathbf{x}_n \leq x_n\} \quad \text{(B.30)}$$

$$f(x_1, x_2, \ldots x_n) = \frac{\partial^n F(x_1, \ldots, x_n)}{\partial x_1, \ldots \partial x_n} \quad \text{(B.31)}$$

A group of random variables $[\mathbf{x}_1, \mathbf{x}_2, \ldots, \mathbf{x}_n]$ in a random vector are independent of another group of random variables $[\mathbf{y}_1, \mathbf{y}_2, \ldots, \mathbf{y}_n]$ if the joint density function can be written as

$$f(x_1, x_2, \ldots x_n, y_1, y_2, \ldots y_n) = f(x_1, x_2, \ldots x_n)f(y_1, y_2, \ldots y_n) \quad \text{(B.32)}$$

The mean of a function of the entries of the random vector X; $g(\mathbf{x}_1, \mathbf{x}_2, \ldots, \mathbf{x}_n)$ equals

$$E[g(\mathbf{x}_1, \mathbf{x}_2, \ldots, \mathbf{x}_n)] = \int_{-\infty}^{\infty} \cdots \int_{-\infty}^{\infty} g(\mathbf{x}_1, \mathbf{x}_2, \ldots, \mathbf{x}_n)f(x_1, \ldots, x_n)dx_1, \ldots, dx_n$$

$$\text{(B.33)}$$

The mean of a random vector is defined as

$$E[X)] = \begin{pmatrix} E[\mathbf{x}_1] \\ E[\mathbf{x}_2] \\ \vdots \\ E[\mathbf{x}_n] \end{pmatrix} = \begin{pmatrix} \eta_1 \\ \eta_2 \\ \vdots \\ \eta_n \end{pmatrix} \tag{B.34}$$

For a random vector we define the covariance matrix as:

$$C = \begin{pmatrix} C_{11} & \cdots & C_{1n} \\ \vdots & \cdots & \vdots \\ C_{n1} & \cdots & C_{nn} \end{pmatrix} \tag{B.35}$$

where $C_{ij} = E[(\mathbf{x}_i - \eta_i)(\mathbf{x}_j - \eta_j)]$ is the covariance of any two random variables in the random vector X. The correlation matrix consists of the correlation of any two random variables:

$$R = \begin{pmatrix} R_{11} & \cdots & R_{1n} \\ \vdots & \cdots & \vdots \\ R_{n1} & \cdots & R_{nn} \end{pmatrix} \tag{B.36}$$

where $R_{ij} = E[\mathbf{x}_i \mathbf{x}_j]$. Note that the correlation matrix can be written in matrix form as:

$$R = E[X^T X] \tag{B.37}$$

where T denotes matrix transpose.

If the group of random variables $[\mathbf{x}_1, \mathbf{x}_2, \ldots, \mathbf{x}_n]$ are pairwise independent then

$$\text{var}(\mathbf{x}_1 + \ldots + \mathbf{x}_n) = E[(\mathbf{x}_1, \ldots, \mathbf{x}_n)^2] \tag{B.38}$$

$$= E[\mathbf{x}_1^2] + \ldots + E[\mathbf{x}_n^2] + \sum_{i \neq j} E[\mathbf{x}_i \mathbf{x}_j] \tag{B.39}$$

$$= \text{var}(\mathbf{x}_1) + \ldots + \text{var}(\mathbf{x}_n) \tag{B.40}$$

B.4 Complex Random Variables

A complex random variable takes the form $\mathbf{z} = \mathbf{x} + j\mathbf{y}$, where $\mathbf{x}$ and $\mathbf{y}$ are real random variables. The statistics of the complex random variable $\mathbf{z}$ are defined by the joint density function $f_{xy}(x, y)$. Similarly, a complex random vector $Z = [\mathbf{z}_1, \mathbf{z}_2, \ldots, \mathbf{z}_n]$ where $\mathbf{z}_i = \mathbf{x}_i + j\mathbf{y}_i$ is characterized by the joint density function $f(x_1, \ldots, x_n, y_1, \ldots, y_n)$ of the $2n$ random variables $\mathbf{x}_i$ and $\mathbf{y}_i$. The set of complex random variables in a random vector are said to be independent if the joint density of $\mathbf{x}_i$ and $\mathbf{y}_i$ can be written as $f(x_1, \ldots, x_n, y_1, \ldots, y_n) = f(x_1, \ldots, x_n) f(y_1, \ldots, y_n)$.

The correlation matrix C_Z is defined in terms of the correlation of any two complex random variables:

$$R_{ij} = E[\mathbf{z}_i \mathbf{z}_j^*] \tag{B.41}$$

where the $*$ indicates complex conjugation. The covariance matrix C_Z is defined in terms of the covariance of any two complex random variables:

$$C_{ij} = E[(\mathbf{z}_i - \eta_i)(\mathbf{z}_j^* - \eta_j^*)]$$

$$= E[\mathbf{z}_i \mathbf{z}_j^*] - \eta_i \eta_j^* = R_{ij} - \eta_i \eta_j^* \tag{B.42}$$

Note that for complex random variables $C_{ij} = C_{ji}^*$ and $R_{ij} = R_{ji}^*$. The variance of a complex random variable $\mathbf{z}_i$ is then defined as:

$$\sigma_i^2 = E[|\mathbf{z}_i - \eta_i|^2]$$

$$= E[|\mathbf{z}_i|^2] - |\eta_i|^2 \tag{B.43}$$

B.5   Gaussian Random Variables

A Gaussian distribution (sometimes called the natural distribution) is the most commonly used distribution in communication system theory because it models noise signals. The importance of the Gaussian distribution stems from the central limit theorem, which states that in the limit, the average of a big number of independent identically distributed random variables approaches the normal distribution. The Gaussian distribution enjoys many interesting properties. It can be shown that

- The sum of independent Gaussian random variables is Gaussian.
- Random variables are jointly Gaussian if an arbitrary linear combination is Gaussian.
- Uncorrelated jointly Gaussian random variables are independent.

B.5.1   Single Gaussian Random Variable

A Gaussian random variable has the density function denoted as $N(\mu, \sigma)$ and defined as:

$$f_x(x) = \frac{1}{\sqrt{2\pi\sigma^2}} e^{-(x-\mu)^2/2\sigma^2} \tag{B.44}$$

where σ and μ are parameters the determine the shape of the function. Note that the Gaussian distribution is determined by its mean and variance. If the Gaussian random variable has zero mean and unity variance then it is called a standard Gaussian random variable and denoted as $N(0, 1)$. The distribution function is given by

$$F_x(x) = \int_{-\infty}^{x} \frac{1}{\sqrt{2\pi\sigma^2}} e^{-(u-\mu)^2/2\sigma^2} du = G\left(\frac{x-\mu}{\sigma}\right) \tag{B.45}$$

where

$$G(x) = \int_{-\infty}^{x} \frac{1}{\sqrt{2\pi}} e^{-u^2/2} du \tag{B.46}$$

B.5.2 Moments of Single Gaussian Random Variable

The moments of a Gaussian random variable have special importance since Gaussian distributions enjoy some interesting characteristics. Consider the Gaussian density function given by Equation B.44 the first moment can be found using Equation B.10 as:

$$\eta = E[\mathbf{x}] = \mu \tag{B.47}$$

The second central moment (the variance) can be found similarly as:

$$\mu_2 = E[(\mathbf{x} - \eta)^2] = \sigma^2 \tag{B.48}$$

For a zero-mean real-valued Gaussian random variable the higher-order moments are given in closed form by:

$$E[\mathbf{x}^n] = \begin{cases} 0, & n = 2k+1; \\ 1 \cdot 3 \cdot 5 \ldots (n-1)\sigma^n, & n = 2k. \end{cases} \tag{B.49}$$

and

$$E[|\mathbf{x}|^n] = \begin{cases} 1 \cdot 3 \cdot 5 \ldots (n-1)\sigma^n, & n = 2k+1; \\ 2^k k! \sigma^{2k+1} \sqrt{2/\pi}, & n = 2k. \end{cases} \tag{B.50}$$

Moments of random variables with other distributions than Gaussian can also be found in closed form as shown in Papoulis (1994).

B.5.3 Jointly Gaussian Random Variables

Two random variables are jointly Gaussian if their joint density function is given by

$$f_{xy}(x, y) = \frac{1}{2\pi\sigma_1\sigma_2\sqrt{1 - r^2}}$$

$$\times \exp\left\{-\frac{1}{1 - r^2}\left(\frac{(x - \eta_1)^2}{\sigma_1^2} - 2r\frac{(x - \eta_1)(y - \eta_2)}{\sigma_1\sigma_2} + \frac{(y - \eta_2)^2}{\sigma_2^2}\right)\right\} \tag{B.51}$$

where η_1 and η_2 are the means and σ_1^2 and σ_2^2 are the variances of the random variables $\mathbf{x}$ and $\mathbf{y}$, respectively. Therefore, the joint density function is denoted as $N(\eta_1, \eta_2, \sigma_1^2, \sigma_2^2, r)$. The parameter r represents the correlation coefficient as defined in Equation (B.29). The marginal densities can be found as in Equation (B.20) where it can be shown that they are still Gaussian with density functions of the form Equation (B.44). Note that when the two random variables are uncorrelated ($r = 0$) then the joint density function can be written as the product of the marginal density functions, which means that the two random variables are independent; an important property of Gaussian random variables. The sum of two jointly Gaussian random variables is also Gaussian.

B.5.4   Price's Theorem

Price's theorem is an important theorem for evaluating higher-order joint moments of functions of two jointly Gaussian random variables (Price, 1958). Price's theorem states that if $g(x, y)f_{xy}(x, y) \to 0$ as $(x, y) \to \infty$ where g is any function of the random variables $\mathbf{x}$ and $\mathbf{y}$, then

$$\frac{\partial^n I(C)}{\partial C^n} = E\left[\frac{\partial^{2n} g(\mathbf{x}, \mathbf{y})}{\partial \mathbf{x}^n \partial \mathbf{y}^n}\right] \tag{B.52}$$

where C is the covariance of $\mathbf{x}$ and $\mathbf{y}$ and I is the expected value:

$$I = E[g(\mathbf{x}, \mathbf{y})] = \int_{-\infty}^{\infty} \int_{-\infty}^{\infty} g(x, y)f_{xy}(x, y)dxdy \tag{B.53}$$

An important result of Price's theorem is that if $\mathbf{x}$ and $\mathbf{y}$ are jointly Gaussian then

$$E[\mathbf{x}^2\mathbf{y}^2] = E[\mathbf{x}^2]E[\mathbf{y}^2] + 2E^2[\mathbf{xy}] \tag{B.54}$$

B.5.5   Multiple Gaussian Random Variable

Random variables in a random vector $X = [\mathbf{x}_1, \mathbf{x}_2, \ldots, \mathbf{x}_n]$ are said to be jointly Gaussian if $\mathbf{a}^T X$ is Gaussian for any vector $\mathbf{a}$, that is they are jointly Gaussian is their linear combination forms a Gaussian random variable. Their joint density function is given by

$$f_X(X) = \frac{1}{\sqrt{(2\pi)^n |C_X|}} \exp\{-\frac{1}{2} X C_X^{-1} X^T\} \tag{B.55}$$

where C_X is the covariance matrix as defined in Equation (B.35) and $|\bullet|$ indicates determinant.

Note that for a complex normal vector Z and if $E[\mathbf{z}_i] = 0$ the covariance matrix of the random vector Z can be written as:

$$C_Z = E[Z^T Z^*] = C_X + C_Y - j(C_{XY} - C_{YX}) \tag{B.56}$$

where C_Z has elements $C_{ij} = E[\mathbf{z}_i \mathbf{z}_j^*]$, C_X has elements $C_{ij} = E[\mathbf{x}_i \mathbf{x}_j]$, C_Y has elements $C_{ij} = E[\mathbf{y}_i \mathbf{y}_j]$, C_{XY} has elements $C_{ij} = E[\mathbf{x}_i \mathbf{y}_j]$ and C_{YX} has elements $C_{ij} = E[\mathbf{y}_i \mathbf{x}_j]$. For the special case when $C_X = C_Y$ and $C_{XY} = C_{YX}$, the joint density function of a complex Gaussian random vector can then be given by

$$f_Z(Z) = \frac{1}{\pi^n |C_Z|} \exp\{-Z^* C_Z^{-1} Z^T\} \tag{B.57}$$

An important result from the properties of jointly Gaussian random variables is the following:

$$E[\mathbf{x}_1\mathbf{x}_2\mathbf{x}_3\mathbf{x}_4] = C_{12}C_{34} + C_{13}C_{24} + C_{14}C_{23} \tag{B.58}$$

where $C_{ij} = E[\mathbf{x}_i\mathbf{x}_j]$. If the random variables are uncorrelated Gaussian then they are independent and hence their joint density function can be written as the product of their marginal density functions. The converse is also true.

Another important result from the properties of jointly Gaussian random variables is Miller's formula on the moments of complex Gaussian random variables (Miller, 1969). Miller formula derives a closed form for the higher-order moments of complex multiple random variables:

$$E[(\mathbf{x}_{i_1}^* - \eta_{i_1}^*) \ldots (\mathbf{x}_{i_n}^* - \eta_{i_n}^*)(\mathbf{x}_{j_1} - \eta_{j_1}) \ldots (\mathbf{x}_{j_m} - \eta_{j_m})] = 0 \tag{B.59}$$

if $m \neq n$ and for $m = n$:

$$E[(\mathbf{x}_{i_1}^* - \eta_{i_1}^*) \ldots (\mathbf{x}_{i_n}^* - \eta_{i_n}^*)(\mathbf{x}_{j_1} - \eta_{j_1}) \ldots (\mathbf{x}_{j_m} - \eta_{j_m})] = \sum C_{j_1 i_{\pi(1)}} C_{j_n i_{\pi(n)}} \tag{B.60}$$

where i_k and j_k, are integers from the set $(1, 2, ..., N)$ and $\pi(n)$, is a permutation of the set of integers $(1, 2, ..., n)$. For a zero-mean random variable:

$$E[\mathbf{x}_1^* \mathbf{x}_2^* \mathbf{x}_3 \mathbf{x}_4] = C_{31} C_{42} + C_{32} C_{41} \tag{B.61}$$

and it follows that

$$E[|\mathbf{x}|^{2n}] = n!(E[|\mathbf{x}|^2])^n$$

$$E[(\mathbf{x}_1^* \mathbf{x}_2)^n] = n!(E[\mathbf{x}_1^* \mathbf{x}_2])^n \tag{B.62}$$

The main advantage of the complex formalism in this application lies in the fact that the moment matrix for the complex process has half the dimensionality of the corresponding real moment matrix for the process when regarded as a joint real process.

B.5.6 Central Limit Theorem

The central limit theorem is one of the weak convergence properties of sequences of random variables. It is well known in probability theory that a sequence of independent identically distributed (iid) random variables converges to a particular distribution. The central limit theorem states that if the random variables have finite variance then the sum of these variables converges to a Gaussian distribution. The mathematical explanation of the CLT is as follows: Let $S_n = \mathbf{x}_1 + \mathbf{x}_2 + \ldots + \mathbf{x}_n$ where $\mathbf{x}_i$ are iid with equal means η and equal variances σ then the variance of S_n is $\sigma_S^2 = n\sigma^2$ and the mean is $\eta_S = \eta$. The central limit theorem states that as n increases the random variable S_n approaches Gaussian distribution with mean $n\mu$ and variance $n\sigma^2$; i.e. $N(n\eta, n\sigma^2)$. In other words let

$$Z_n = \frac{S_n - n\eta}{\sqrt{n}\sigma} \tag{B.63}$$

then

$$\lim_{n \to \infty} P(Z_n < z) = G(z) = \int_{-\infty}^{z} \frac{1}{\sqrt{2\pi}} e^{-u^2/2} du \tag{B.64}$$

which means that $F_z(z) \to G(z)$ as $n \to \infty$

B.6 Random Processes

In a definition of random variables, we assign a number $\mathbf{x}$ to an outcome ζ of an experiment S. If ζ designates the outcome of the random experiment at the time t, then the collection of random variables $X = x(t; \zeta)$, where $t \in T$ and T designates the set of times is called a random process. Therefore a random process is a mapping from a sample space to an ensemble of time functions. When T is countable, then $x(t)$ is a discrete-time random process and when T is an interval (possibly infinite), then $x(t)$ is called a continuous-time random process.

A random process can therefore be characterized by a finite vector of random variables with joint CDF $F(X)$ and joint PDF $f_X(X)$ of any finite set of random variables $x(t_1), ..., x(t_n)$. Therefore, the CDF of a random process is called a finite-dimensional distribution function of a random process and is defined as $F(x_1, ...x_n, t_1, ..., t_n) = P(x(t_1) \leq x_1, ..., x(t_n) \leq x_n)$. The PDF of a random process can then be defined as:

$$f(x_1, ..., x_n, t_1, ..., t_n) = \frac{\partial^n F(x_1, .., x_n, t_1, ..., t_n)}{\partial x_1 ... \partial x_1} \tag{B.65}$$

Note that since $X = x(t, \zeta)$ and t and ζ are variables then X represents a family of time functions for every ζ, which altogether are called an ensemble. If t is fixed then X represents a random variable.

A common example of a random process is an electric signal with random amplitude and phase

$$x(t) = \mathbf{v}\cos(\omega t + \phi) \tag{B.66}$$

here $\mathbf{v}$ and ϕ are random variables that may or may not be correlated. The time function $x(t)$ represent a continuous random process.

The mean of a random process is the time function defined as

$$\eta(t) = E[x(t)] = \int_{-\infty}^{\infty} x f(x, t) dx \tag{B.67}$$

The correlation of two random variables $x(t_1)$ and $x(t_2)$ $R(t_1, t_2)$ derived from the process $x(t)$ at times t_1 and t_2 is called the autocorrelation function of the process and is defined as

$$R(t_1, t_2) = E[x(t_1)x(t_2)] = \int_{-\infty}^{\infty} \int_{-\infty}^{\infty} x_1 x_2 f(x_1, x_2, t_1, t_2) dx_1 dx_2 \tag{B.68}$$

By setting $t_1 = t$ and $t_2 = t + \tau$, where $\tau = t_2 - t_1$ the autocorrelation function can be written as $R(t, t + \tau) = E[x(t)x(t + \tau)]$. Note that $R(t, t) = E[x(t)^2]$, which represents the average power of the process that is a function of time. The autocovariance of the process is defined as

$$C(t_1, t_2) = E[x(t_1)x(t_2)] - E[x(t_1)]E[x(t_2)] = R(t_1, t_2) - \eta(t_1)\eta(t_2) \tag{B.69}$$

A complex random process $Z(t) = x(t) + jy(t)$ is characterized by the joint statistics of the real processes $x(t)$ and $y(t)$. The autocorrelation function is defined as $R(t_1, t_2) = E[x(t_1)x^*(t_2)]$ and the autocovariance as $C(t_1, t_2) = E[x(t_1)x^*(t_2)] - \eta_1\eta_2^*$.

B.6.1 Stationarity

Stationarity is one of the important properties of random processes. It states that the statistics of the process do not change over time. This means that the nth-order density function $f(x_1, ..., x_n, t_1, ..., t_n)$ does not vary with a time shift; that is $f(x_1, ..., x_n,$ $t_1, ..., t_n) = f(x_1, ..., x_n, t_1 + \tau, ..., t_n + \tau)$. This means that the process has the same distribution for any time shift τ. In other words, the joint statistics of X of all orders are unaffected by a shift in time.

Stationarity in the density function is called Strict Sense Stationarity (SSS). Another form of stationarity is the wide sense stationarity. A random process is called WSS if the following hold:

1. The mean $\eta(t)$ is constant.
2. The autocorrelation function is a function of time shift and not the absolute time:
   $R(t, t + \tau) = R(\tau) = E[x(t)x(t + \tau)]$.

This implies that by fixing the time at different time instants, the resulting random variables have the same mean regardless of the absolute time. Moreover, property 2 implies that the random variables that result from fixing the time at two time instants t_1 and t_2 have the same correlation as the random variables resulting from fixing the time at time instants $t_1 + \tau$ and $t_2 + \tau$. Note that if the process is SSS then this implies that it is WSS. However, the converse is not true where WSS does not imply SSS. A result of the WSS property of a random process that represents an electric signal is that the average power of the process is independent of time; i.e. $E[|x(t)|^2] = R(0)$ =constant.

B.6.2 Ergodicity

Ergodicity is an important property of a random process that enables the statistics of the process to be estimated from real data realizations. Roughly speaking, a stochastic process is ergodic if the statistics of the process do not depend on the realization of the process. That is, one realization of the process is representative of the ensemble of the process.

A process is called ergodic is the statistical averages are equal to the time averages of the process. Since the time averages are constants the equality with the statistical averages requires that the process be stationary. Ergodicity therefore implies stationarity, however, the converse is not true since there are processes that are stationary but not ergodic. For example, the statistical mean of the process can be found from the time average of a single realization of the process:

$$\eta = E[x(t)] = \int_{-\infty}^{\infty} x f(x, t) dx = \lim_{T \to \infty} \frac{1}{2T} \int_{-T}^{T} x(t) dt \qquad (B.70)$$

and the autocorrelation of an ergodic process can be found as

$$R(\tau) = E[x(t)x(t + \tau)] = \int_{-\infty}^{\infty} \int_{-\infty}^{\infty} x_1 x_2 f(x_1, x_2, \tau) dx$$

$$= \lim_{T \to \infty} \frac{1}{2T} \int_{-T}^{T} x(t)x(t + \tau) dt \qquad (B.71)$$

B.6.3 White Processes

A random process $x(t)$ is called white if its samples (the random variables resulting from fixing the time) at any arbitrary time instants t_i and t_j ($t_i \neq t_j$) are uncorrelated.

This implies that the autocovariance function takes the form $C(t_1, t_2) = q(t_1)\delta(t_1 - t_2)$. Furthermore, if $x(t)$ is stationary then $q(t) = q$ and the autocovariance can be written as:

$$C(\tau) = q\delta(\tau) \tag{B.72}$$

A white process can have any distribution, however, the whiteness of the process depends on the form of its autocovariance function, which must be a delta function. A common assumption for noise in communication channels is the white Gaussian noise where the samples of the noise process have Gaussian distribution, while the autocorrelation of the process is a delta function.

The significance of a white noise process stems from the fact that passing a white noise process through a linear filter, the resulting process has an autocorrelation function that takes the shape of the impulse response of the filter, as will be shown in the next sections.

B.6.4 Gaussian Processes

A Gaussian process $x(t)$ is the process whose samples $x(t_1), ..., x(t_n)$ are jointly Gaussian for any n and $t_1, ..., t_n$. A Gaussian process has a density function given by Equation (B.55) with $X_i = x(t_i)$ and $C_{X_i X_j} = C(x(t_i)x(t_j)) = C(t_i, t_j)$.

Since the statistics a Gaussian random vector are uniquely determined by its covariance matrix and the vector of mean values, the statistics a Gaussian random process are uniquely determined by its autocovariance function. Therefore, given any arbitrary mean function $\eta(t)$ autocovariance function $C(t_1, t_2)$ we can construct a Gaussian process. One common example is the AWGN process whose autocovariance function is given by a delta function.

If a Gaussian process is WSS then it is SSS. This results from the fact that for any $t_1, t_2, ..., t_n$, the random vector $[x(t_1 + \tau), x(t_2 + \tau), ..., x(t_n + \tau)]^T)$ is Gaussian with mean $[\mu, \mu, ..., \mu]^T$ and covariance matrix with ijth entry $C_x((t_i + \tau) - (t_j + \tau)) = C_x(t_i - t_j)$. This means that the mean and covariance matrix do not depend on τ. Thus, the distribution of the vector does not depend on τ. Therefore, $x(t)$ is stationary. In summary, if $x(t)$ is stationary then $x(t)$ is WSS, and if $x(t)$ is both Gaussian and WSS, then $x(t)$ is stationary.

B.7   The Power Spectrum

The spectrum of a deterministic signal is defined as the Fourier transform of the signal that gives the voltage density against frequency. The existence of the Fourier transform requires that the signal has finite energy. Finite power signals can also have Fourier transforms. For random processes, a voltage Fourier transform is not defined since random processes may have infinite energy. The spectrum of a random process is defined as the Fourier transform of the average power of the process that is defined by the autocorrelation function. The average power of a random process is usually finite, and hence a Fourier transform exists. Furthermore, since the power spectrum is the Fourier transform of an average then it is a deterministic quantity and it is real.

Therefore, given a WSS random process $x(t)$, the power spectrum (denoted by $S_x(\omega)$) is defined as the Fourier transform of its correlation function R_{xx} is denoted by S_{xx}.

$$S_{xx}(\omega) = \int_{-\infty}^{\infty} R_x(\tau)e^{-j\omega\tau} d\tau \tag{B.73}$$

The power spectrum is a real quantity regardless of the process being real or complex. This results from the definition of the power spectrum and the fact that $R_{xx}(-\tau) = R_{xx}^*(\tau)$.

Similarly, if $y(t)$ and $x(t)$ are jointly WSS, then the cross power spectrum of $y(t)$ and $x(t)$ is defined as the Fourier transform of their cross-correlation function

$$S_{yx}(\omega) = \int_{-\infty}^{\infty} R_{yx}(\tau)e^{-j\omega\tau} d\tau \tag{B.74}$$

The inverse Fourier inversion of $S_{xx}(\omega)$ yields the autocorrelation function

$$R_{xx}(\tau) = \frac{1}{2\pi} \int_{-\infty}^{\infty} S_{xx}(\omega)e^{j\omega\tau} d\omega \tag{B.75}$$

In particular, the average power of the process is defined in terms of the power spectrum as

$$E[|x(t)|^2] = R_{xx}(0) = \frac{1}{2\pi} \int_{-\infty}^{\infty} S_{xx}(\omega)d\omega \tag{B.76}$$

Note that this quantity is always positive and it follows by the Wiener-Khinchin theorem that the power spectrum $S_{xx}(\omega)$ is always positive.

The average power of the process within a frequency band $B = [\omega_1, \omega_2]$ is defined as

$$P_B = \frac{1}{2\pi} \int_{-\omega_1}^{\omega_2} S_{xx}(\omega)d\omega \tag{B.77}$$

B.7.1   White Noise Processes

The power spectrum of a stationary white noise process is the Fourier transform of its autocorrelation function, which is defined by Equation (B.72) as $R(\tau) = N_0/2\delta(\tau)$. Therefore, the power spectrum of a white noise process is

$$S_{xx}(\omega) = N_0/2 \tag{B.78}$$

which defines a constant power spectrum over all frequencies.

A band limited white noise process is the noise process whose spectrum is nonzero and constant over a finite frequency band W and zero everywhere else. Thus, the power spectrum of a bandlimited white noise is given by (Peebles, 1987):

$$S_{xx}(\omega) = \begin{cases} \frac{P\pi}{W}, & -W < \omega < W \\ 0, & \text{elsewhere.} \end{cases} \tag{B.79}$$

where P is the noise power. The autocorrelation function of this process is found from the inverse Fourier transform of the power spectrum:

$$R_{xx}(\tau) = P\frac{\sin(W\tau)}{W\tau} \tag{B.80}$$

B.7.2 Narrowband Processes

Narrowband process refers to processes with a power spectrum confined to a narrowband of frequencies W much smaller than some frequency ω_0 where the spectrum is at its maximum. That is, if $x(t)$ is a WSS real process then $x(t)$ is a narrowband process if its power spectrum satisfies

$$S_{xx} \neq 0, \, 0 < \omega_0 - W_1 < |\omega| < \omega_0 - W_1 + W;$$

$$S_{xx} = 0, \qquad\qquad\qquad \text{elsewhere.} \qquad\qquad (B.81)$$

where W and W_1 are real positive constants.

A narrowband random process can be represented as a baseband random process that is modulated by a deterministic sinusoid. Hence, complex random processes naturally arise as baseband equivalent processes for real-valued narrowband random processes. A NB process is defined in the time domain as follows: Let $u(t)$ and $v(t)$ be jointly WSS real-valued baseband random processes, and let $x(t)$ be defined as (Hajek, 2006):

$$x(t) = u(t)\cos(\omega_c t) - v(t)\sin(\omega_c t) \qquad\qquad (B.82)$$

and can be written as

$$x(t) = A(t)\cos(\omega_0 t + \Phi(t)) \qquad\qquad (B.83)$$

where $A(t) = \sqrt{u^2(t) + v^2(t)}$ and $\Phi(t) = \tan^{-1}[u(t)/v(t)]$.

Equivalently, defining the complex envelope of $x(t)$ by $\tilde{x}(t) = u(t) + jV(t)$, then $x(t) = \textbf{Re}\{\tilde{x}(t)e^{j\omega_c t}\}$. $x(t)$ is then a NB process However, such an $x(t)$ need not even be WSS. For $x(t)$ to be WSS the following must hold:

- $\eta_u(t) = \eta_v(t) = 0$;
- $R_u(\tau) = R_v(\tau)$;
- $R_{uv}(\tau) = -R_{vu}(\tau)$.

Equivalently, $x(t)$ is WSS if and only if $\tilde{x}(t) = u(t) + jV(t)$ has a zero mean and $E[\tilde{x}(t_1)\tilde{x}(t_2)] = 0$ for all t_1, t_2. The autocorrelation function of the NB process $x(t)$ can then be written as

$$R_{xx}(\tau) = R_{uu}(\tau)\cos(\omega_0 \tau) + R_{uv}(\tau)\sin(\omega_0 \tau) \qquad\qquad (B.84)$$

and the autocorrelation function of the complex envelope of $x(t)$ can be written as

$$R_{xx}(\tau) = \textbf{Re}\{R_{\tilde{x}\tilde{x}}(\tau)e^{j\omega_0 \tau}\} \qquad\qquad (B.85)$$

where $R_{\tilde{x}\tilde{x}}(\tau) = E[\tilde{x}(t)\tilde{x}^*(t + \tau)]$.

The power spectrum of a NB process can be found using Equation (B.81) and Equation (B.82) as:

$$S_{xx}(\omega) = \frac{1}{2}\left[S_{uu}(\omega - \omega_0) + S_{uu}(\omega + \omega_0) - jS_{uv}(\omega - \omega_0) + jS_{uv}(\omega + \omega_0)\right] \qquad (B.86)$$

Narrowband White Noise (NBWN) is defined as a narrowband process whose spectrum is nonzero and constant over a finite frequency band and zero everywhere else. Therefore such a spectrum takes the form (Peebles, 1987):

$$S_{xx}(\omega) = \begin{cases} \frac{P\pi}{W}, & \omega_0 - W/2 < |\omega| < \omega_0 + W/2 \\ 0, & \text{elsewhere.} \end{cases} \qquad \text{(B.87)}$$

and the autocorrelation function is then expressed as

$$R_{xx}(\tau) = P \frac{\sin(W\tau/2)}{W\tau/2} \cos(\omega_0 \tau) \qquad \text{(B.88)}$$

Note that a Gaussian process $x(t)$ can be a white noise process provided that it has a power spectrum as defined by Equation (B.87). In this case the $x(t)$ can be written as in Equation (B.83) and the processes $A(t)$ and $\Phi(t)$ are Rayleigh and uniform in $[0, 2\pi]$, respectively, and the process is called the Narrowband Gaussian Noise (NBGN) process.

Appendix C

Introduction to MATLAB®

MATLAB® is an efficient simulation tool for various math and engineering disciplines. In addition, MATLAB® is a language for technical computing, where scripts can be written using built-in functions and used for a given simulation problem. Therefore, MATLAB® performs computation, programming and visualization using a user-friendly interface that enables complex design, modeling and simulation problems to be handled efficiently.

The basic element in MATLAB® is the array. Therefore, it handles computations using matrix-based programming where variables, functions or mathematical expressions are dealt with as matrices. Mathematical operations are applied to these matrices and yield new matrices. MATLAB® can perform different types of operations including arithmetic and logical operations, mathematical functions, graphical functions, and input/output operations (Czylwik, 2005). Mathematical operations can be performed using MATLAB® expressions that are based on variables, numbers, operators and functions. These expressions involve matrices and are used to perform computations for a given problem.

In this appendix, an introduction to MATLAB® including the basic MATLAB® tools needed for the simulation of communications systems is presented. These include deterministic and random signals generation and Fourier analysis. The basics of the use of MATLAB® commands and their functionalities will not be discussed here and the readers are referred to different MATLAB® handbooks, MATLAB® manuals and MATLAB® help which are all available on the web site of TheMathWorks™ (MATLAB® Documentation, 2009).

C.1 MATLAB® Scripts

MATLAB® can perform a wide variety of mathematical operations using its commands, which are built-in functions. These functions can be used in the main command menu or by using M-files. Simple problems can be solved efficiently in the command menu but if the problem involves the use of a large number of commands, then an M-file needs to be used. M-files can either be scripts or functions. A script performs a number of commands one-by-one in a similar way to the main command window. On the other

Nonlinear Distortion in Wireless Systems: Modeling and Simulation with MATLAB®, First Edition.
Khaled M. Gharaibeh.
© 2012 John Wiley & Sons, Ltd. Published 2012 by John Wiley & Sons, Ltd.

hand, a function accepts a set of input arguments and returns another set of output arguments. The main difference between scripts and functions is that the variables in a function are local to that function and do not appear on the workspace, while all the variables in a script can be recalled in the workspace. On the other hand, the first line in a function needs to start with the declaration of the function, its input arguments and its output arguments. This line appears as

```
function [a,b,c] = MyFunction(x, y, z)
```

where the `MyFunction` is the name of the function, `a,b,c` are the output arguments and `x, y, z` are the input arguments.

C.2 MATLAB® Structures

A structure in MATLAB® is a group of fields that are accessed by the structure name and the field name. For example, the commands:

```
Circuit.InputResistance=50
Circuit.OutputResistance=1e3
Circuit.Capacitance=1e-12
```

establish a MATLAB® struct under the name `Circuit` that has the fields `InputResistance`, `OutputResistance`, and `Capacitance`. When this struct is called in the command menu, the fields are displayed as:

```
>> Circuit
Circuit =
     InputResistance: 50
    OutputResistance: 1000
         Capacitance: 1.0000e-012
```

Stucts are useful in simulations of communication systems because a number of parameters of a system or a process can be grouped in one variable. For example, objects used to measure the performance of a communication system such as `commmeasure` are structs.

C.3 MATLAB® Graphics

MATLAB® is a very efficient tool for visualization of vectors and matrices through its various two- and three-dimensional plotting functions. For example, the `plot` function provides cartesian plots of a given vector with the ability to specify a color, axes, labels, titles, markers, etc. Other 2D plotting functions include `bar`, `stem`, `stairs`, `polar`. On the other hand, 3D plotting can be performed through the various 3D plotting functions such as `mesh`, `surf`, `surfl` and `contour`. These functions provide 3D plots defined by four matrix arguments (X, Y, Z, C) of 2D functions of the form $Z = f(X, Y)$ where C defines the color map of the plot.

C.4 Random Number Generators

Uniform random numbers can be generated in MATLAB® using the command `rand(1,n)`, which generates an $1 \times n$ vector of uniformly distributed random numbers in [0,1]. The default random numbers have a zero mean and a unity variance. For example, the command

```
x=rand(1,100)
```

generates a random vector x of size 100 whose elements are uniformly distributed within the range [0,1]. To generate random numbers that are uniformly distributed in an arbitrary interval [a,b], a simple linear transformation of the form $y = (a - b)x + b$ can be used and then the random vector y is uniform in [a,b].

```
x=rand(1,100); %Uniform in (0,1)
a=2;
b=5;
y=-3*x+5;  % Uniform in (2,5)
```

To generate discrete uniformly distributed random numbers, such as random bit sequences, the functions: `floor`, `ceil`, `sign` can be used. For example, the command:

```
a=floor(N*rand(1,n)+1);
```

generates an n by 1 vector elements that are uniformly distributed integer numbers within the range [1,...,N]. The command

```
a=sign(2*rand(1,n)-1);
```

generates an n by 1 vector elements that are uniformly distributed integer numbers that take the values of ±1 only; i.e. the values ±1 are generated with equal probability. Another example is a random bit generator that takes the values 1 or 0:

```
a=ceil(rand(1,100)-.5)
```

Gaussian random numbers can be generated using the function `randn(n,m)` which generates an n by m matrix of Gaussian distributed random numbers with zero mean and unity variance in a similar way to uniform random numbers.

To generate random numbers of a given distribution, the theory transformation of random variables can be used (Papoulis, 1994). Using the theory of inverse transformation of random variables, a random variable y with density function $f_y(y)$ can be generated from a uniformly distributed random variable u in [0,1] using the transformation $y = F_y^{-1}(u)$ where $F_y(y)$ is the distribution function of y.

For example, to generate an exponentially distributed random variable with $f_y(y) = e^{-x}U(x)$ where $U(x)$ is a unit step function, we transform a uniformly distributed random variable u according to $y = F_y^{-1}(u)$. Therefore, $F_y(y) = 1 - e^{-y}$ and hence, $y = F_y^{-1}(u) = -\ln 1 - u$. The following script generates an exponentially distributed random variable.

```
% Generate a uniformly distributed random vector
x=rand(1,1000);
% Apply transformation to produce an exponentially
% distributed random variable
y=-log(1-x);
% Plot the histogram of x and y
```

```
hist(x,50)
hist(y,50)
```

MATLAB® also has built-in functions in The Statistics Toolbox (The Statistics Toolbox, 2009) that supports the generation of most of the well-known distributions.

C.5 Moments and Correlation Functions of Random Sequences

Table C.1 shows the commands used to compute the moments and correlation functions of random vectors.

C.6 Fourier Transformation

Fourier transform is implemented in MATLAB® in its discrete version, the DFT, which can be computed using the FFT algorithm. Consider a time-domain signal given by an n by 1 vector x, its FFT is computed using the function FFT(x,N) which computes the N-point FFT, padded with zeros if n<N and truncated if n>N. The function FFT-SHIFT(FFT(x)) shifts zero-frequency component to the center of the spectrum producing a centered spectrum.

The following script computes the FFT of a signal and plots the spectrum versus frequency.

```
clear all
% Frequency of single tone
f=100;
% Sampling frequency
fs=20*f;
% Sampling time
Ts=1/fs;
% Define a time vector
t=0:Ts:2e4*Ts;
% Single tone signal
xt=cos(2*pi*f*t);
% Find FFT
len=length(xt);
Xf=fftshift(fft(hanning(len)'.*xt/len));
len=length(Xf);
% Define frequency resolution
delf=fs/len;
% Set the frequency vector
m=len/2;
z=-(m):(m-1);
f=delf*z;
% Plot FFT
plot(f,Xf)
```

The PSD of a random process can be found in a similar way from the FFT of the autocorrelation function.

Table C.1   Moments and correlation functions in MATLAB®

Command	Description
mean(x)	mean value of x- or nth power of x
var(x)	variance
std(x)	standard deviation
cov(x,y)	covariance of x and y
corrcoef(x,y)	Correlation coefficient of x and y
xcorr(x,y)	Autocorrelation of x or cross correlation of x and y
xcov(x,y)	Autocovariance of x or cross covariance of x and y

C.7   MATLAB® Toolboxes

In addition to the built-in functions in the MATLAB® environment that represent general-purpose functions, MATLAB® has a number of application-specific packages of functions called "Toolboxes". MATLAB® Toolboxes are software packages that provide specialized functions, plots, simulation algorithms and many other extra capabilities to the built-in components in the MATLAB® environment that serve different fields of math, statistics and engineering. Table C.2 lists the main toolboxes available from TheMathWorks™. In the context of the subject of this book, the most commonly used toolboxes are the RF, Communications and Signal Processing toolboxes.

Table C.2   MATLAB® Toolboxes (MATLAB® Documentation, 2009)

Toolbox	Areas Covered
Communication	Communication System Design
RF	RF Circuit Design
Signal Processing	Signal Processing algorithms and systems
Filter Design	Filter Design
Partial Differential Equations, Statistics	Math & Statistics
System Identification, Optimization, Curve Fitting Statistics Wavelet Spline	System Optimization and Identification
Genetic Algorithm, Neural Network, Fuzzy Logic, Fixed Point	Search Algorithms
Control System, Instrument Control, Robust Control, Model Predictive Control	Control System
Data Base, Data Feed	Data Management
Embedded, Parallel Computing	Computer Engineering
Image Processing Image Acquisition	Image Processing

C.7.1 The Communication Toolbox

The Communication Toolbox provides a large number of functions, plots, graphical user interfaces that serve as tools for the efficient simulation of communication systems. There are a number of features that make this toolbox an efficient tool in the design and analysis of communication system, among which are the following categories (The Communication Toolbox, 2009):

- Libraries of functions for the generation of various communication signals using random data, Table C.3.
- Libraries of functions for simulation of communication channel models that include the commonly used channel models, Table C.4.
- Libraries of communication channel equalizers that aids the design of equalizers to combat the effects of channel impairments, Table C.5.
- Libraries of functions for evaluation of communication system performance, Table C.6.

  Furthermore, the toolbox contains a graphical user interface, the BERTool, which can be used for simulation of BER and comparing simulated values with analytical and semianalytical results of most modulation formats.

C.7.2 The RF Toolbox

The RF toolbox provides tools for creation, simulation and visualization of passive and active RF circuits/networks and their characteristics. Table C.7 shows the RF components and networks and the corresponding data and visualization objects supported by the toolbox.

Table C.3 Communication signal generation in The Communication Toolbox (The Communication Toolbox, 2009)

Sublibrary	Functionality	Commonly Used Functions
Signal Sources	Generation of WGN, Random Symbols, Random Integers, BER Patterns	wgn, randsrc, randint, randerr
Source Coding	Quantization, Companding, Huffman Coding, PCM	quantiz compand dpcmenco dpcmdeco
Error Detection and Correction	Channel Coding: Block, Convolutional, CRC Codes	encode, decode, gen2par, syndtable hammgen rsgenpoly bchenc, bchdec, rsenc, rsdec
Special Filters	Pulse Shaping: RC filters	rcosflt
Modulation & Demodulation	Analogue Modulation: AM, FM; Digital Modulation: ASK, MPSK, MQAM, DQPSK, OOPSK, MFSK, MSK, GMSK	ammod, pmmod, fmmod; modem object

Table C.4 Channel models in The Communication Toolbox (The Communication Toolbox, 2009)

Channel	Commonly Used Functions
AWGN	`awgn`
MIMO	`mimochan` object
Fading	`rayleighchan`, `ricianchan`
Binary Symmetric Channel	`bsc rcosflt`

Table C.5 Equalization in The Communication Toolbox (The Communication Toolbox, 2009)

Equalization Technique	Adaptive Algorithm	Commonly Used Functions
Linear	CMA, LMS, RLS	`lineareq`, `equalize`, `lms`, `rls`, `cma`
Decision Feedback	CMA, LMS, RLS	`dfe`, `equalize`, `lms`, `rls`, `cma`
MLSE (Maximum-Likelihood Sequence Estimation)	CMA, LMS, RLS	`mlseeq`, `equalize`, `lms`, `rls`, `cma`

Table C.6 Evaluation of communication system performance in The Communication Toolbox (The Communication Toolbox, 2009)

Performance Metric	Method	Commonly Used Functions
BER	Simulation	`symerr`, `biterr`, `berfit`, `berconfint`
BER	Semianalytic	`semianalytic`
BER	Theoretical	`berawgn`, `bercoding`, `berfading`, `bersync`
Communication Scopes	Eye Diagram, Scatter Plot	`commscope.eyediagram`, `commscope.ScatterPlot`
Communication Measurements	EVM, Modulation Error Rate (MER)	`commmeasure.EVM`, `commmeasure.MER`

The toolbox also contains the RF Analysis GUI that provides a visual interface for creation, analysis and simulation of RF components and networks and enables the visualization of their parameters. Figure C.1 shows the layout of the RF GUI.

C.8 Simulink®

Simulink® has been a very useful tool for analysis, modeling and simulation of dynamic systems in various disciplines of engineering. Through its graphical user interface (GUI), Simulink® enables models of linear and nonlinear systems to be built as block diagrams and simulated in continuous, discrete or hybrid times.

Table C.7   RF circuit analysis, simulation and visualization in The RF Toolbox (The RF Toolbox, 2009)

RF Circuit/Network	Passive networks, Amplifiers and mixers, Amplifiers and mixers, Transmission lines, LC ladder filters, Networks
Creation Object	rfckt
Circuit Analysis	analyse, calculate, extract, listformat, listparam
Visualization	plotyy, smith
RF Data Object	rfdata
Data Extraction	extract, read, write getdata
Type of Data	Network parameters, Spot noise, Noise figure, Third-order intercept point (IP3)
Data Format	SnP, YnP, ZnP, HnP, AMP

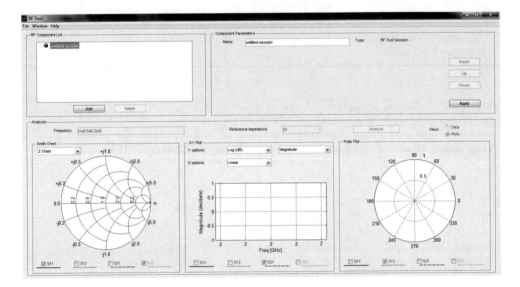

Figure C.1   RF GUI in the RF toolbox (The RF Toolbox, 2009).

Design and modeling with Simulink® is based on model-based design. Complex system models can be built in Simulink® using its comprehensive libraries of blocks which represent submodels and components. These libraries contain a wide variety of sources, sinks, components and connectors. Simulink® blocks are hierarchical, which means that depending on the level of detail required by the design, modeling or simulation problem, these models can be modified by changing their internal design, which can be viewed by looking under their masks. On the other hand, a number of block can be grouped in one block with a certain number of inputs and outputs. New blocks can be built using existing blocks in the library to fit a user requirement for a given design that makes Simulink® unlimited in covering the various engineering problems.

Table C.8 Simulink® Blocksets and specialized software (Simulink® User's Guide, 2009)

Blockset/Toolbox/Software	Areas Covered
Communication	Communication system design
RF	RF Ccircuit/system design
Signal Processing	Signal processing algorithms and systems
Video and Image Processing	Image processing
SimPowerSystems	Power system design
SimElectronics	Electronic system design
SimHydraulics®, SimMechanics, SimDriveline	Mechanical systems
Simscape	Physical systems design and simulation
SimEvents®	Discrete-event simulation
Design Optimization	Model parameter numerical optimization
Real − TimeWorkshop®	Embedded system design and verification
Control Design	Control systems and models

Simulink® software is integrated with the MATLAB® environment where it could use MATLAB® for a number of tasks such as storing model output, calling MATLAB functions within a block, loading a stored signal, etc.

In a similar way to MATLAB®, Simulink® has a number of "Blocksets" and software that contain specialized blocks for various disciplines. Table C.8 shows the main blocksets supported by Simulink®.

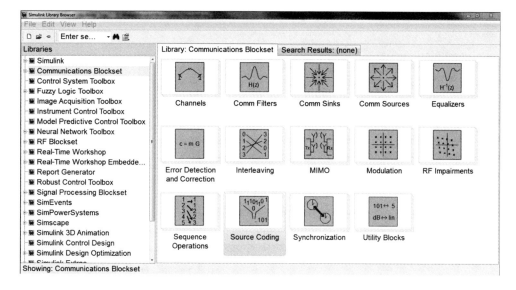

Figure C.2 Sublibraries in the Communication Blockset (The Communication Blockset, 2009).

Table C.9   Sublibraries in the Communication Blockset (The Communication Blockset, 2009)

Sub-Library	Functionality	Blocks
Communication Sources	Random Data Generation	Noise, Sequences BER Patterns
Communication Sinks	Visualization	Error statistics, Scopes
Source Coding	Quantization, Companding	Quantization, Companding
Error Detection and Correction	Channel Coding	Block, Convolutional, CRC Codes
Interleaving	Interleaving	Block, Convolutional
Communication Filters	Pulse Shaping	RC filters
Modulation & Demodulation	Analogue/Digital Modulation	AM, FM; Digital Modulation: ASK, MPSK, MQAM, DQPSK, OOPSK, MFSK, MSK, GMSK
Channels	Channel Models	AWGN, Fading, Binary Symmetric
MIMO	MIMO Systems	Space-Time Coding
RF Impairments	RF Impairments	Nonlinearity and I/Q imbalance, phase/frequency offsets, phase noise, thermal noise and path loss
Equalizers	Adaptive Equalizers	Linear equalizers, Decision-feedback equalizers, MLSE
Sequence Operation	Operations on sequences	Scrambling, puncturing
Synchronization	Timing, Phase Recovery	Timing, Phase and Carrier Recovery, voltage-controlled oscillator (VCO), phase-locked loop (PLL)

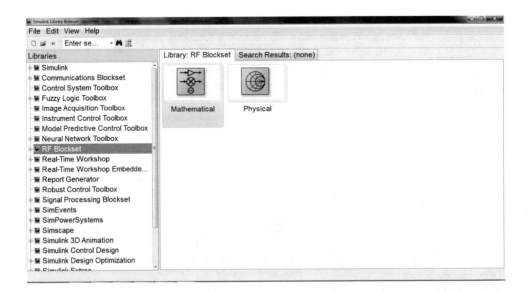

Figure C.3   Sublibraries in the RF Blockset (The RF Blockset, 2009).

Table C.10 Sublibraries in the RF Blockset (The RF Blockset, 2009)

Sub-Library	Blocks
Physical	Amplifiers, Black Box Element, I/O Ports, Mixers, Ladder Filters, RLC Circuits, Transmission Lines
Mathematical	Amplifiers, Mixers, Filters (BPF, BSF, HPF, LPF),

The most important Blocksets in the context of topics covered in this book are The Communication Blockset, The RF Blockset and The Signal Processing Blockset.

C.8.1 The Communication Blockset

The Communication Blockset contains a number of sublibraries that cover various aspects of communication system design and simulation. Figure C.2 shows the sublibraries within The Communication Blockset and Table C.9 shows their description.

C.8.2 The RF Blockset

The RF Blockset contains two sublibraries that cover various aspects of RF system design and simulation. The mathematical sublibrary contains blocks with models based on mathematical representations of the amplifier, mixer, and filter blocks. The physical sublibrary contains models for physical and electrical components and are based on specifying physical properties of components or on importing measured data. Figure C.3 shows the sublibraries within the RF Blockset and Table C.10 shows their description.

References

Agilent Technologies, Inc. 2006 *Harmonic Balance Simulation*, Agilent.

Arabi E and Sadiq A 2008 *Behavioral Modeling of RF front end devices in* Simulink®. Master thesis, Chalmers University of Technology Göteborg, Sweden.

Akhtar S 2010 2G-4G Networks: Evolution of Technologies, Standards, and Deployment. White paper, College of Information Technology, UAE University.

Aparin V 2001 Analysis of CDMA signal spectral regrowth and waveform quality *IEEE Trans. Microwave Theory Techn.* **49**, 2306–2314.

Aparin V 2003 Analysis and reduction of cross-modulation distortion in CDMA receivers *IEEE Trans. Microwave Theory Techn.* **51**, 1591–1602.

Aparin V and Larson L 2004 Analysis of Cross Modulation in W-CDMA Receivers, 787–790.

Baer H 1982 Interference effects of hard limiting in PN spread-spectrum systems. *IEEE Trans. Comm.* **30**, 1010–1017.

Banelli P 2003 Theoretical analysis and performance of OFDM signals in nonlinear fading channels. *IEEE Trans. Wireless Comm.* **2**, 284–293.

Bennett D W, Kenington P B and Wilkinson R J 1997 Distortion effects of multicarrier envelope limiting. *IEE Proc. Comm.* **144**, 349–356.

Bendat J S, 1990 *Nonlinear System Analysis and Identification*, John Wiley and Sons, UK.

Berenji K, Abdipour A, and Mohammadi A 2006 A New Circuit Envelope Simulation Technique for Analysis of Microwave Circuits. *Iran. J. Elect. Comput. Eng.* **5**.

Boyd S and Chua L 2000 Uniqueness of a basic nonlinear structure. *IEEE Trans. Circuits Systems* **30**, 648–651.

Blachman N 1968 The Signal × Signal, Noise × Noise, and Signal × Noise Output of a Nonlinearity. *IEEE Trans. Inf. Theory* **1**, 21–27.

Blachman N, 1968 The Uncorrelated Output Components of a Nonlinearity. *IEEE Trans. Inf. Theory* **3**, 250–255.

Boccuzzi J 2008 *Signal Processing for Wireless Communications* McGraw-Hill.

Briffa M 1996 *Linearization of RF Power Amplifiers*. PhD dissertation, Victoria University of Technology, Australia.

Bussgang J 1952 *Cross-correlation functions of amplitude-distorted Gaussian signals* MIT Research Laboratory for Electronic Technology Report **216**, 1–14.

Chabrak K 2006 *System Simulation of the Analog and Digital Front-End for Reconfigurable Multi-Standard Wireless Receivers* Ph.D. dissertation, Universität Erlangen-Nrnberg, Germany.

Chang C R 1990 *Computer Aided Analysis of Nonlinear Analog Microwave Circuits using Frequency Domain Spectral Balance*, Ph.D. Dissertation, North Carolina State University, Raleigh, NC.

Chen H, Jacobson L, Gaska J and Pollen D 1989 Structural classification of multi-input biological nonlinear systems. In *IEEE Conf. Systems, Man and Cybernetics* **3**, pp. 903–908.

Nonlinear Distortion in Wireless Systems: Modeling and Simulation with MATLAB®, First Edition.
Khaled M. Gharaibeh.
© 2012 John Wiley & Sons, Ltd. Published 2012 by John Wiley & Sons, Ltd.

Chen H, 1995 Modeling and identification of parallel nonlinear systems: structural classification and parameter estimation methods. *Proc. IEEE* **83**, 39–66.

Conti Dardari D, Tralli V and Vaccari A 2000 A theoretical characterization of nonlinear distortion effects in OFDM systems. *IEEE Trans. Comm.* **48**, 1755–1764.

Cripps S 2000 *RF Power Amplifiers for Wireless Communications* Artech House Inc.

Cooper G and McGillem C 1986 *Modern Communications and Spread Spectrum* McGraw-Hill, NY.

Cylan N 2005 *Linearization of Power Amplifiers by Means of Digital Predistortion*. PhD Thesis, Der Technischen Fakultat der Universitat Erlangen-Nurnberg Erlangen, Germany.

Czylwik A 2005 *Matlab for Communications*, NTS seminar online.

3GPP 2005 *Technical Specification Group GSM/EDGE Radio Access Network, Multiplexing and Multiple Access on the Radio Path* 2005 Tech. Rep. TS 45.002V6.9.0, 3rd Generation Partnership Project.

Dardari D Tralli V and Vaccari A 2000 A theoretical characterization of nonlinear distortion effects in OFDM systems. *IEEE Trans. Comm.* **48**, 1755–1764.

Ding L Qian H Chen N and Zhou G T 2004 A memory polynomial predistorter implemented using TMS320C67xx. In *Proc. Texas Instruments Developer Conference*, pp 1–7.

Eslami M and Shafiee H 2002 Performance of OFDM receivers in presence of nonlinear poweramplifiers. In *Int. Conf. Comm. Sys.* **1**, pp. 25–28.

Fenton R 1977 Noise-loading analysis for a general memoryless system. *IEEE Trans. Inst. Meas.* **26**, 61–64.

Franz O, Scholkopf B 2003 *Implicit Wiener Series*, Technical Report No. TR-114.

Frevert R, Haase J, Jancke R 2005 *Modeling and Simulation for RF System Design* Springer.

Furuskar A Mazur S Muller F and Olofsson H 1999 EDGE: enhanced data rates for GSM and TDMA/136 evolution. *IEEE Wireless Commun. Mag.* **6**, 56–66.

Gard K Larson L E and Steer M B 2005 The impact of RF front-end characteristics on the spectral regrowth of communications signals. *IEEE Trans. Microwave Theory Techn.* **53**, 2179–2186.

Gard K Gutierrez H and Steer M B 1999 Characterization of spectral regrowth in microwave amplifiers based on the nonlinear transformation of a complex Gaussian process. *IEEE Trans. Microwave Theory Tech.* **47**, 1059–1069.

Gardner W A and Archer T L 1993 Exploitation of cyclostationarity for identifying the Volterra kernels of nonlinear systems. In *IEEE Trans. Inf. Theory*, 535–542.

Gharaibeh K M Gard K G and Steer, M B 2006 In-Band Distortion of Multisines. *IEEE Trans. Microwave Theory Techn.* **8**, 3227–3236.

Gharaibeh K M Gard K and Steer M B 2004 The impact of nonlinear amplification on the performance of CDMA systems. In *2004 IEEE Radio and Wireless Conf.*, pp. 83–86.

Gharaibeh K M Gard K and Steer M B 2005 Estimation of in-band distortion in digital communication system. In *Int. Mic. Symp.*, 1963–1966.

Gharaibeh K M Gard K and Steer M B 2006 Characterization of in-band distortion in RF front-ends using multisine excitation. In *2006 IEEE Radio Wireless Symp.* **pp** 487–490.

Gharaibeh K M 2010 On the relationship between the noise-to-power ratio (NPR) and the effective in-band distortion of WCDMA signals. *Int. J. Electron. Commun.* **64**, 273–279.

Gharaibeh K M 2009 Simulation of Noise Figure of Nonlinear Amplifiers Using the Orthogonalization of the Nonlinear Model. *International J. RF Microwave Comput-Eng.* **19**, 502–511.

Gharaibeh K M Gard K and Steer M B 2007 Estimation of Co-Channel Nonlinear Distortion and SNDR in Wireless System. *IET Microwave Antenna and Propagation* **1**, 1078–1085.

Gharaibeh K M and Steer M B 2005 Modeling distortion in multichannel communication systems. *IEEE Trans. Microwave Theory and Tech.* **53**, 1682–1692.

Gharaibeh K M and Steer M B 2003 Characterization of cross modulation in multichannel amplifiers using a statistically based behavioral modeling technique. *IEEE Trans. Microwave Theory Techn.* **51**, 2434–2444.

Gharaibeh K M 2004 *Design Methodology for Multichannel Communication Systems*. PhD dissertation, North Carolina State University, Raleigh NC.

Gharaibeh K, Gard K and Steer M B 2004 Accurate Estimation of Digital Communication System Metrics – SNR, EVM and Rho in a Nonlinear Amplifier Environment. In *Proc. of the 42 Automatic RF Techniques Group (ARFTG)* Orlando, FL, pp. 41–44.

Ghosh A Wolter D R Andrews J G and Chen R 2005 Broadband wireless access with WiMax/802.16: current performance benchmarks and future potential. *IEEE Communications Magazine* **43** 129–136.

Gilmore R and Steer M B 1991 Nonlinear Circuit Analysis Using the Method of Harmonic Balance-A Review of the Art. Part I. Introductory Concepts. *Int. J. Microw. Millimeter-Wave Comput.-Aided Eng.* **1**, 22–37.

Goldsmith A 2005 *Wireless Communications* Cambridge, MA,Cambridge University Press, MA (USA).

Graham J and Ehrmen L 1973 *Nonlinear System Modeling and Analysis with Application to Communication Receivers*, Rome Air Development Center, Rome, N.Y.

Gross R and Veeneman D 1994 SNR and spectral properties for a clipped DMT ADSL signal. In *IEEE Conf. Comm.* **2**, pp. 843–847.

Gross R and Veeneman D 1993 Clipping distortion in DMT ADSL systems. *Electron. Lett.* **29**, 2080–2081.

Hasan M 2007 Performance Evaluation of WiMAX/IEEE 802.16 OFDM Physical Layer. Masters thesis, Helsinki University of Tchnology, Helsinki, Finland.

Hajek B 2006 *An Exploration of Random Processes for Engineers* Lecture notes, The University of Illinois at Urbana-Champaign.

Hart F, Stephenson D, Chang CR, Gharaibeh K, Johnson R and Steer M B 2003 Mathematical foundations of frequency domain modeling of nonlinear circuits and systems using the arithmetic operator method. *Int. J. RF Microw. Comput.-Aided Eng.* **13**, 473–495.

IEEE 802.162004 2004 *IEEE Standard for Local and Metropolitan Area Networks Part 16: Air Interface for Fixed Broadband Wireless Access Systems*.

IS95 Standard 1993 *Mobile Station-Base Station Compatibility Standard for Dual-Mode Wideband Spread-Spectrum Cellular Systems* TIA/EIA Standard IS-95.

IThink 2009 *http://www.systems-thinking.org/modsim/modsim.htm*

Jeruchem M, Balaban P and Shanmugan S, 2000 Simulation of Communication Systems, Kluwer Academic/Plenum Publishers, NY(USA).

Jones J 1963 Hard-limiting of two signals in random noise. *IEEE Trans. Information Theory* **9**, 34–42.

Kashyap K, Wada T, Katayama M, Yamazato T and Ogawa A 1996 The performance of CDMA system using p/4-shift QPSK and $\pi/2$-shift BPSK with the nonlinearity of HPA. In *IEEE Int. Symp. Personal, Indoor and Mobile Radio Comm., PIMRC 96* **2**, 492–496.

Kenington P B 2000 *High Linearity Amplifier Design* Artech House Inc, Norwood, MA.

Kim S B and Powers E J 1993 Orthogonalized Frequency Domain Volterra Model for Non-Gaussian Inputs. *IEEE Proc. (Special Issue on Applications of Higher-Order Statistics)* **40**, 402–409.

Kim S B and Powers E J 1993 Orthogonal Development of a Discrete Frequency- Domain Third-Order Volterra Model. In *Proc. Acoustics, Speech, and Signal Processing Conf.* , pp. 484–487.

Kirlin R L 1977 Performance of polarity correlators or limiters with pseudorandom additive input noise. In *IEEE Trans. Information Theory* **23**, 265–267.

Korenberg M, 1991 Parallel cascade identification and kernel estimation for nonlinear systems. Annals Biomed. Eng. **19**, 429–455.

Koch R 1971 Random signal method of nonlinear amplifier distortion measurement. *IEEE Trans. Inst. Meas.* **20**, 95–98.

Kokkeler A 2005 Modeling Power Amplifiers using Memory Polynomials. In: *Proceedings of the 12th IEEE Benelux Symposium on Communications and Vehicular Technology in the Benelux (SCVT2005)* **3**, pp. 1–6.

Kundert K 2003 Introduction to RF Simulation and its Application. *IEEE J. Solid-State Circuits* **34**, 1298–1319.

Kundert K, White J and Vincentelli A 1989 A Mixed Frequency-Time Approach for Distortion Analysis of Switching Filter Circuits. *IEEE J. Solid-State Circuits* **24**, 443–451.

Kuo Y 1973 Noise loading analysis of a memoryless nonlinearity characterized by a Taylor series of finite order. *IEEE Trans. Inst. Meas.* **22**, 246–249.

Lavrador P M de Carvalho N B and Pedro J C 2001 Evaluation of signal-to-noise and distortion ratio degradation in nonlinear systems. In *IEEE Trans. Microwave Theory Tech.* **52**, 813–822.

Lunsford P J 1993 *The frequency domain behavioral modleing and simulation of nonlinear analoge circuits and systems*, Ph.D. dissertation, North Carolina State University.

M. Maqusi Maqusi M 1985 Characterization of nonlinear distortion in HRC multiplexed cable television systems. *IEEE Trans. Circuits Systems* **32**, 605–609.

Matthias F O, Scholkopf B 2006 A Unifying View of Wiener and Volterra Theory and Polynomial Kernel Regression. *Neural Computation* **18**, 3097–3118.

Mathews V J 1995 Orthogonalization of correlated Gaussian signals for Volterra system identification. *IEEE Signal Proc. Let*. **2**, 188–190.

Maurer L Chabrak K Weigel R 2004 Design of mobile radio transceiver RFICs: current status and future trends. In *International Symposium on Control, Communications and Signal Processing* **1**, 53–56.

Mayaram K, Lee D, Moinian S, Rich D, and Roychowdhury J 2000 Computer-Aided Circuit Analysis Tools for RFIC Simulation: Algorithms, Features, and Limitations. *IEEE Trans. Circuits Systems II: Analog Digital Signal Process*. **47**, 274–286.

Max J 1970 Envelope Fluctuations in the Output of a Bandpass Limiter. *IEEE Trans. Comm*. **18**, 597–605.

McGuffin B 1992 The effect of bandpass limiting on a signal in wideband noise. In *IEEE Military Comm. Conf*. **1**, pp. 206–210.

McGee W F 1971 Complex Gaussian Noise Moments. *IEEE Trans. Inf. Th*. **17**, 149–157.

Morgan D R, Ma Z, Kim J, Zierdt M G, Pastalan J 2006 A Generalized Memory Polynomial Model for Digital Predistortion of RF Power Amplifiers. *IEEE Trans. Signal Proces*. **54**, 3852–3860.

Miller K S 1969 Complex Gaussian Processes. *Siam Review* **11**, 544–567.

Muha M, Clark C, Moulthrop A and Silva C 1999 Validation of power amplifier nonlinear block models. In *IEEE MTT-S Int. Microwave. Symp. Digest* **2**, 759–762.

Nuttal A 1958 *Theory and application of the separable class of random processes* Ph.D. Dissertation, Massachusetts Institute of Technology.

Papoulis A 1994 *Random Variables, and Stochastic Processes* New York,McGraw-Hill.

Peebles P 1987 *Probability, Random Variables and Random Signal Principles* McGraw-Hill.

Pedro J C and de Carvalho N B 2001 Evaluating co-channel distortion ratio in microwave power amplifiers. *IEEE Trans. Microwave Theory Techn*. **49**, 1777–1784.

Pedro J C and de Carvalho N B 2001 On the use of multitone techniques for assessing RF components' intermodulation distortion. *IEEE Trans. Microwave Theory Techn*. **47**, 2393–2402.

Pedro J C and de Carvalho N B 2001 Characterizing nonlinear RF circuits for their in-band signal distortion. *IEEE Trans. Microwave Theory Techn*. **51**, 420–426.

Price R 1958 A useful theorem for nonlinear devices having Gaussian inputs. In *IEEE Trans. Informat. Theory* **4**, 69–72.

Pickholtz R L, Milstein L B., and Schilling D L 1991 Spread Spectrum for Mobile Communications. *IEEE Trans. Vehic. Technol*. **40** 313–322.

Raich R, Qian, H, and Zhou G T 2005 Optimization of SNDR for amplitude-lmited nonlinearities. *IEEE Trans. Comm*. **53**, 1964–1972.

Raich, R. Qian, H., and Zhou, G. T. Raich, R., Qian, H., and Zhou, G. T., 2004 Orthogonal polynomials for power amplifier modeling and predistorter design. In *IEEE Trans. Vehic. Tech*. **53**, 1468–1479.

El-Rabaie S, Fusco V, and Stewart C 1988 Harmonic Balance Evaluation of Nonlinear Microwave Circuits-A Tutorial Approach. *IEEE Trans. Educat*. **31**, 181.

Rapp C 1991 Effects of HPA-Nonlinearity on a 4-DPSK/OFDM-Signal for a Digital Sound Broadcasting System. In *Proceedings of the Second European Conference on Satellite Communications*, pp. 179–184.

Rappaport T 2000 *Wireless Communications: Principles and Practice* Prentice Hall, NJ.

Razavi B, 1998 *RF Microelectronics*, New Jersy Prentice Hall.

Recommended Minimum Performance Standards for Dual-Mode Spread Spectrum Mobile Stations 1999 Tech. Rep. TIA/EIA-98-C, TIA/EIA Standards.

Reis G, 1976 Further results in the noise-loading analysis of a memoryless nonlinearity characterized by a Taylor series of finite order. *IEEE Trans.Inst. Meas*. **25**, 28–33.

Roberts J, Tsui E, Watson D 1979 Signal-to-noise ratio evaluations for nonlinear amplifiers. *IEEE Trans. Comm*. **27**, 197–201.

Rizzoli V, Neri N, Mastri F, and Lipparini A 1999 A Krylov-subspace technique for the simulation of integrated RF/microwave subsystems driven by digitally modulated carriers. *Int. J. RF Microw. Comput.-Aided Eng*. **9**, 490–505.

Rudko M and Wiener D 1978 Volterra Systems with Random Inputs: A Formalized Approach. *IEEE Trans. Comm*. **COM-26**, 217–226.

Rugini L, Banelli P and Cacopardi S 2002 Performance analysis of the decorrelating multiuser detector for nonlinear amplified DS-CDMA signals. In *IEEE Conf. Comm.* **3**, pp. 1466–1470.

Rhyne G W and Steer M B 1987 Generalized power series analysis of intermodulation distortion in a MESFET amplifier: simulation and experiment. *IEEE Trans. Microwave Theory Techn.* **35**, 1248–1255.

Rhyne G W 1988 *Nonlinear Analysis of Microwave Circuits*, Ph.D. Dissertation, North Carolina State University, Raleigh, NC.

Rumney M 2008 IMT-Advanced: 4G Wireless Takes Shape in an Olympic Year. *Agilent Measurement Journal*.

Saleh A 1981 Frequency-independent and frequency-dependent nonlinear models of TWT amplifiers. *IEEE Trans. Comm.* **29**, 1715–1720.

Scott I and Mulgrew B Nonlinear system identification and prediction using orthogonal functions. In *IEEE Transactions Signal Processing* **45**, 1842–1853.

Schetzen M, 1981 Nonlinear System Modeling based on the Wiener theory. *Proc. IEEE* **69**, 1557–1573.

Sevy J 1966 The Effect of Multiple CW and FM Signals Passed Through a Hard Limiter or TWT. *IEEE Trans. Comm.* **14**, 568–578.

Sheng L and Larson L 2003 An SiSiGe BiCMOS Direct-Conversion Mixer With Second-Order and Third-Order Nonlinearity Cancelation for WCDMA Applications. *IEEE Trans. Microwave Theory Techn.* **51**, 2211–2220.

Shiryaev, A N 1996 *Probability*, New York,Springer.

Shi J and Sun H 1990 Nonlinear System Identification for Cascaded Block Model: An Application to Electrode Polarization Impedance. *IEEE Trans. Biomedical Eng.* **37**, 574–587.

Shi J and Sun H 1990 Nonlinear System Identification Via Parallel Cascaded Structure. In *Proceedings Int. Conf. IEEE Engin. Medicine Biology* , pp. 1897–1898.

Silva C, Moulthrop A and Muha M 2002 Polyspectral techniques for nonlinear system modeling and distortion compensation. In *IEEE Int. Vacuum Electronics Conf., IVEC 2002* pp. 314–315.

Silva C P, Moulthrop A A, Muha M S and Clark C J 2001 Application of polyspectral techniques to nonlinear modeling and compensation. In *IEEE MTT-S Int. Microwave Symp. Digest* **1**, 13–16.

Silva C, Clark C, Moulthrop A and Muha M 2000 Optimal-filter approach for nonlinear power amplifier modeling and equalization. In *IEEE MTT-S Int. Microwave Symp. Digest* **1**, pp. 437–440.

Silva C, Moulthrop A, Muha M and Clark C 2001 "Application of polyspectral techniques to nonlinear modeling and compensation. In *IEEE MTT-S Int. Microwave Symp. Digest* **1**, 13–16.

Stanley G 2000 Wiener kernel estimation for neural systems with natural inputs. In *10th Annual Computational Neuroscience Meeting*, Monterey, CA.

Strang G. 1988 *Linear Algebra and its Applications* Harcourt, Brace, Jovanovich, Publishers, CA(USA).

Terrovitis M and Meyer R 2000 Intermodulation Distortion in Current-Commutating CMOS Mixers. *IEEE J. Solid-State Circuits* **35**, 1461–1473.

TheMathWorks™ Inc. 2009 MATLAB® help.

TheMathWorks™ Inc. 2011 *www.mathworks.com/matlabcentral*.

TheMathWorks™ Inc. 2009 Simulink® *v7.3 (R2009a) User's Guide*, online.

TheMathWorks™ Inc. 2009 *The Statistics Toolbox 4 User's Guide*, online.

TheMathWorks™ Inc. 2009 *The Signal Processing Toolbox 6.12 User's Guide*, online.

TheMathWorks™ Inc. 2009 *The Signal Processing Blockset Users' Guide*, online.

TheMathWorks™ Inc. 2009 *The RF Blockset v. 2 Users' Guide*, Online.

TheMathWorks™ Inc. 2009 *The RF Toolbox v. 2.5 Users' Guide*, Online.

TheMathWorks™ Inc. 2009 *The Communication Blockset Users' Guide*, online.

TheMathWorks™ Inc. 2009 *The Communication Toolbox 4 Users' Guide*, online.

Vandermot K, Van Moer W, Schoukens J and Rolain Y 2006 Understanding the Nonlinearity of a Mixer Using Multisine Excitations. In *Instrumentation and Measurement Technology Conference* **1**, pp. 24–27.

Vanhoenacker K, Dobrowiecki T, and Schoukens J 2001 Design of multisine excitations to characterize the nonlinear distortions during FRF-measurements. *IEEE Trans. Instrum. Meas.* **50**, 1097–1102.

Westwick D and Kearney R 2000 Identification of a Hammerstein model of the stretch reflex EMG using separable least squares. In *22nd Annual Int. Conf. IEEE Engineering in Medicine and Biology Society* **3**, pp. 1901–1904.

Zhou G T, 2000 Analysis of spectral regrowth of weakly nonlinear power amplifiers. In *IEEE Conf. Acoustics, Speech, and Signal Processing* **5**, pp. 2737–2740.

Zhou G T, Qian H, Ding L, and R Raich 2005 On the Baseband Representation of a Bandpass Nonlinearity. *IEEE TRANSACTIONS ON SIGNAL PROCESSING* **53**, 2953–2957.

Yap H S 1997 Designing to Digital Wireless Specifications Using Circuit Envelope Simulation. In *Microwave Conference Proceedings* **1**, pp. 173–176.

Index

Nonlinear Distortion in Wireless Systems: Modeling and Simulation with MATLAB®, First Edition.
Khaled M. Gharaibeh.
© 2012 John Wiley & Sons, Ltd. Published 2012 by John Wiley & Sons, Ltd.